Methods in Physiology

Published on behalf of The American Physiological
Society by Springer

Methods in Physiology

This book series is published on behalf of the American Physiological Society (APS) by Springer. Access to APS books published with Springer is free to APS members.
APS publishes three book series in partnership with Springer: *Physiology in Health and Disease* (formerly *Clinical Physiology*), *Methods in Physiology*, and *Perspectives in Physiology* (formerly *People and Ideas*), as well as general titles.

More information about this series at http://www.springer.com/series/11781

Susan R. Hopkins • Peter D. Wagner

The Multiple Inert Gas Elimination Technique (MIGET)

Springer

American Physiological Society

Susan R. Hopkins
Division of Physiology, Department of Medicine
University of California, San Diego
La Jolla, CA, USA

Department of Radiology
University of California, San Diego
La Jolla, CA, USA

Peter D. Wagner
Division of Physiology, Department of Medicine
University of California, San Diego
La Jolla, CA, USA

Methods in Physiology
ISBN 978-1-4939-7440-5 ISBN 978-1-4939-7441-2 (eBook)
https://doi.org/10.1007/978-1-4939-7441-2

Library of Congress Control Number: 2017957063

Printed on acid-free paper

This Springer imprint is published by Springer Nature
The registered company is Springer Science+Business Media, LLC
The registered company address is: 233 Spring Street, New York, NY 10013, U.S.A.

Foreword

It is a special pleasure to write a foreword for this important book by my colleagues Susan Hopkins and Peter Wagner. There are several reasons for this. One is that Peter joined me at UCSD in 1970 and thus has been a colleague for 47 years. This may not be a record, but it has been an unusually rewarding relationship, and, especially in the early days, we cooperated on many projects. Susan has also been a valuable colleague and incidentally is a talented artist. Perhaps I should add that I have not seen quite so much of Peter recently because he has taken up multitasking and now runs a vineyard that produces grapes for what are expected to be exceptional wines.

But the main reason why this book gives me so much pleasure is that it describes an outstanding advance in understanding ventilation-perfusion ($\dot{V}_A/\dot{Q}$) inequality in diseased lungs. $\dot{V}_A/\dot{Q}$ was my first love in respiratory physiology and still remains dear to me. The first paper that I ever published (in 1957) was on the measurement of $\dot{V}_A/\dot{Q}$ inequality from a single expiration, and oddly enough only a few years ago, we used the same technique to make measurements on astronauts in space. In the period during and after World War II, pulmonary gas exchange was a particularly exciting research topic. Two groups of investigators, Fenn, Rahn, and Otis in Rochester, NY, and Riley and his colleagues at Johns Hopkins, made extraordinary advances. I was so enamored that I spent a sabbatical year with Rahn. In those days, the only possible method of analysis was graphical, and Rahn and Fenn produced a softcover booklet with foldout graphs that was the bible for acolytes. The breakthrough came with the application of digital computing, and this volume is a splendid example of what this can do.

$\dot{V}_A/\dot{Q}$ inequality is a demanding area of analysis. The basic physiology is very simple: the gas exchange that occurs in any single lung unit depends solely on the ratio of ventilation to blood flow. But once one starts to consider $\dot{V}_A/\dot{Q}$ distributions, which typically occur in lung disease, the situation quickly becomes very complicated. The development of the multiple inert gas elimination technique has clarified many aspects of the abnormal gas exchange that occurs in patients with lung disease and also in normal subjects under unusual physiological conditions. The book is not bedtime reading, but the serious student of pulmonary gas exchange will find it enormously stimulating. I believe that it will be a valuable reference book for many years to come.

La Jolla, CA, USA John B. West

Preface

The multiple inert gas elimination technique (MIGET) was born in the early 1970s in John West's UCSD laboratory. Some 45 years later, we have thought to bring together in one place as much detail as we can about the method. Why now? It is the sobering thought that we are not immortal and that a lot of the technical information is hard to find, scattered as it is in the literature. It certainly requires a lot of searching, and even then, those not mathematically inclined would be in doubt as to the completeness of the descriptions. Our vision has therefore been to create a reference source that present and future users of MIGET can utilize to enhance their understanding of its theoretical and practical aspects. We also undertook the daunting task of reading and summarizing the world's literature on all aspects of the method, especially concerning its physiological and clinical applications, or at least the literature in the English language, but even there, we may have missed articles for which we apologize in advance.

While the book is certainly not light reading, each chapter has been written with the reader in mind—so that even if the detailed mathematics are set aside, the principles and approaches used hopefully make sense, especially coupled to the liberal use of figures. We begin with a short history and then move to the basic physiological principles underlying MIGET. These are the very same principles used some 70 years ago by our pioneering gas exchange predecessors—the principle that alveolar gas exchange for any gas follows simple laws of mass conservation. What follows are chapters on the mathematical equations required to make MIGET work: how we convert inert gas exchange data into the ventilation-perfusion distribution. A consideration of helpful physiological constraints that are integrated into the mathematics, a consideration of experimental error, and then a description of the physiological information content of MIGET are included.

The direction then changes to cover the practical laboratory aspects including a number of seemingly minor but yet very important technical issues associated with the core technology—gas chromatography. Coupled to this is a discussion of the data entry process which enables the inert gas concentrations to be processed to yield the ventilation-perfusion distribution.

Finally, we discuss how MIGET has been applied in hundreds of studies around the world to throw light on complex gas exchange processes in health and disease. To keep this section under control, we have grouped outcomes into logical topic areas and confined our editorial opinion on the findings in each case to major observations of likely mechanistic importance. The domain of research here has been extremely diverse—small and large animals (mammals, birds, reptiles), human health and disease, and exercise and environmental stresses—posing us a major challenge to be complete yet not endless.

While this final chapter shows that a great many people have contributed to the development and implementation of MIGET over the past 45 years, there are four individuals in particular without whom MIGET likely would never have been imagined, developed, or applied, and this book is dedicated to them: John West, who supported the development of MIGET for many years; Herb Saltzman (chief of the Duke University hyperbaric chambers), who spent a sabbatical with John in the early 1970s and whose discussions kick-started thinking about the use of inert gases to investigate ventilation-perfusion inequality; John Evans, a UCSD mathematics professor (and physician!) who understood our goals and was absolutely instrumental in writing computer programs that converted our amateurish data processing code into a rigorous, understandable procedure; and Harrieth Wagner, whose decades-long technical skill with the gas chromatograph made the difference between success and failure.

La Jolla, CA, USA

Susan R. Hopkins
Peter D. Wagner

Contents

Chapter 1
Introduction and History

Abstract After Rahn and Fenn published *A Graphical Analysis of the Respiratory Gas Exchange* in 1955, understanding of the importance of ventilation-perfusion matching to pulmonary gas exchange grew rapidly. Rahn and Fenn highlighted that alveolar gas concentrations were intimately related to the associated ventilation-perfusion ($\dot{V}_A/\dot{Q}$) ratio. Riley and Cournand expanded this into the three-compartment lung model with shunt, ideal, and deadspace compartments. This early work was important since the multiple inert gas technique (MIGET) uses the same principles of mass conservation. MIGET also built on the work of Kety, who showed that the fractional retention of an inert gas in arterial blood was function of solubility and $\dot{V}_A/\dot{Q}$ ratio. Later, Farhi and Yokoyama devised a method for measuring retention of inert gases during washout, from which a two-compartment model of $\dot{V}_A/\dot{Q}$ inequality was derived. Lenfant and Okubo exploited the fact that arterial partial pressure of oxygen (P_{O_2}) response to inspired O_2 depended on the $\dot{V}_A/\dot{Q}$ ratio and modeled a continuous distribution of $\dot{V}_A/\dot{Q}$ ratios to compute the underlying $\dot{V}_A/\dot{Q}$ ratios explaining the response. Using computer algorithms to solve the gas exchange equations of these pioneering investigators and to quantify the O_2 and CO_2 dissociation curves, Kelman, Olszowka, and West studied complex gas exchange behavior by taking theoretical $\dot{V}_A/\dot{Q}$ distributions and evaluating the effect on O_2 and CO_2 exchange. Wagner and West working with Saltzman realized that by using the inert gas approach of Farhi and Yokoyama, and measuring the elimination, simultaneously, of several inert gases of widely varying solubility, continuous distributions of $\dot{V}_A/\dot{Q}$ could be measured. This is what gave rise to the multiple inert gas elimination technique.

1.1 Rahn, Fenn, Riley, and Cournand (et al.)

The concept that ventilation ($\dot{V}_A$), perfusion ($\dot{Q}$), and their ratio ($\dot{V}_A/\dot{Q}$) might be important in pulmonary gas exchange was raised in the first part of the twentieth century, but little was done in the area until around and soon after the World War II. In 1955, Rahn and Fenn published their seminal monograph, *A Graphical Analysis of the Respiratory Gas Exchange* [10]. This analysis, which preceded the computer age by several years, painstakingly provided the physiological and mathematical basis for understanding both the qualitative and quantitative aspects

S.R. Hopkins, P.D. Wagner, *The Multiple Inert Gas Elimination Technique (MIGET)*, Methods in Physiology, https://doi.org/10.1007/978-1-4939-7441-2_1

of alveolar exchange of gases. Perhaps the key concept from their work was the intimate, dependent relationship between alveolar gas concentrations (for any gas) and the associated $\dot{V}_A/\dot{Q}$ ratio. The latter determined the former and did so according to rather simple principles of conservation of mass.

For O_2, it runs like this:

The amount of O_2 taken up from the inspired air (I) each minute ($\dot{V}_{O_2}$) can be written down as:

$$\dot{V}_{O_2} = \left(\dot{V}_I \times F_{IO_2}\right) - \left(\dot{V}_E \times F_{EO_2}\right) \tag{1.1}$$

Here $\dot{V}_I$ and $\dot{V}_E$ are inspired and expired total minute ventilation, respectively; F_{IO_2} and F_{EO_2} are inspired and mixed expired fractional concentrations of O_2, respectively. This is nothing more than expressing mass conservation of O_2. It implies no specific transport mechanism. While an approximation, we often simplify $\dot{V}_I$ to equal $\dot{V}_E$, strictly true only when the respiratory exchange ratio (R) is 1.00, but even when $R = 0.8$, the difference between $\dot{V}_I$ and $\dot{V}_E$ is no more than 1%. In addition, we replace $\dot{V}_E$ and F_{EO_2} by $\dot{V}_A$ and F_{AO_2} (where $\dot{V}_A$ is alveolar, not minute ventilation, and F_{AO_2} is exhaled alveolar O_2 concentration rather than mixed expired). This works because the deadspace is not involved in gas exchange (see below), such that:

$$\dot{V}_E \times F_{EO_2} = \dot{V}_A \times F_{AO_2} \tag{1.2}$$

Thus, Eq. (1.1) is simplified as follows:

$$\dot{V}_{O_2} = \dot{V}_A \times \left[F_{IO_2} - F_{AO_2}\right] \tag{1.3}$$

The next step realizes that the O_2 taken out of the gas as expressed above must appear in the blood, which again espouses only the principle of mass conservation for O_2 (and no particular transport mechanism). This allows a blood-side equivalent equation to be written:

$$\dot{V}_{O_2} = \left(\dot{Q}_T \times C_{aO_2}\right) - \left(\dot{Q}_T \times C_{\bar{v}O_2}\right) \tag{1.4}$$

This mass conservation equation, known as the Fick principle, states that the amount of O_2 taken up per minute equals the difference between the amount leaving the lungs for the left atrium $\left(\dot{Q}_T \times C_{aO_2}\right)$ and that entering the lungs in the mixed venous blood $\left(\dot{Q}_T \times C_{\bar{v}O_2}\right)$. Here, $\dot{Q}_T$ is cardiac output, while C_{aO_2} and $C_{\bar{v}O_2}$ are the concentrations of O_2 in the systemic arterial and mixed venous bloodstreams, respectively. Here we assume that there is no significant change in $[O_2]$ from when a red cell leaves the pulmonary microcirculation and when it appears in the systemic arterial blood, also reasonable. Textbooks indicate that a small amount of coronary venous blood drains directly into the left ventricular cavity, perhaps 1% or less of the cardiac output, but here this is not considered because when it comes time to apply the same concepts to inert gas exchange, this process is not relevant.

The next step is to recognize that these two expressions for $\dot{V}_{O_2}$ depict the same O_2 flux (mass conservation) and are thus equal (as long as the lungs are in a steady state of gas exchange). Hence:

$$\dot{V}_{O_2} = \dot{V}_A \times [F_{IO_2} - F_{AO_2}] = \dot{Q}_T \times [C_{aO_2} - C_{\bar{v}O_2}] \quad (1.5)$$

which is readily rearranged into the famous $\dot{V}_A/\dot{Q}$ equation:

$$\dot{V}_A/\dot{Q} = [C_{aO_2} - C_{\bar{v}O_2}]/[F_{IO_2} - F_{AO_2}] \quad (1.6)$$

We like to convert from fractional gas concentration F to partial pressure P, which are proportional to each other via Dalton's Law of partial pressures. Thus:

$$\dot{V}_A/\dot{Q}_T = k \times [C_{aO_2} - C_{\bar{v}O_2}]/[P_{IO_2} - P_{AO_2}] \quad (1.7)$$

k equals 8.63 at 37 °C. (given by the formula $k = 7.60 \times (273 + T)/273$, where T is body temp, deg centigrade. This stems from the convention that gas-side variables are expressed BTPS, while blood-side variables are expressed STPD (don't shoot the messenger).

This equation is valid regionally in a single homogeneous unit of the lung (where the ratio of ventilation to blood flow is written more generally as $\dot{V}_A/\dot{Q}$, with $\dot{V}_A$ and $\dot{Q}$ each representing the ventilation and blood flow of that single unit) and also applies globally to the lung as a whole (where $\dot{V}_A$ and $\dot{Q}_T$ are then total lung values of alveolar ventilation and blood flow).

Applied to single homogeneous units, the assumption is made that alveolar and endcapillary P_{O_2} values are equal (i.e., there is no diffusion limitation). This means that in Eq. (1.7), if one knows P_{AO_2} one knows C_{aO_2} and vice versa. This statement implies knowledge of the O_2 Hb dissociation curve relating P_{O_2} to O_2 concentration. *Thus, for a given set of values for* $\mathbf{C_{\bar{v}O_2}}$ *and* $\mathbf{P_{IO_2}}$*, there is a unique relationship between* $\dot{\mathbf{V}}_\mathbf{A}/\dot{\mathbf{Q}}$ *and* $\mathbf{P_{AO_2}}$*: one* $\dot{\mathbf{V}}_\mathbf{A}/\dot{\mathbf{Q}}$ *ratio corresponds to one and only one alveolar* $\mathbf{P_{O_2}}$.

This consequence of a single, simple, physical principle is remarkable for what it tells us about gas exchange: In words, it is this. The ventilation perfusion ($\dot{V}_A/\dot{Q}$) ratio is the unique determinant of the alveolar P_{O_2} under a given set of boundary conditions (i.e., mixed venous and inspired O_2 concentrations and O_2 Hb dissociation curve). The identical equation works for CO_2, and the principles indeed apply for all gases.

Riley and Cournand [11] had done similar work to that of Rahn and Fenn but with an emphasis on using gas exchange data and the principles laid out as in Rahn and Fenn to learn about and quantify gas exchange abnormalities rather than lay out fundamental principles themselves. They conceived the famous three-compartment lung: A diseased lung, however complex, could be portrayed *as if* it consisted of

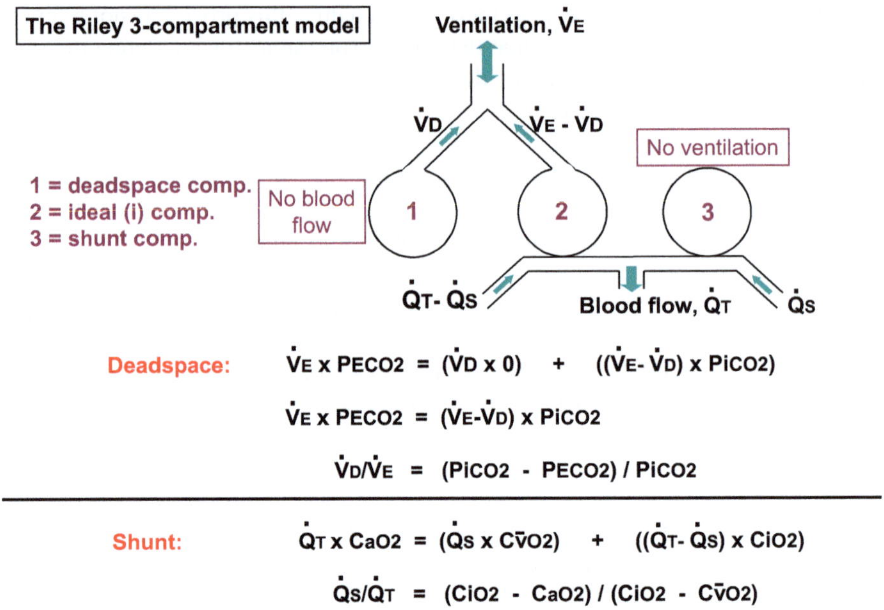

Fig. 1.1 The Riley three-compartment model of ventilation/perfusion heterogeneity. The three compartments are deadspace (ventilated but unperfused); ideal (ventilated and perfused); shunt (unventilated but perfused). The equations in the lower part of the figure show how deadspace and shunt fractions are calculated from measured mixed expired and arterial O_2 and CO_2 levels, with the ideal levels computed according to the respiratory exchange ratio that defines the ideal compartment

only three compartments, which were labeled (a) shunt, (b) ideal, and (c) deadspace (Fig. 1.1). This was seen to be of potentially great clinical value.

The ideal compartment was the "normal" compartment. More specifically, ideal meant a compartment that operated at the same respiratory exchange ratio (i.e., $\dot{V}_{CO_2}/\dot{V}_{O_2}$) as the subject's actual lungs. The other two compartments represented extremes in which no gas exchange took place. Shunt meant a compartment with blood flow but no ventilation. In this compartment, blood emerged into the pulmonary vein unchanged in composition from that of mixed venous blood. Symmetrically, deadspace meant the converse—a compartment with ventilation but no blood flow. Thus, the O_2 concentration in such alveoli and airways would be that of the inspired air, since due to lack of blood flow, there could be no gas exchange in those regions. Healthy subjects should display essentially no shunt and only that deadspace corresponding to the volume of gas in the conducting airways from the mouth to the respiratory bronchioles. Importantly, it must be remembered that this construct was not imagined to accurately portray how ventilation and perfusion were actually distributed. After all, there are some 500 million alveoli clustered into perhaps 100,000 acini or gas exchange units, and depicting these as falling into just one of three functional categories was acknowledged to be unrealistic. The construct was advanced rather as a simple tool to measure how well or poorly an individual subject's lung was exchanging gas.

The three-compartment model worked like this: Using the actual, measured, systemic arterial (a) and pulmonary arterial (mixed venous, $\bar{v}$) P_{O_2}'s (and the corresponding O_2 concentrations, C), arterial blood was modeled as derived from a mixture of the blood from the ideal (i) and shunt (s) compartments according to the same mass conservation principles underlying the work of Rahn and Fenn. The mixture of course had to reflect the relative amounts of the blood perfusing the two compartments and thus appears as a weighted average:

$$C_{aO_2} = \left[\dot{Q}_i \times C_{iO_2} + \dot{Q}_s \times C_{\bar{v}O_2}\right] / \left[\dot{Q}_i + \dot{Q}_s\right] \quad (1.8)$$

where $\dot{Q}_i + \dot{Q}_s = \dot{Q}_{total}$ (i.e., cardiac output, $\dot{Q}_T$). Therefore:

$$C_{aO_2} = \left[\left(\dot{Q}_T - \dot{Q}_S\right) \times C_{iO_2} + \dot{Q}_s \times C_{\bar{v}O_2}\right] / \dot{Q}_T \quad (1.9)$$

Rearranging this equation yields the desired variable, the ratio $\dot{Q}_s/\dot{Q}_T$:

$$\dot{Q}_s/\dot{Q}_T = \left[C_{iO_2} - C_{aO_2}\right] / \left[C_{iO_2} - C_{\bar{v}O_2}\right] \quad (1.10)$$

To use this formula requires not only direct measurements of arterial and mixed venous O_2 concentrations but also knowledge of the O_2 concentration in the ideal compartment. Unfortunately, the ideal compartment is a virtual concept, and thus its gas tensions are not actually measureable. But, using the Rahn and Fenn analysis, C_{iO_2} can be calculated if the respiratory exchange ratio is known.

The outcome is a number for the magnitude (as a fraction of the cardiac output) that a shunt would have to be to fully explain the measured arterial $\mathbf{P_{O_2}}$ *in a given subject.*

Deadspace reflects a very symmetrical concept and set of calculations but applied to the gas side rather than the blood side and on the basis of concentrations (F) of CO_2 rather than O_2. But the principle of mass conservation remains its underpinning. Thus, the CO_2 in the mixed expired (E) gas collected from a subject was modeled as coming from just two functional compartments: the ideal compartment (i) and the unperfused but ventilated (deadspace, D) compartment in which the CO_2 concentration must be that of the inspired gas, I, usually essentially (and taken to be) zero. Again, flow weighting the two gas streams must be done to assure mass conservation, and the equation thus looks like this:

$$F_{ECO_2} = \left[\dot{V}_A \times F_{iCO_2} + \dot{V}_D \times F_{ICO_2}\right] / \left[\dot{V}_A + \dot{V}_D\right] \quad (1.11)$$

where $\dot{V}_A + \dot{V}_D =$ total ventilation, $\left(\dot{V}_E\right)$. Therefore:

$$F_{ECO_2} = \left[\left(\dot{V}_E - \dot{V}_D\right) \times F_{iCO_2} + \dot{V}_D \times F_{ICO_2}\right]/\dot{V}_E \tag{1.12}$$

Rearranging this equation yields the desired variable, the ratio $\dot{V}_D/\dot{V}_E$:

$$\dot{V}_D/\dot{V}_E = [F_{iCO_2} - F_{ECO_2}]/[F_{iCO_2} - F_{ICO_2}] \tag{1.13}$$

Because in the gas phase, concentration F and partial pressure P are linearly related (Dalton's Law of partial pressures), this equation can be written:

$$\dot{V}_D/\dot{V}_E = [P_{iCO_2} - P_{ECO_2}]/[P_{iCO_2} - P_{ICO_2}] \tag{1.14}$$

or, as is more common, because $P_{ICO_2} \sim 0$:

$$\dot{V}_D/\dot{V}_E = [P_{iCO_2} - P_{ECO_2}]/P_{iCO_2} \tag{1.15}$$

$\dot{V}_D/\dot{V}_E$ is thus also a fraction, of total ventilation, associated with unperfused alveoli and airways that if present would fully explain the mixed expired $\mathbf{P_{CO_2}}$*—that is, why it was lower than that exhaled from the ideal compartment, "i."*

You may wonder why so much space has been given to these seemingly ancient, established, and "simple" concepts that do not even address MIGET.

Two major reasons:

1. MIGET uses exactly the same principles of mass conservation as does the Rahn and Fenn and Riley and Cournand work summarized above. Nothing is different, just that the same concepts are applied to exchange of foreign (to the body) inert gases.
2. The three-compartment model raises awareness of the concept of the distribution of ventilation, of blood flow, and of their ratio $\dot{V}_A/\dot{Q}$.

Indeed, the three-compartment model can be nicely depicted on the basic diagram that represents the output of the MIGET: the $\dot{V}_A/\dot{Q}$ distribution. This is done in Fig. 1.2, and if you wish to understand MIGET fully, you need to become comfortable with this way of summarizing $\dot{V}_A/\dot{Q}$ inequality, both qualitatively and quantitatively. Here is the bottom line for Fig. 1.2: The most concise and useful visual depiction of $\dot{V}_A/\dot{Q}$ inequality is to imagine the lung as made up of separate compartments each with their own distinct $\dot{V}_A/\dot{Q}$ ratio. The question then becomes, how much of the blood flow and how much of the ventilation are distributed to compartments of different $\dot{V}_A/\dot{Q}$ ratio? The Riley three-compartment model, while greatly oversimplifying the real lung, fits into this construct well as Fig. 1.2 shows—there are just three $\dot{V}_A/\dot{Q}$ ratios corresponding to shunt ($\dot{V}_A/\dot{Q} = 0$); the ideal compartment ($\dot{V}_A/\dot{Q}$ based on the respiratory exchange ratio and usually having a value of about 1); and the deadspace compartment ($\dot{V}_A/\dot{Q}$ = infinitely great). Thus the abscissa is the $\dot{V}_A/\dot{Q}$ ratio, and the two ordinates are ventilation and blood flow in each compartment.

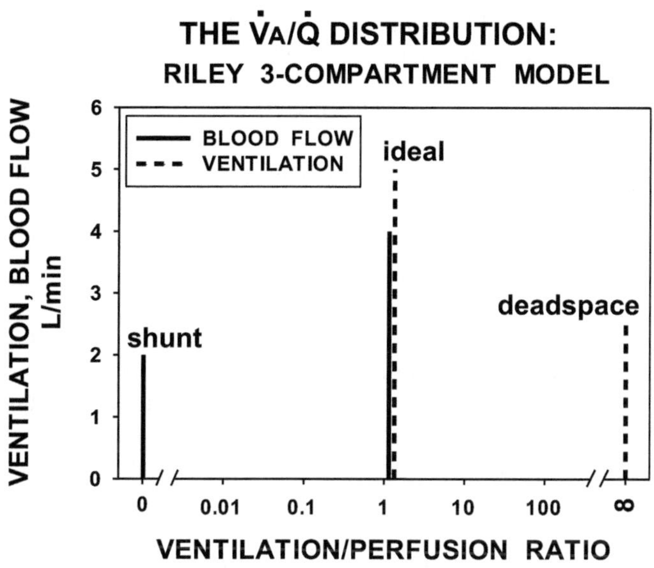

Fig. 1.2 The Riley three-compartment model depicted as a ventilation/perfusion ratio ($\dot{V}_A/\dot{Q}$) distribution. Shunt appears at $\dot{V}_A/\dot{Q} = 0$; deadspace at $\dot{V}_A/\dot{Q} =$ infinite; the ideal point at $\dot{V}_A/\dot{Q}$ corresponding to the overall respiratory exchange ratio. Ordinate values indicate the amounts of ventilation and blood flow in each compartment

1.2 Seymour Kety

In 1951, Seymour Kety, a pharmacologist, published a remarkable, huge dissertation focused on inert gases and their exchange in, but not confined to, the lungs [5]. In effect, he identified the famous "kernel" of the MIGET—an algebraic expression for fractional retention (in the systemic arterial blood) of an inert gas being eliminated from the body by the lungs in a steady state. He showed that this expression was a simple function of the inert gas' solubility in the blood and the $\dot{V}_A/\dot{Q}$ ratio of the lung, as follows:

As stated, Eq. (1.7) above applies to any gas. If we drop the O_2 subscript and generalize, Eq. (1.7) looks like this ($\dot{V}_A$ and $\dot{Q}_T$ here are whole lung values):

$$\dot{V}_A/\dot{Q}_T = k \times \left[C_a - C_{\bar{v}}\right]/[P_I - P_A] \tag{1.16}$$

Now, if inspired concentration of the gas is zero (thus, the gas is undergoing elimination from the blood), it becomes:

$$\dot{V}_A/\dot{Q}_T = k \times \left[C_{\bar{v}} - C_a\right]/P_A \tag{1.17}$$

The beauty of inert gases, defined here as gases carried in the blood only in physical solution and thus not chemically combined with any molecule in the blood, is that concentration C and partial pressure P are linearly related (Henry's Law). In fact the relationship has a constant equal to the solubility, $\boldsymbol{\beta}$:

$$C = \beta \times P \tag{1.18}$$

We usually express C in ml/dl; P in mm Hg; and β therefore in ml/dl per mm Hg. Accordingly, this can be substituted into Eq. (1.17) as follows:

$$\dot{V}_A/\dot{Q}_T = k \times \beta \times \left[P_{\bar{v}} - P_a\right]/P_A \tag{1.19a}$$

Next, $\boldsymbol{k} \times \boldsymbol{\beta}$ turns out to be the dimensionless variable λ, the blood/gas partition coefficient of the inert gas. In words, λ is the ratio of concentrations of the gas in blood and in gas when the gas and blood have the same partial pressure (i.e., are in physical-chemical equilibrium). We therefore simplify $\boldsymbol{k} \times \boldsymbol{\beta}$ to $\boldsymbol{\lambda}$.

We then make the same assumption as for O_2 above: that $P_A = P_a$ in Eq. (1.19a), implying the absence of any diffusion limitation.

$$\dot{V}_A/\dot{Q}_T = \lambda \times \left[P_{\bar{v}} - P_a\right]/P_a \tag{1.19b}$$

Rearranging Eq. (1.19b) and now considering that $\dot{V}_A/\dot{Q}$ applies either to the whole lung when it is perfectly homogeneous, or to a homogeneous lung unit within the lung, gives:

$$P_A/P_{\bar{v}} = P_a/P_{\bar{v}} = \lambda/\left[\lambda + \dot{V}_A/\dot{Q}_T\right] \tag{1.20}$$

Here, when applied to a homogeneous whole lung, P_A refers to alveolar gas; P_a to systemic arterial blood. For a single homogeneous lung unit, P_A refers to the alveolar gas of just that lung unit and P_a refers to the endcapillary blood leaving that same lung unit headed toward the left atrium.

This simple, elegant equation is the basis of MIGET. In words, the fractional retention in blood (P_a/P_v) of any inert gas being eliminated by the lungs in a steady state is given by the ratio of λ to the sum of λ and $\dot{V}_A/\dot{Q}$ ratio. This is applied to each and every compartment that may be present in the lungs—from a $\dot{V}_A/\dot{Q}$ of zero (shunt) to one of infinity (deadspace)—and all values in between. We have Kety to thank for first detailing these relationships systematically, more than 50 years ago. There will be much more on this later in the next chapter.

1.3 Yokoyama and Farhi

Tetsuro Yokoyama and Leon Farhi took Eq. (1.20) and in the 1960s devised a feasible method for measuring retention of three inert gases (methane, ethane, and nitrous oxide) during the late phases of their washout after having previously inhaled them for several minutes [14]. They considered the late phase of elimination to be "steady state"—which cannot really be true, when gas is continually being lost from the body. Note that all of the preceding 20 equations explicitly cover *only* the steady state and have to be thrown out under nonsteady state conditions. More importantly, Yokoyama and Farhi chose to apply the Riley and Cournand three-compartment model concept to the retention data from their three gases. They did not constrain the compartments to be shunt, ideal, and deadspace. In fact, it was a two-compartment model of the complex lung. Their paper seems to show physiologically sensible results, but there were the above limitations to their approach. A simplifying feature of their approach was that with three data points (one each from three gases), one could easily deduce the $\dot{V}_A/\dot{Q}$ values of two compartments. Their approach did not catch on clinically but certainly stimulated work in San Diego. We became fairly evident gas exchange competitors at national meetings in the ensuing years.

1.4 Lenfant and Okubo

Claude Lenfant, who would later become the long-time director of the NIH NHLBI, and his colleague Takao Okubo then became involved in the quest for a method to measure the $\dot{V}_A/\dot{Q}$ distribution [6]. They took a very different approach and recognized that as inspired P_{O_2} (P_{IO_2}) was raised progressively in a given subject, the amount of increase in arterial P_{O_2} at any P_{IO_2} depended on the $\dot{V}_A/\dot{Q}$ ratios present in the lung. The principle is illustrated in Fig. 1.3 where for several single, homogeneous $\dot{V}_A/\dot{Q}$ compartments, their alveolar P_{O_2} is plotted as a function of P_{IO_2}. Where do these data come from? From calculations made directly by solving Eq. (1.7) repeatedly as P_{IO_2} is altered. What is shown is that the lower the $\dot{V}_A/\dot{Q}$ ratio, the less P_{AO_2} rises with P_{IO_2}, until finally at high P_{IO_2}, P_{AO_2} takes off. Breathing pure O_2, P_{AO_2} becomes essentially independent of P_{IO_2}, converging on a value in the mid-600 mm Hg range. Thus, a normal lung without low $\dot{V}_A/\dot{Q}$ areas would show a

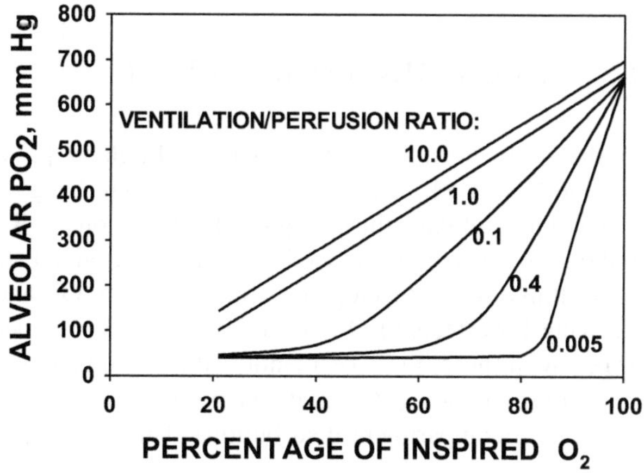

Fig. 1.3 Increase in alveolar P_{O_2}, as F_{IO_2} is increased, for gas exchange units of indicated $\dot{V}_A/\dot{Q}$ ratio from 10 to 0.005. Note in particular how alveolar P_{O_2} fails to increase substantially when $\dot{V}_A/\dot{Q}$ is very low until F_{IO_2} approaches 1.0, while when $\dot{V}_A/\dot{Q}$ is normal (~1.0) or greater, alveolar P_{O_2} rises linearly with F_{IO_2}

Table 1.1 Limitations to the use of progressive increase in F_{IO_2} to estimate $\dot{V}_A/\dot{Q}$ inequality

– Time-consuming sequential exposures of many minutes needed at each F_{IO_2} to establish a good steady state at each
– Potential for the $\dot{V}_A/\dot{Q}$ distribution, total ventilation, and cardiac output to change randomly or systematically during the procedure
– Absorption atelectasis developing as F_{IO_2} is raised in areas of low $\dot{V}_A/\dot{Q}$ ratio, increasing shunt as the measurements are carried out
– Release of hypoxic pulmonary vasoconstriction, especially in low $\dot{V}_A/\dot{Q}$ regions, changing the $\dot{V}_A/\dot{Q}$ distribution as it is measured
– Difficulty in accurate measurement of arterial P_{O_2} at high F_{IO_2}, underestimating that P_{O_2} and thus overestimating shunt
– Need to know mixed venous P_{O_2} at every F_{IO_2}
– Gives no information about high $\dot{V}_A/\dot{Q}$ areas

linear response to P_{IO_2}, while a lung with lots of low $\dot{V}_A/\dot{Q}$ regions would show little increase in arterial P_{O_2} until high F_{IO_2} gas was inhaled, and then arterial P_{O_2} would rise to the mid-600 mm Hg range on 100% O_2.

What Lenfant and Okubo realized is that because the actual pattern of arterial P_{O_2} response to P_{IO_2} depended on the $\dot{V}_A/\dot{Q}$ ratios present, they could use that P_{O_2} response to compute what distribution of $\dot{V}_A/\dot{Q}$ ratios had to be present to explain the actual response. This was very clever, especially because it used a naturally occurring gas and because they were able to use a mathematical method (LaPlace transform) that gave them a continuous distribution of $\dot{V}_A/\dot{Q}$ ratios—not just a three-compartment virtual analog of the lung [6].

There were however several physiological limitations to this approach, which functionally "forces" the lungs to exchange O_2 across several F_{IO_2} values, and they are listed in Table 1.1. They should be self-explanatory to anyone interested enough in gas exchange to have read this far!

This method also failed to catch on, likely both because the mathematics was obscure to most workers and because of the problems listed in Table 1.1.

1.5 Kelman, Olszowka, and West

In the mid-1960s, George Kelman in England [2–4], Albert Olszowka [7] in Buffalo New York, and John West [13] in San Diego more or less simultaneously realized that all the painstaking hand calculations of Rahn, Fenn, Riley, Cournand, and others to solve the gas exchange equations given above could be computerized. Algorithms were developed to quantify the O_2 and CO_2 dissociation curves allowing for effects of O_2/CO_2 interaction, pH, temperature, Hb, and its p50, and armed with those, the aforementioned investigators set about (separately, not as a team) to solve the $\dot{V}_A/\dot{Q}$ equations, not just for a single $\dot{V}_A/\dot{Q}$ compartment but for entire multicompartment distributions. This opened an era in which complex gas exchange behavior of entire $\dot{V}_A/\dot{Q}$ distributions could be studied theoretically, leading to many insights not otherwise apparent [7, 12, 13].

Note that all of these ventures took a chosen theoretical $\dot{V}_A/\dot{Q}$ distribution and studied how that particular case affected O_2 and CO_2 exchange. Such calculations may be termed the *forward* problem—start with a $\dot{V}_A/\dot{Q}$ distribution and ask how it affects gas exchange. This was not yet a direction of study that would enable movement in the opposite direction—the use of measured gas exchange behavior to determine what kind of $\dot{V}_A/\dot{Q}$ distribution had to be present (what we call the *inverse* problem, which was starting to be addressed, as above, by the Farhi and Lenfant groups). As you know, this is what MIGET was expressly designed to do—solve the inverse problem. Much like the Lenfant and Okubo approach for the O_2 approach mentioned above, but without many of the drawbacks listed in Table 1.1.

In the late 1960s, West published his well-known "LOGNOR" program [13] which started with logarithmically normal $\dot{V}_A/\dot{Q}$ curves, incorporated Kelman's subroutines for the O_2 and CO_2 dissociation curves [2–4] and then analyzed how $\dot{V}_A/\dot{Q}$ inequality compromised the exchange of gases—O_2 and CO_2 in particular [12, 13]. This program, slightly modernized from its original form to speed it up, is still used inside the MIGET software in order to predict what the arterial P_{O_2} and P_{CO_2} ought to be in the presence of a particular measured $\dot{V}_A/\dot{Q}$ distribution. Thus, instead of inputting a theoretical lognormal $\dot{V}_A/\dot{Q}$ distribution, the MIGET version takes the MIGET-measured $\dot{V}_A/\dot{Q}$ distribution (which may or may not be lognormal) as input.

1.6 Saltzman, Kylstra, Cohen, and Overfield

In 1972, Herb Saltzman from Duke University spent a sabbatical with us in La Jolla. Herb was director of the diving chambers at Duke and expressed an interest in what increased barometric pressure would do to respiratory gas exchange. He wanted to apply "LOGNOR" to the problem. His chambers also went the other way—altitude simulation by hypobaria—and he produced some papers, then just published by people in his unit (Kylstra, Cohen, Overfield, [1, 8, 9] where they had tried to separate out the causes of the alveolar-arterial P_{O_2} difference in health and disease by clever strategies that changed both F_{IO_2} and barometric pressure at the same time. It was those conversations with Herb that stimulated thinking about the Lenfant and Okubo approach whereby F_{IO_2} was varied and the distribution computed based on the arterial P_{O_2} response (see above). But we also had seen the inert gas approach of Yokoyama and Farhi (also described above) and soon came to realize that we could go after continuous distributions in a manner similar to that of Lenfant, but do so by using inert gas elimination rather than varying F_{IO_2}, thus bypassing the limitations listed in Table 1.1. The rest is history, and a lot of it!!

References

1. Cohen R, Overfield EM, Kylstra JA. Diffusion component of alveolar-arterial oxygen pressure difference in man. J Appl Physiol. 1971;31:223–6.
2. Kelman GR. Computer programs for the production of O_2-CO_2 diagrams. Resp Physiol. 1968;4:260–9.
3. Kelman GR. Digital computer procedure for the conversion of oxygen tension into saturation. J Appl Physiol. 1966;21:1375–6.
4. Kelman GR. Digital subroutines for the conversion of PCO_2 into blood CO_2 content. Resp Physiol. 1967;3:111–6.
5. Kety SS. The theory and applications of the exchange of inert gas at the lungs and tissues. Pharmacol Rev. 1951;3:1–41.
6. Lenfant C, Okubo T. Distribution function of pulmonary blood flow and ventilation-perfusion ratio in man. J Appl Physiol. 1968;24:668–77.
7. Olszowka AJ, Farhi LE. A system of digital computer subroutines for blood gas calculations. Respir Physiol. 1968;4:270–80.
8. Overfield EM, Kylstra JA. Change in (A-a)DO 2 as a function of FI O2 at a constant PI O2 in normal men. J Appl Physiol. 1971;31:581–5.
9. Overfield EM, Kylstra JA. Distribution component of alveolar-arterial oxygen pressure difference in man. J Appl Physiol. 1969;27:634–6.
10. Rahn H, Fenn WO. Graphical analysis of the respiratory gas exchange. Bethseda, MD: American Physiological Society; 1955.
11. Riley RL, Cournand A. Ideal alveolar air and the analysis of ventilation-perfusion relationships in the lungs. J Appl Physiol. 1949;1:825–47.
12. West JB. Causes of carbon dioxide retention in lung disease. N Engl J Med. 1971;284:1232–6.
13. West JB. Ventilation-perfusion inequality and overall gas exchange in computer models of the lung. Respir Physiol. 1969;7:88–110.
14. Yokoyama T, Farhi LE. Study of ventilation-perfusion ratio distribution in the anesthetized dog by multiple inert gas washout. Respir Physiol. 1967;3:166–76.

Chapter 2
MIGET: Basic Physiological Principles

Abstract This chapter builds on the historical concepts outlined in the previous chapter and expands them for a general understanding of the underlying theory of MIGET. The technique is based on simple principles of mass balance: inert gas retention (i.e., the amount of each gas retained in arterial blood as a fraction of the amount delivered to the lung in the pulmonary arterial blood) is a function of the inert gas partition coefficient (λ) and the ventilation ($\dot{V}_A/\dot{Q}$) ratio distribution. Thus, the $\dot{V}_A/\dot{Q}$ distribution can be determined by measuring λ and retention during steady-state inert gas elimination. By evaluating data from gases of varying solubility (λ between 0.001 and 100), the overall distribution can be recovered. Low-solubility gases will be largely eliminated, high-solubility gases will be almost completely retained, and gases of medium solubility will have intermediate retention in blood. In a lung with $\dot{V}_A/\dot{Q}$ inequality, the retention of any gas in arterial blood is the blood flow-weighted average of the retention values from all regions of the lung, which may have potentially differing $\dot{V}_A/\dot{Q}$ ratio. The $\dot{V}_A/\dot{Q}$ distribution can be represented by plotting ventilation and blood flow separately against $\dot{V}_A/\dot{Q}$ ratio as a frequency distribution, represented by 48 $\dot{V}_A/\dot{Q}$ compartments with additional compartments for each of the extremes: shunt and deadspace. Parameters used to describe the distribution quantitatively include the first moment (mean position of the distribution along the $\dot{V}_A/\dot{Q}$ axis) and the second moment about the mean, expressing heterogeneity. The second moments of the ventilation and perfusion distributions are called LogSDV_A and LogSDQ, respectively. The higher the value of these, the greater the extent of $\dot{V}_A/\dot{Q}$ mismatch.

2.1 Inert Gas Retention at a Given $\dot{V}_A/\dot{Q}$ Ratio

As already mentioned in reference to Eq. (1.20) above, reproduced here, MIGET is based on the same mass conservation principles as used by Rahn, Fenn, Riley, Cournand, Kety, Farhi, and others in the middle of the twentieth century.

$$P_A/P_{\bar{v}} = P_a/P_{\bar{v}} = \lambda/\left[\lambda + \dot{V}_A/\dot{Q}\right] \tag{1.20}$$

S.R. Hopkins, P.D. Wagner, *The Multiple Inert Gas Elimination Technique (MIGET)*, Methods in Physiology, https://doi.org/10.1007/978-1-4939-7441-2_2

Applied to a homogeneous whole lung, P_A, P_a, and $P_{\bar{v}}$ are, respectively, alveolar, arterial, and mixed venous inert gas partial pressures. λ is the blood/gas partition coefficient of the inert gas, and $\dot{V}_A/\dot{Q}$ is the ratio of total alveolar ventilation to cardiac output. Applied to a single homogeneous gas exchange lung unit, P_A, P_a, and $P_{\bar{v}}$ are, respectively, local alveolar, local endcapillary, and mixed venous inert gas partial pressures. λ is the blood/gas partition coefficient of the inert gas, and $\dot{V}_A/\dot{Q}$ is the ratio of local ventilation to blood flow.

The basic idea underlying MIGET is simple: if inert gas retention ($P_a/P_{\bar{v}}$) is a simple function of the inert gas partition coefficient λ and the $\dot{V}_A/\dot{Q}$ ratio, as Eq. (1.20) shows, then $\dot{V}_A/\dot{Q}$ can be measured if one knows λ and measures retention during steady-state inert gas elimination; λ can be measured from a blood sample, and inert gases can be dissolved in saline or dextrose and infused intravenously so that when they reach the lungs in the mixed venous blood, they can diffuse into the alveolar gas and be eliminated (in part). Then, the systemic arterial and pulmonary arterial blood can both be sampled and the inert gas levels measured from each to yield their ratio, retention.

This concept however is just the beginning. As presented above, this would be a method for measuring overall $\dot{V}_A/\dot{Q}$ ratio in a homogeneous lung, but what we really want is to measure $\dot{V}_A/\dot{Q}$ inequality. Note that $\dot{V}_A/\dot{Q}$ inequality is that state that exists when the local $\dot{V}_A/\dot{Q}$ ratios throughout the lung are not everywhere the same.

This then begs the question of what the retention-solubility curves would look like in compartments of different $\dot{V}_A/\dot{Q}$ ratio. Figure 2.1 shows the retention-solubility curve for a single compartment of normal $\dot{V}_A/\dot{Q}$ ratio (here taken to be

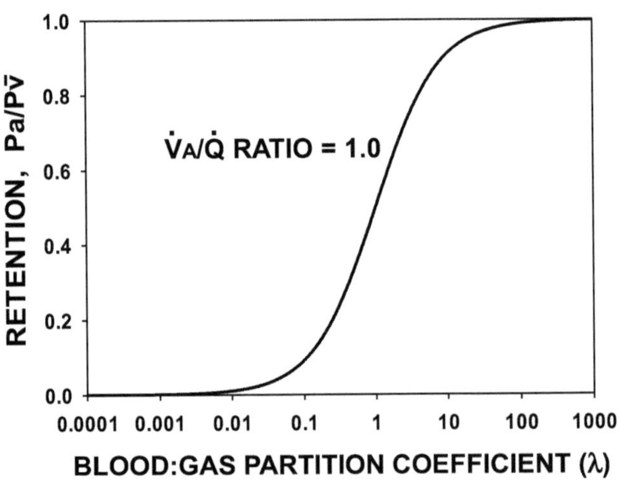

Fig. 2.1 Inert gas retention ($P_a/P_{\bar{v}}$) as a function of inert gas partition coefficient (λ, log scale) during steady-state gas elimination in a homogeneous lung of $\dot{V}_A/\dot{Q} = 1.0$. Low-solubility gases are essentially completely eliminated from the blood; high-solubility gases are essentially completely retained in the arterial blood. The curve expresses the basic mass conservation equation for inert gases: $P_a/P_{\bar{v}} = \lambda/\left[\lambda + \dot{V}_A/\dot{Q}\right]$

exactly 1.0 for simplicity). This curve is nothing more than the calculated value of $P_a/P_{\bar{v}}$ from Eq. (1.20) over a wide range of λ and for a single $\dot{V}_A/\dot{Q}$ ratio, here equal to 1.0. Thus, $Y = X/(1 + X)$. The ordinate (Y) is retention ($P_a/P_{\bar{v}}$), and the abscissa (X) is λ. This curve has some remarkable properties other than the obvious.

The obvious property is that low-solubility gases (low λ) will be almost completely eliminated from the pulmonary capillary blood (by ventilation), while high-solubility gases will be almost completely retained in the blood. Gases of medium solubility will be partly retained and partly excreted. This should be intuitively evident without the need for any equations.

The less obvious property is that the curve is smooth and monotonic. In plain English, this means that moving from low to high solubility, the curve can only increase and never decrease. To prove this, one simply uses calculus: The first derivative of Eq. (1.20) with respect to λ (i.e., the slope of the line in Fig. 2.1) cannot be zero unless λ is zero, which cannot occur in real life—all gases have some solubility no matter how low. That means the retention-solubility curve has to be monotonic. The second derivative and in fact *all* subsequent derivatives have that same property—they cannot be zero. That means that the curve is highly constrained by its mathematical form to a smooth, continually rising function. *In turn, that means that you do not need to make measurements of hundreds of gases of different λ to experimentally define the curve, because for values of λ in between those actually measured, the curve has very limited room to "wiggle." Thus, a small number of well-chosen (in terms of λ) gases will allow the entire retention-solubility curve to be well-identified. That is the bottom line.*

Here is an analogy: MIGET, as you likely know, is based on retention measurements of six different gases. Suppose you throw 50 separate tennis balls onto a tennis court at random. They will come to rest in 50 different locations that could be anywhere on the court. It will take 50 measurements to locate them all. Now, repeat the whole experiment with one difference: the 50 balls are *pre-connected by a stiff string* (like a string of pearls), and you therefore toss the 50 balls as a single entity. You have dramatically reduced the complexity of the possible ball location pattern. If you measure the location of only 6 well-chosen balls among the 50, you can have a good idea of where the other 44 must be, thanks to the string. To choose the six balls "well" means they would be equally spaced along the string from start to end (e.g., rather than clustered all at one end). The mathematical properties of Eq. (1.20) described above in essence mimic the string and imply that the string scenario exists for inert gases.

2.2 The Retention-Solubility Curve and Different $\dot{V}_A/\dot{Q}$ Ratios

Let us now take the next step and add a different curve computed from Eq. (1.20) in addition to that already shown in Fig. 2.1 for $\dot{V}_A/\dot{Q} = 1$. Let's use a $\dot{V}_A/\dot{Q}$ of 0.01 for this additional curve. This is shown in Fig. 2.2, where we now have two

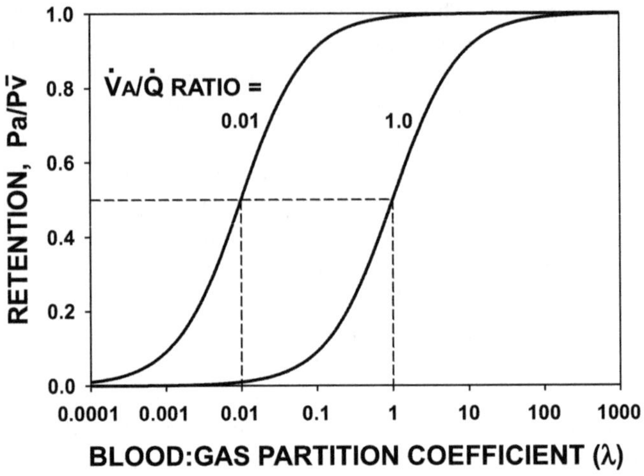

Fig. 2.2 Inert gas retention ($P_a/P_{\bar{v}}$) as a function of inert gas partition coefficient (λ, log scale) during steady-state gas elimination for two homogeneous lung units, one of $\dot{V}_A/\dot{Q} = 1.0$ and the other of $\dot{V}_A/\dot{Q} = 0.01$. The two retention curves are of identical shape, but that for $\dot{V}_A/\dot{Q} = 0.01$ is shifted two decades to the left (because the $\dot{V}_A/\dot{Q}$ ratio is two decades lower). The dashed lines indicate that 50% retention in each unit will occur for a gas whose $\lambda = \dot{V}_A/\dot{Q}$ of the unit

retention-solubility curves, one for $\dot{V}_A/\dot{Q} = 1$ and one for $\dot{V}_A/\dot{Q} = 0.01$. As Eq. (1.20) predicts, the curve for $\dot{V}_A/\dot{Q} = 0.01$ has the exact same shape as that for $\dot{V}_A/\dot{Q} = 1$, but it is left-shifted by exactly two decades on the λ axis (because the $\dot{V}_A/\dot{Q}$ ratio is two decades less). Note that the "P_{50}" for each curve—the value of λ at which retention is 0.5—occurs when $\lambda = \dot{V}_A/\dot{Q}$.

One can draw any number of similar curves, each for a different $\dot{V}_A/\dot{Q}$ ratio, and they will all reflect the same outcome: left- or right-shifted with a P_{50} wherever $\lambda = \dot{V}_A/\dot{Q}$, but they will all have the same shape (i.e., can be completely superimposed by sliding them left or right).

What these curves are really saying is that there is a unique shape to the retention-solubility curve for a homogeneous lung or lung unit no matter what the $\dot{V}_A/\dot{Q}$ ratio and that the $\dot{V}_A/\dot{Q}$ ratio of any such homogeneous lung (unit) equals the value of λ at which retention ($P_a/P_{\bar{v}}$) is exactly 0.5.

2.2.1 *The Retention-Solubility Curve and $\dot{V}_A/\dot{Q}$ Inequality*

The next step is to consider a lung with $\dot{V}_A/\dot{Q}$ inequality. The simplest would be a two-compartment lung with each compartment having a different $\dot{V}_A/\dot{Q}$ ratio. Suppose we continued from Fig. 2.2 by taking the example of a lung with two compartments, one of $\dot{V}_A/\dot{Q}$ ratio 1 and the other with $\dot{V}_A/\dot{Q}$ ratio 0.01. The latter ratio could be the result of severe but not complete (otherwise $\dot{V}_A/\dot{Q}$ would equal zero) airway obstruction leading to reduced ventilation and thus a low $\dot{V}_A/\dot{Q}$ ratio, such as might result from an inhaled foreign object; the former ratio would represent the remaining unobstructed lung. Further, let us allocate equal blood flows to the two compartments—think of them as the right and left lungs with unilateral main stem bronchial obstruction on one side, if you wish. The normal side would produce a retention curve as shown in Fig. 2.2 for $\dot{V}_A/\dot{Q} = 1$, while the obstructed side would produce a retention curve as shown in Fig. 2.2 for the other compartment with $\dot{V}_A/\dot{Q} = 0.01$.

The key concept is that the effluent blood from the two lungs will be ducted back to the left atrium and will mix in proportion to their blood flows to produce a mixed arterial blood retention curve that is in this case simply the equally weighted average of the two curves (equally weighted because the blood flows are equal). Figure 2.3 shows what this equally weighted average curve will look like. *The critical observation is that the shape of this composite curve no long resembles the shape of the individual $\dot{V}_A/\dot{Q}$ compartment curves of any one $\dot{V}_A/\dot{Q}$ ratio.* This change in shape can mean only one thing—$\dot{V}_A/\dot{Q}$ inequality is present.

It was easy to perform the forward calculation shown in Fig. 2.3 given the $\dot{V}_A/\dot{Q}$ distribution (in this case simply two compartments of different $\dot{V}_A/\dot{Q}$ but equal blood flow), to determine the composite retention curve.

What MIGET is all about is measuring the composite curve and performing the inverse calculation to come up with the responsible $\dot{V}_A/\dot{Q}$ distribution.

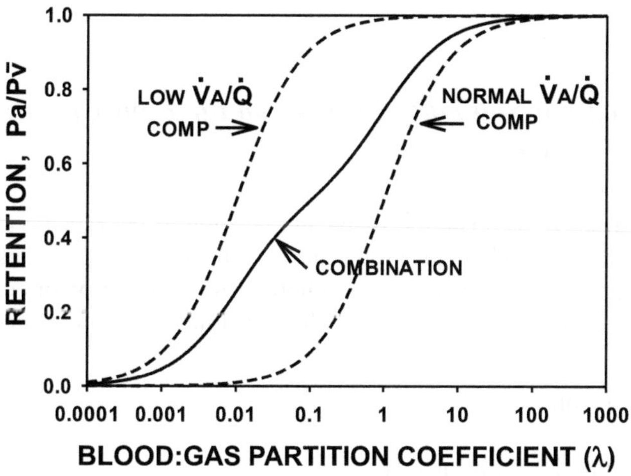

Fig. 2.3 Retention curve (solid line) for a two-compartment lung in which the two $\dot{V}_A/\dot{Q}$ ratios are 1.0 and 0.01 and where there is equal perfusion of each compartment. The retention curve is the equally weighted average of the individual retention curves of the two compartments (dashed lines). The important outcome is that the combined curve has a very different shape compared to either of the curves that contribute to it

2.2.2 *Choice of Gases to Use in Measuring the Retention Curve*

Figure 2.2 helps to identify the optimal gases we would want to use to best "tie down" the whole retention-solubility curve. Thus, the entire curve can be adequately described by a handful of gases selected to have well-spaced out values of λ between about 0.001 and 100. This is depicted in Fig. 2.4a by the six dots that indicate the values of λ of the six gases we have come to use in MIGET. Figure 2.4b shows on the same diagram where the three gases of Yokoyama and Farhi lie (methane, ethane, and nitrous oxide; see Sect. 1.3). It is clear that these three gases fail to span the entire range of the retention domain.

There are of course other factors that go into the choice of gases. These include (a) lack of toxicity (even though the gases are present in the blood in only parts per million concentration), (b) availability, and (c) ease of laboratory measurement. They should also not react chemically with any elements in the blood or lung tissues.

The gases we use are, in order of increasing solubility, sulfur hexafluoride (SF_6), ethane, cyclopropane, enflurane (or, more recently, isoflurane), diethyl ether, and acetone. They have solubilities corresponding to the points illustrated in Fig. 2.4, are available, have not shown any toxicity in the parts per million concentrations used in MIGET, and are easily measured, even when present together, by gas chromatography at these low concentrations (see later).

2.2.3 *Retention-Solubility Curves and the Four Causes of Hypoxemia*

It is very instructive to relate the shape and position of the retention curve to the type of gas exchange disturbance (i.e., causes of hypoxemia) that could be present when the lungs are not functioning as a homogeneous organ with normal overall $\dot{V}_A/\dot{Q}$. This is conveniently done by recalling that there are four primary causes of hypoxemia:

1. Hypoventilation
2. Shunt
3. Ventilation/perfusion inequality
4. Diffusion limitation

Assuming that each of the above four causes occurs singly, Fig. 2.5 shows, against the backdrop of the normal retention curve, how each disturbance would affect the retention curve. The outcomes are a translational shift, a change in asymptote (as solubility approaches zero), a change in shape, and no effect, respectively. These remarkably and uniquely different changes are readily

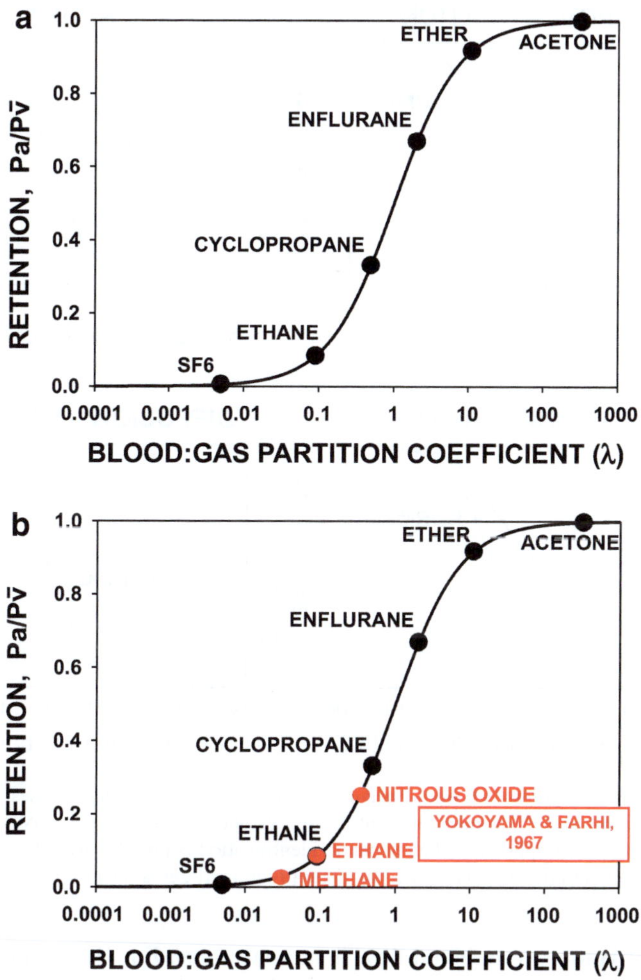

Fig. 2.4 (**a**) Retention curve for a homogeneous lung with $\dot{V}_A/\dot{Q} = 1.0$, with the retention and partition coefficient values for the six gases commonly used in MIGET indicated by the solid points. Note how the six gases capture essentially the entire range of the retention curve by virtue of their partition coefficient values. (**b**) Superposition of the three gases used by Yokoyama and Farhi in 1967 to estimate parameters of the $\dot{V}_A/\dot{Q}$ distribution. They are positioned to capture only a small portion of the entire retention curve

explained simply by inserting appropriate $\dot{V}_A/\dot{Q}$ values into Eq. (1.20) and solving (assuming no diffusion limitation for inert gases, to be discussed later).

Hypoventilation alone (i.e., in the absence of any other disturbance, thus implying a homogeneous lung) will simply *shift* the retention curve leftward. This reflects the logic and analysis surrounding Fig. 2.2 above, which will therefore not be repeated. All that has happened is that total $\dot{V}_A$ is below normal by definition

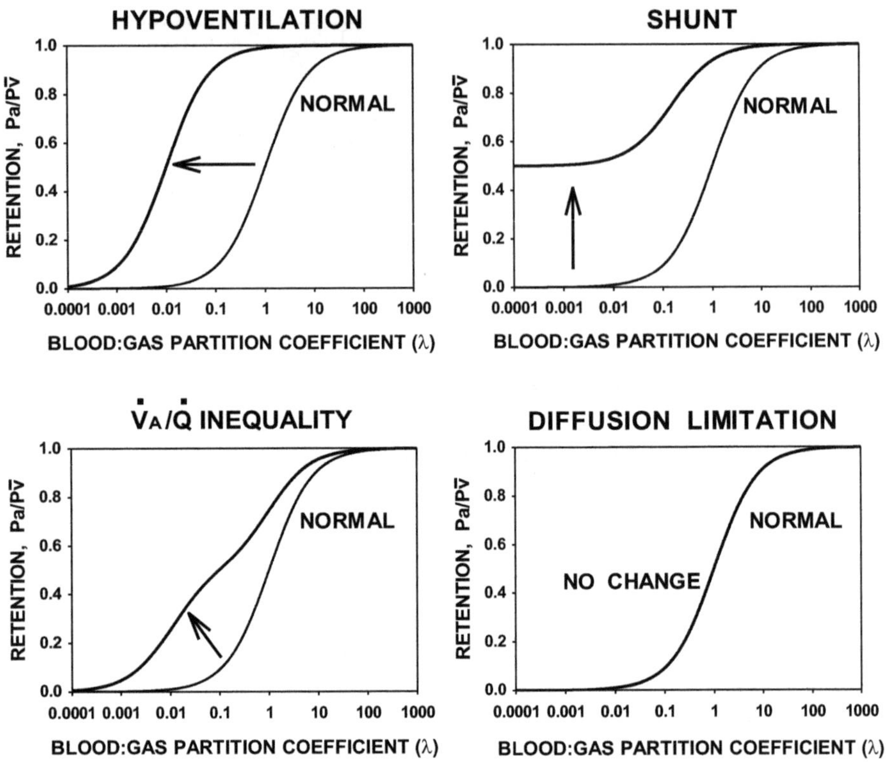

Fig. 2.5 Retention curves that would occur in each of the four classical causes of hypoxemia indicated in each panel. Note how different the shapes and positions of the curves are across all four conditions. This implies that their separate identification is possible by MIGET if the whole domain of the retention curves is defined by appropriate inert gases (as in Fig. 2.4a)

of hypoventilation, while blood flow is normal. Thus, Eq. (1.20) applies as for a homogeneous lung, but one in which overall $\dot{V}_A/\dot{Q}$ is reduced. The *shape* of the retention curve thus remains unaffected, and the *asymptote*, as solubility approaches zero, is zero, also as in the homogeneous lung.

Shunt results in compression of the retention curve upward but *no change in shape* or *shift* compared to normal. The *asymptote*, as solubility approaches zero, is the fractional shunt. In other words, suppose shunts were 30% of the cardiac output, retention would asymptotically approach 0.3 as solubility approached zero. Why does this happen? Reexamine Eq. (1.20). A shunt is a compartment without ventilation—that is with a $\dot{V}_A/\dot{Q}$ ratio of zero. From Eq. (1.20), retention will be complete (i.e., equal to 1.0) for *all* gases in such a unit. Suppose the lung contains just two units—a normal unit and a shunt compartment—and suppose that the blood flow through the shunt is 50% of the cardiac output. When the effluent blood from the two compartments flow together into the left atrium, the left atrial blood will display retention values for each gas that are the simple weighted average of their retentions in the two contributing blood streams. This can be set out in another equation as follows, abbreviating $P_a/P_{\bar{v}}$ to R (for retention):

$$R(\text{left atrium}) = 50\% \times R(\text{shunt}) + 50\% \times R(\text{normal}) \quad (2.1)$$

This equation simply expresses mass conservation when two streams mix. Thus, if the $\dot{V}_A/\dot{Q}$ ratio in the normal unit were 1.0,

$$R(\text{left atrium}) = 0.5 \times 1.0 + 0.5 \times \lambda/[\lambda + 1.0] \quad (2.2)$$

Thus, R (left atrium) equals a constant (0.5 in this example) plus a λ-dependent component that has the normal configuration of Eq. (1.20) but contributes only 50% of the weight to the relationship. It should be clear that the lowest value of R in the left atrium will occur as λ approaches zero and cannot therefore be less than 0.5 (in this example). As λ approaches infinitely high values, it should be clear that R (left atrium) approaches 1.0, which is exactly the same as in a homogeneous lung. The bottom line for shunt is that the retention curve will asymptote at a value equal to the shunt fraction as λ approaches zero but have the same shape as the retention curve in a homogeneous lung.

Ventilation/perfusion inequality will *change the shape* of the retention curve but not cause a *shift* or change in *asymptote*, as already described in Fig. 2.3. The particular shape change will depend on the quantitative nature of the pattern of $\dot{V}_A/\dot{Q}$ inequality. Equation (1.20) again underlies the concept as follows: the simplest pattern of $\dot{V}_A/\dot{Q}$ inequality is a two-compartment system, and the definition of $\dot{V}_A/\dot{Q}$ inequality is simply the existence of compartments with different $\dot{V}_A/\dot{Q}$ ratios. Suppose a two-compartment lung has 50% of its blood flow distributed to each compartment, but that one compartment has only 1% of the ventilation while the other therefore receives 99%. Further suppose that total ventilation and total lung blood flow each equal 6 l/min. In this particular example, the poorly ventilated compartment has ventilation of 0.06 l/min, 3 l/min blood flow and thus a $\dot{V}_A/\dot{Q}$ ratio of 0.06/3, or 0.02. The other compartment has ventilation of 5.94 l/min, blood flow of 3 l/min, and a $\dot{V}_A/\dot{Q}$ ratio of 5.94/3, or 1.98. We use the identical concept as for shunt above: the left atrial blood is a weighted average of blood draining from the two compartments. Based on Eq. (1.20), we have:

$$\begin{aligned} R(\text{left atrium}) = 0.5 \times R(&\text{poorly ventilated unit}) + 0.5 \\ &\times R(\text{well ventilated unit}) \end{aligned} \quad (2.3)$$

thus:

$$R(\text{left atrium}) = 0.5 \times \lambda/[\lambda + 0.02] + 0.5 \times \lambda/[\lambda + 1.98] \quad (2.4)$$

Figure 2.5 is very instructive in pointing out the *huge* differences in the retention curve between lungs having a shunt compared to lungs having very low $\dot{V}_A/\dot{Q}$ areas.

Table 2.1 Effects of gas exchange disturbances on retention curves

Gas exchange disturbance	Curve shift	Shape change	Asymptote value
Hypoventilation	+	–	Zero
Shunt	–	–	Increased
$\dot{V}_A/\dot{Q}$ inequality	–	+	Zero
Diffusion limitation	–	–	Zero

Importantly, O_2 does *not* display such differences—P_{O_2} in low $\dot{V}_A/\dot{Q}$ areas and P_{O_2} of shunted blood are very similar (breathing room air).

The general equation for retention emanating from such a two-compartment model of $\dot{V}_A/\dot{Q}$ mismatch is therefore:

$$R(\text{left atrium}) = \left[(\dot{Q}_1 \times R_1) + (\dot{Q}_2 \times R_2)\right] / \left[(\dot{Q}_1 + \dot{Q}_2)\right] \quad (2.5)$$

where $R_1 = \lambda/\left[\lambda + \dot{V}_{A_1}/\dot{Q}_1\right]$ and $R_2 = \lambda/\left[\lambda + \dot{V}_{A_2}/\dot{Q}_2\right]$

Diffusion limitation (affecting O_2) *will not change shape or asymptote* and *will not cause a shift* in the retention curve—diffusion limitation is without effect on inert gases. It is possible to think of diffusion limitation severe enough to affect inert gases, but because inert gases are an order of magnitude less vulnerable to diffusion limitation than is O_2, a person would be fatally hypoxic from such a severe level of diffusion limitation. Thus, in practice, inert gases are not measurably affected by diffusion limitation, even when severe enough to affect O_2.

The qualitative effects on retention curves that result from the four basic gas exchange disturbances are summarized in Table 2.1.

2.2.4 Retention and, Now, Excretion

Up to this point, the discussion has been limited to the concept of retention. Retention was defined as the ratio of systemic arterial inert gas partial pressure to pulmonary arterial (mixed venous) partial pressure: $P_a/P_{\bar{v}}$. When the lung is homogeneous such that there is but a single $\dot{V}_A/\dot{Q}$ ratio present, the alveolar/mixed venous partial pressure ratio ($P_A/P_{\bar{v}}$) equals the systemic arterial/mixed venous partial pressure ratio ($P_a/P_{\bar{v}}$) simply because under the assumption of no diffusion limitation for inert gases, $P_A = P_a$. In such a homogeneous lung, if expired alveolar gas was collected and its partial pressure, P_A, measured, we could accordingly use the resulting ratio $P_A/P_{\bar{v}}$ instead of the more difficult to measure ratio $P_a/P_{\bar{v}}$ which requires an arterial blood sample.

Just as $P_a/P_{\bar{v}}$ is termed "retention," $P_A/P_{\bar{v}}$ is termed "excretion." However, since we collect mixed expired gas when applying the MIGET and not expired alveolar (end-tidal) gas (because the latter is problematic when $\dot{V}_A/\dot{Q}$ mismatch exists), *we* *have come to use the term "excretion" to reflect the ratio of mixed expired gas partial pressure (P_E) to that in mixed venous blood: $P_E/P_{\bar{v}}$.*

In real lungs, where there is always some $\dot{V}_A/\dot{Q}$ inequality, P_A does not equal P_a, and so the strategy of replacing $P_a/P_{\bar{v}}$ by $P_A/P_{\bar{v}}$ will not work. Why is P_A not equal to P_a in lungs with $\dot{V}_A/\dot{Q}$ inequality? The explanation is as follows:

Retention, being the arterial blood/mixed venous ratio, reflects a mixture of all the blood draining from the many different alveoli and is therefore a *blood flow-weighted* average of the individual alveolar gas tensions. Excretion on the other hand, being the mixed expired/mixed venous ratio, reflects a mixture of all the exhaled gas coming from the many different alveoli—the very same alveoli as for blood flow of course—but in this case, the resulting mixed expired gas is a *ventilation-weighted* average of the individual alveolar gas tensions. Thus, mixed arterial inert gas partial pressures will always exceed those of mixed expired gas. This is analogous to arterial P_{CO_2} exceeding mixed expired P_{CO_2}.

In any lung (homogeneous or one with $\dot{V}_A/\dot{Q}$ inequality), retention and excretion are related mathematically by the laws of mass conservation, commonly referred to as the Fick principle. For inert gases undergoing elimination, the Fick principle looks like this:

$$\dot{V}_E \times P_E = \lambda \times \dot{Q}_T \times \left[P_{\bar{v}} - P_a\right] \tag{2.6}$$

Here, $\dot{V}_E$ is minute ventilation and $\dot{Q}_T$ is cardiac output. P_E is mixed expired, P_a is mixed systemic arterial, and $P_{\bar{v}}$ is mixed venous partial pressures of the inert gas. Remember, this mass conservation equation is true no matter whether the lungs are homogeneous or not. Diving throughout by $P_{\bar{v}}$ gives:

$$\dot{V}_E \times P_E/P_{\bar{v}} = \lambda \times \dot{Q}_T \times \left[1 - P_a/P_{\bar{v}}\right] \tag{2.7}$$

$$\dot{V}_E \times \text{Excretion} = \lambda \times \dot{Q}_T \times [1 - \text{Retention}] \tag{2.8}$$

Note several things here:

1. Excretion is NOT simply 1 − Retention. Only in the special case that $\lambda = \dot{V}_E/\dot{Q}_T$ will excretion equal 1 − Retention. For all other λ, that is, for the general case, excretion does not equal 1 − Retention.
2. If you measure Retention, λ, $\dot{V}_E$ and $\dot{Q}_T$ you can calculate Excretion, and vice versa if you know Excretion, you can calculate Retention. Thus they do NOT give independent information.
3. If you measure BOTH Retention and Excretion together, you have in effect two duplicate sources of information about inert gas exchange—one from the vascular side and one from the airway side. Duplicates reduce variance and improve data quality.
4. While these two sources in **theory** each hold the same information about gas exchange, in **reality**, this is not the case because of random experimental error. As will be described later in much detail, Retention data are better suited to

assessing areas of low $\dot{V}_A/\dot{Q}$ ratio, while Excretion data are better for assessing areas of high $\dot{V}_A/\dot{Q}$ ratio. This is related to how random experimental error affects the variance of retention and excretion differently.

What this means in practice is that the best and most complete information in MIGET comes from simultaneously measuring both Retention and Excretion, even though in theory this gives duplicate information.

The preceding discussion has provided several examples of retention curves. But what do excretion curves then look like?

In the **homogeneous lung**, one would expect that the excretion curve should be exactly the same as the retention curve, because $P_A = P_a$. This would be true if there was no such thing as anatomical deadspace. However, because of deadspace, the mixed expired gas is an air-diluted sample of the alveolar gas (for each of the six gases). This is the case because the deadspace is assumed not to take part in exchange of any gases with the blood and, thus after each inspiration, contains only fresh inspired air that is free of inert gases. Thus, just as for CO_2 elimination, mixed expired gas is diluted with room air. Thus, the dilution has the same quantitative effect on every gas.

As an example, suppose the tidal volume is 500 ml and the anatomic deadspace volume is 150 ml, both normal values for healthy humans. Exhaled alveolar gas is therefore 350 ml in volume, and the mixed expired gas is a volume-weighted average of the two:

$$P_E = [P_A \times 350 + \text{Pdeadspace} \times 150]/500 \tag{2.9}$$

As stated, Pdeadspace is zero (the inhaled gas contains no inert gases). Thus:

$$P_E = [P_A \times 350]/500 = 0.7 \times P_A \tag{2.10}$$

This is true for every gas, as stated above, so that the excretion curve has the same shape as the retention curve, but for every gas it is situated 30% lower, as in Fig. 2.6. If the deadspace were 25% of tidal volume, the excretion curve would asymptote at a value of 0.75; if the deadspace were 35%, the excretion curve would asymptote at 0.65, etc.

Note the symmetry between deadspace and shunt: both are regions that exhibit zero gas exchange, and both result in a "squeezing" of the inert gas curves [upward squeezing of retention for shunt; downward squeezing of excretion for deadspace, in each case simply representing dilution of the alveolar gas value by the incoming gas stream (mixed venous for retention, inspired for deadspace)]. For both, there is no fundamental shape change to the curves compared to normal.

The one point of asymmetry of course is that shunt does not normally exist but deadspace is normal, due to the tidal nature of ventilation combined with existence of the conducting airway tree.

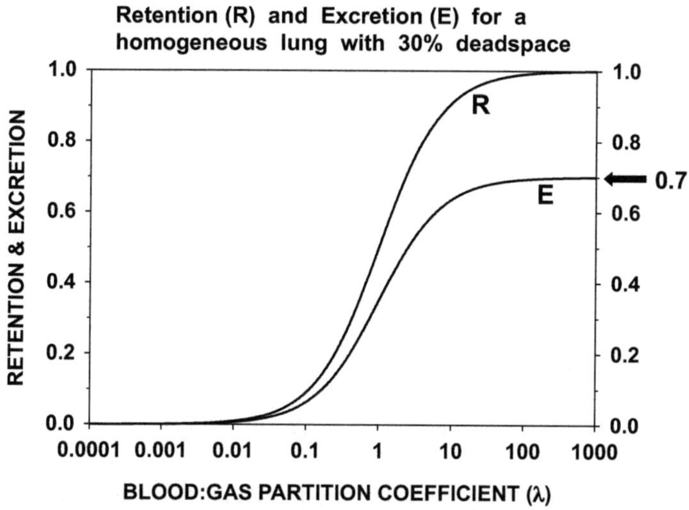

Fig. 2.6 Addition of the excretion curve plot to that of retention. Whereas retention is the blood flow-weighted average of all alveolar inert gas tensions (defined as systemic arterial/mixed venous tension ratio), excretion is the ventilation-weighted average of the same set of all alveolar gas tensions and is defined as the mixed expired/mixed venous tension ratio. In this figure, the diluting effect of conducting airway deadspace ventilation (identical for all six gases) is evident for the particular value of deadspace (30% of tidal volume), reducing excretion values to 70% of retention values in this example of a homogeneous lung of $\dot{V}_A/\dot{Q} = 1.0$

Finally, we can apply the same rules to calculation of excretion, E, in a nonhomogeneous lung as we did for retention in Eq. (2.5) above. In the simplest example of a two-compartment lung:

$$E = \left[\left(\dot{V}_{A_1} \times E_1\right) + \left(\dot{V}_{A_2} \times E_2\right)\right] / \left[\dot{V}_{A_1} + \dot{V}_{A_2}\right] \tag{2.11}$$

where $E_1 = \lambda/\left[\lambda + \dot{V}_{A_1}/\dot{Q}_1\right]$ and $E_2 = \lambda/\left[\lambda + \dot{V}_{A_2}/\dot{Q}_2\right]$.

In words, excretion is the ventilation-weighted mean of the alveolar gas tensions of the two compartments; retention is the blood flow-weighted mean of the endcapillary gas tensions of the same two compartments; remember that endcapillary tension equals alveolar in each compartment because of lack of diffusion limitation (such that $R_1 = E_1$ and $R_2 = E_2$).

2.2.5 The R–E Difference

Note: $R_1 = E_1$; $R_2 = E_2$; but R does not equal E. That conclusion comes from looking at Eqs. (2.5) and (2.11) and comparing them. Except for the (theoretical) homogeneous lung *without* anatomical deadspace, R exceeds E during inert gas elimination, much as alveolar P_{O_2} exceeds arterial P_{O_2} and creates an alveolar-arterial gas tension difference.

We are going to get a bit mathematical here, but an interesting result is seen on further analysis of *R–E* from Eqs. (2.5) and (2.11). Recall from Eq. (2.5):

$$R = \left[\dot{Q}_1 \times \lambda/\left(\lambda + \dot{V}_{A_1}/\dot{Q}_1\right) + \dot{Q}_2 \times \lambda/\left(\lambda + \dot{V}_{A_2}/\dot{Q}_2\right)\right]/\left[\dot{Q}_1 + \dot{Q}_2\right]$$

While from Eq. (2.11), we have:

$$E = \left[\dot{V}_{A_1} \times \lambda/\left(\lambda + \dot{V}_{A_1}/\dot{Q}_1\right) + \dot{V}_{A_2} \times \lambda/\left(\lambda + \dot{V}_{A_2}/\dot{Q}_2\right)\right]/\left[\dot{V}_{A_1} + \dot{V}_{A_2}\right]$$

If you subtract *E* from *R* and differentiate ($R-E$) with respect to λ, after much algebra, you get the interesting result that the retention-excretion difference is *greatest* for a gas whose:

$$\lambda = \text{square root}\left[\dot{V}_{A_1}/\dot{Q}_1 \times \dot{V}_{A_2}/\dot{Q}_2\right] \quad (2.12)$$

What is this saying? It says that for any particular two-compartment lung, *the gas whose exchange is most impaired in that lung is a gas whose blood/gas partition coefficient equals the geometric mean of the* $\dot{V}_A/\dot{Q}$ *ratios of the two compartments.* By implication, different gases are differently impaired in any given lung. *Gases of both higher and lower* λ *than given by Eq. (2.12) will be less affected.*

Suppose a two-compartment lung had one normal $\dot{V}_A/\dot{Q}$ ratio ($\dot{V}_{A_1}/\dot{Q}_1$) and one abnormal $\dot{V}_A/\dot{Q}$ ratio ($\dot{V}_{A_2}/\dot{Q}_2$). For simplicity, let us suppose the normal $\dot{V}_A/\dot{Q}$ ratio equals 1.0, which approximates the usual case. Then, from Eq. (2.12):

$$\lambda, \text{most affected gas} = \text{square root}\left[V_{A_2}/\dot{Q}_2\right] \quad (2.13)$$

which can be turned around to state that if the value of λ **for the most affected gas can be established experimentally,** *the square of* λ *is numerically equal to the* $\dot{V}_A/\dot{Q}$ *ratio of the abnormal* $\dot{V}_A/\dot{Q}$ *compartment—in a two-compartment lung!*

Figure 2.7 shows this relationship (i.e., *R–E*) for a two-compartment lung whose $\dot{V}_A/\dot{Q}$ ratios are 0.01 and 1.0 and which are equally perfused, the same model as shown in Fig. 2.3. The top panel repeats Fig. 2.3 and shows retention as a 50:50 average of the retention values for each compartment. The middle panel shows excretion which is a 1:99 average of the same two retention values. Why the very different weighting? Because if perfusion is equally distributed, ventilation of the low $\dot{V}_A/\dot{Q}$ compartment must be very low, while that of the other compartment must be relatively high. Specifically, to have a $\dot{V}_A/\dot{Q}$ ratio of 0.01 in the low $\dot{V}_A/\dot{Q}$ unit requires that ventilation be 1/100th of its blood flow. In the other unit, ventilation equals blood flow and is thus 100 times greater than in the first unit. Thus, the excretion curve of the two-compartment lung virtually overlies the individual excretion curve of the normal $\dot{V}_A/\dot{Q}$ unit as the middle panel shows, because 99% of the exhaled signal comes from the normal unit's curve. Finally, the retention-excretion difference is plotted in the lower panel to show that indeed, R-E is greatest for a gas whose $\lambda = 0.1$, the geometric mean of 0.01 and 1.0. The $\dot{V}_A/\dot{Q}$ ratio of the abnormal (low) $\dot{V}_A/\dot{Q}$ compartment is given by the square of this

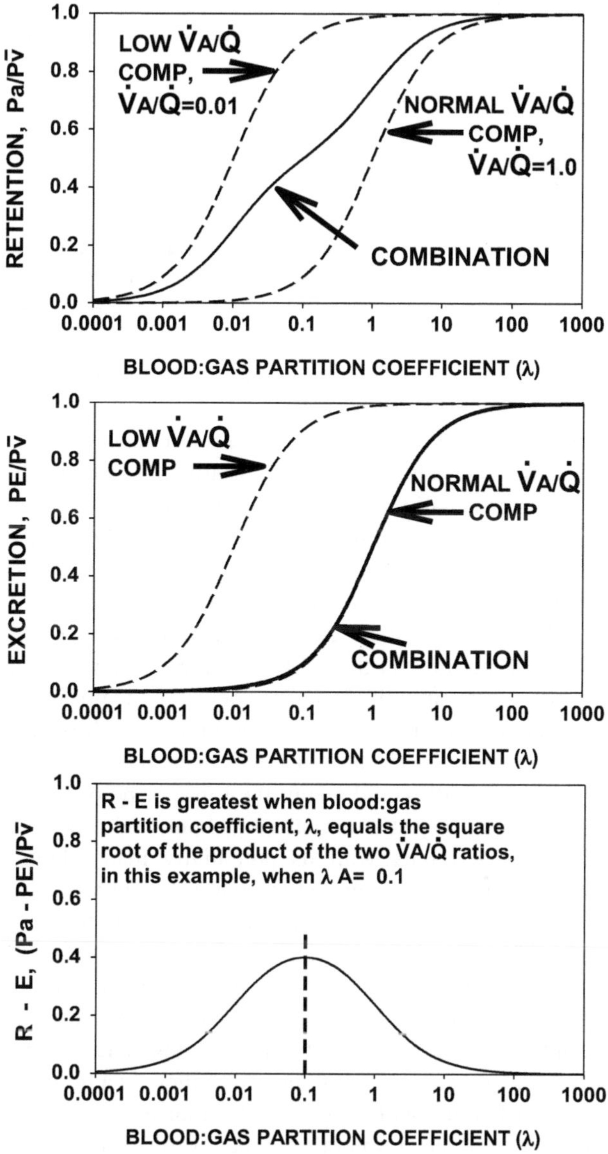

Fig. 2.7 The two-compartment lung from Fig. 2.6 (retention shown in top panel) now incorporates excretion curves from the same lung (middle panel). Excretion curves are dominated by the normal $\dot{V}_A/\dot{Q}$ compartment values since $\dot{V}_A$ is so low in the low $\dot{V}_A/\dot{Q}$ unit, making the combined excretion curve almost overlie that of the normal $\dot{V}_A/\dot{Q}$ curve. The difference between retention and excretion at each partition coefficient along the X axis (lower panel) is analogous to the alveolar-arterial P_{O_2} difference. Note that in this example, gases of partition coefficient = 0.1 have the greatest R–E difference (i.e., show the greatest interference to their elimination) while gases of both lower and higher solubility show lower differences

value of λ and in this case comes to 0.01. Note that in this case, anatomic deadspace is taken out of the calculations to keep things simple.

2.3 The $\dot{V}_A/\dot{Q}$ Distribution

Figure 1.2 introduced the concept of the $\dot{V}_A/\dot{Q}$ distribution as a way to represent the lung by plotting ventilation ($\dot{V}_A$) and blood flow ($\dot{Q}$) separately against $\dot{V}_A/\dot{Q}$ ratio. Why is this the most useful way of depicting the distribution? Equations (2.5) and (2.11) provide the answer by showing how retention and excretion for any gas of known λ are determined only by compartmental $\dot{V}_A$, $\dot{Q}$ and their ratio $\dot{V}_A/\dot{Q}$. Thus, gas exchange in general (be it for inert gases, O_2 or CO_2) is determined by local lung unit ("compartment") $\dot{V}_A/\dot{Q}$ ratios and the amounts of ventilation and blood flow associated with each such lung unit. Pictorially, this can best be shown in the format of Fig. 1.2, because all three variables—$\dot{V}_A$, $\dot{Q}$, and their ratio $\dot{V}_A/\dot{Q}$—are drawn into a single pair of functions: $\dot{V}_A$ against $\dot{V}_A/\dot{Q}$ and $\dot{Q}$ against $\dot{V}_A/\dot{Q}$.

Figure 1.2 was used to picture the Riley three-compartment model, i.e., where a "normal" $\dot{V}_A/\dot{Q}$ unit, an unventilated ("shunt") unit, and an unperfused ("deadspace") unit together explain arterial P_{O_2} and mixed expired P_{CO_2}. The variables in this construct are (1) the split of ventilation between the normal and deadspace units and (2) the split of blood flow between the normal and shunt units.

The real lung may contain thousands of such lung units. It is inconceivable that each would have the same $\dot{V}_A/\dot{Q}$ ratio just as it is inconceivable that in a class of many thousands, students all would weigh the same or be of the same height. Or that the real lung consists of just a normal unit, shunt, and deadspace. So how can we generalize to the real lung with its potentially many thousands of units? Consider the more familiar example: distribution of height within a student class.

The most efficient way to present the array of heights in this large class is by a frequency distribution exactly similar to Fig. 1.2. Numbers of students in each of many discrete height ranges would be added up and the total number of students in each such range plotted on the Y axis against the mean height in that range on the X axis. This would be done across the entire height range encountered from shortest to tallest. When the class size is several thousand, there is a high probability that a great many different heights will be seen and, more importantly, that with these large numbers there will be a smooth frequency distribution across the range encountered. The expected shape of this distribution might be normal or skewed, but it would be smooth, with most people being of similar "average" height and progressively fewer being very much shorter or taller than average.

This is precisely how we think of the $\dot{V}_A/\dot{Q}$ distribution: a very large number of units of different $\dot{V}_A/\dot{Q}$ ratio (=height) might exist, most near the mean and fewer on either side. With thousands of units potentially existing, we expect a smooth curve just as in the height example above. As with the height analogy, we would sum the blood flows in all units having a $\dot{V}_A/\dot{Q}$ ratio within a narrow interval and plot that sum at the mean $\dot{V}_A/\dot{Q}$ ratio of that interval. That summation process would be done across the entire range of $\dot{V}_A/\dot{Q}$ ratios from zero (i.e., shunt) to infinite (i.e., deadspace). In practice, as we shall see, the MIGET can separate shunt from regions whose $\dot{V}_A/\dot{Q}$ ratio is as low as 0.005, but regions with lower $\dot{V}_A/\dot{Q}$ ratio cannot be distinguished from shunt and would be called shunt. Correspondingly, any units whose $\dot{V}_A/\dot{Q}$ ratio is greater than 100 cannot be separated from units that have no blood flow (i.e., deadspace) and would be called deadspace.

It turns out that in the present case of looking at $\dot{V}_A/\dot{Q}$ ratios, using a logarithmic $\dot{V}_A/\dot{Q}$ axis is more useful than using a linear axis for $\dot{V}_A/\dot{Q}$. It also turns out that chopping the $\dot{V}_A/\dot{Q}$ axis between the lower and upper limits defined above (0.005 and 100) into 50 intervals works well. So too would having 60 or 40 intervals, etc. Some choice has to be made, and years of experience have settled on 50. The problem with fewer than 50 is that the intervals become too coarse and information can be lost. The problem with too many intervals is simply technical, not physiological: the greater the number of intervals, the greater is the computing requirement of the MIGET in both number crunching and array storage. The way we put it, we are approximating a continuous depiction of the $\dot{V}_A/\dot{Q}$ axis by having a large enough number of fine but discrete intervals ($\dot{V}_A/\dot{Q}$ ratios) to be able to simulate a continuous curve but without requiring large arrays or presupposing algebraically specified truly continuous functions that may or may not be appropriate.

Figure 2.8 shows a $\dot{V}_A/\dot{Q}$ distribution using 50 compartments. Note that the 50 compartments are represented by the circles. The solid lines joining the circles are best-fit spline functions to better visualize the shape and position of the data. It

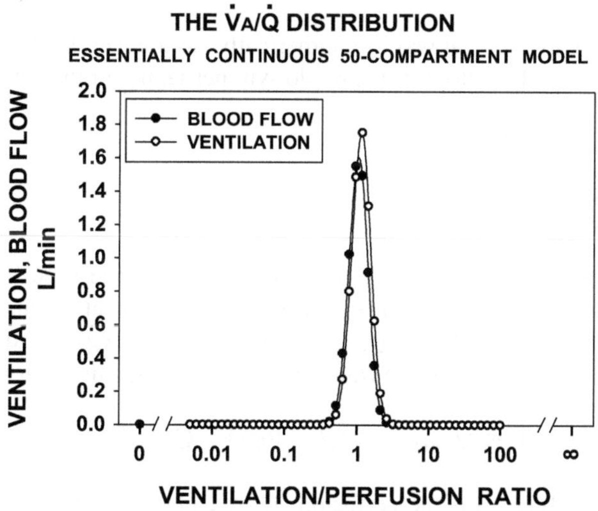

Fig. 2.8 The $\dot{V}_A/\dot{Q}$ distribution plot. Here, both ventilation $(\dot{V}_A)$ and blood flow $(\dot{Q})$ are plotted on the ordinate against their corresponding ratio, $\dot{V}_A/\dot{Q}$, to provide a frequency distribution of how $\dot{V}_A$ and $\dot{Q}$ are distributed across the entire $\dot{V}_A/\dot{Q}$ domain from $\dot{V}_A/\dot{Q} = 0$ to $\dot{V}_A/\dot{Q} =$ infinity. In this example, which comes from a normal lung, the distribution is narrow, with no $\dot{V}_A/\dot{Q}$ ratio units of less than about 0.5 or greater than about 3 appearing. There is no shunt. Deadspace (ventilation of unperfused lung, $\dot{V}_A/\dot{Q} =$ infinite) is too large to appear without compressing the appearance of the rest of the distribution and has been omitted

is of course apparent that a $\dot{V}_A/\dot{Q}$ distribution is in fact TWO distributions—one for ventilation (open circles) and one for blood flow (closed circles). However, since $\dot{V}_A/\dot{Q} = \dot{V}_A$ divided by $\dot{Q}$, we can write:

$$\dot{V}_A = \dot{Q} \times \dot{V}_A/\dot{Q} \tag{2.14}$$

What this means is that if we know the distribution of $\dot{Q}$ against $\dot{V}_A/\dot{Q}$, we must also know the distribution of $\dot{V}_A$ against $\dot{V}_A/\dot{Q}$ from Eq. (2.14). Thus, the two curves in Fig. 2.8 are really a single curve in terms of information—each can be used to compute the other when one is known. Well, almost true: we cannot compute deadspace from the blood flow curve, or, symmetrically, shunt from the ventilation curve. Note that from Eq. (2.14), $\dot{V}_A = \dot{Q}$ only when $\dot{V}_A/\dot{Q} = 1.0$. Thus, the two curves will cross at that $\dot{V}_A/\dot{Q}$ ratio as the figure shows. In addition, at $\dot{V}_A/\dot{Q}$ ratios less than 1.0, $\dot{Q} > \dot{V}_A$ and at $\dot{V}_A/\dot{Q}$ ratios greater than 1.0, $\dot{V}_A > \dot{Q}$, as the figure also shows.

The pictorial representation of the $\dot{V}_A/\dot{Q}$ distribution is important and useful. However, we need to find ways to parameterize these curves by assigning numbers to the basic characteristics of the curves. These are now discussed.

2.4 Parameters of the $\dot{V}_A/\dot{Q}$ Distribution

We can, and do, use several ways to numerically summarize the distributions, and what we do for the blood flow curve, we do symmetrically for the ventilation curve.

2.4.1 Moments

We calculate moments, continuing to employ the log scale used in the pictorial form in Fig. 2.8. The *first moment* is simply the $\dot{V}_A/\dot{Q}$ ratio at the mean position of the curve, and we have called it Q_{mean} for the blood flow curve and VA_{mean} for the ventilation curve. The formula to calculate it for the blood flow curve is:

$$\text{Log}\left[Q_{\text{mean}}\right] = \Sigma\left[\dot{Q} \times \log\left(\dot{V}_A/\dot{Q}\right)\right]/\Sigma\left[\dot{Q}\right] \tag{2.15}$$

In words, Log[Q_{mean}] is the perfusion ($\dot{Q}$)-weighted mean of the logarithm of each of the 48 $\dot{V}_A/\dot{Q}$ ratios other than shunt and deadspace. We cannot include shunt $\left(\dot{V}_A/\dot{Q} = 0\right)$ and deadspace ($\dot{V}_A/\dot{Q}$ infinity) since their logarithms are indeterminate. Then, one just takes the antilog of the result, and that number equals the value of the mean $\dot{V}_A/\dot{Q}$ ratio of the perfusion distribution. If the blood flow distribution is perfectly symmetrical (still on a log scale), the value of Q_{mean} coincides with the peak of the distribution curve. If the curve is not symmetrical, but rather is skewed left or right, the value of Q_{mean} will lie left or right of the peak of the curve but is still a useful outcome parameter for comparison purposes.

For ventilation, we use the same equation except the weighting is by ventilation, again using only the 48 compartments excluding shunt and deadspace as above:

$$\text{Log}\left[V_{A_{mean}}\right] = \Sigma\left[\dot{V}_A \times \log(\dot{V}_A/\dot{Q})\right]/\Sigma\left[\dot{V}_A\right] \tag{2.16}$$

Having computed the first moments, Q_{mean} and $V_{A_{mean}}$, we are positioned to compute the *second moments about the mean.* We have called these moments LogSD,Q and LogSD,V_A, respectively, for the blood flow and ventilation curves. These parameters indicate dispersion either side of the mean, hence the abbreviation "SD" in the terms, implying standard deviation. The higher the value of these two parameters, the greater the dispersion, or $\dot{V}_A/\dot{Q}$ mismatch. If the curves are symmetrical about the mean (on a log scale), the number obtained has the usual meaning of standard deviation. In addition, when perfectly symmetrical, LogSD,Q = LogSD,V_A. If the curve is not symmetrical, the value obtained is still very useful in characterizing the amount of $\dot{V}_A/\dot{Q}$ mismatch, but it is technically wrong to label it "standard deviation."

The formula for the second moment (M_2) for the blood flow curve is:

$$MQ = \Sigma\left[\dot{Q} \times \left[\log(Q_{mean}) - \log(\dot{V}_A/\dot{Q})\right]^2\right]/\Sigma\left[\dot{Q}\right] \tag{2.17}$$

And then:

$$\text{LogSD},Q = \text{Square root of } MQ$$

In words, MQ is the blood flow ($\dot{Q}$)-weighted average of the 48 squared differences between the logs of the mean $\dot{V}_A/\dot{Q}$ (Q_{mean}) and each of the 48 $\dot{V}_A/\dot{Q}$ ratios. Again, we exclude shunt and deadspace for the same reasons as above. Except for the use of logarithms, Eq. (2.17) is essentially the same equation we all use in statistics to compute the standard deviation about a mean.

For ventilation, the process is identical:

$$MV_A = \Sigma\left[\dot{V}_A \times \left[\log(V_{A_{mean}}) - \log(\dot{V}_A/\dot{Q})\right]^2\right]/\Sigma\left[\dot{V}_A\right] \tag{2.18}$$

And then:

$$\text{LogSD},V_A = \text{Square root of } MV_A$$

Finally, we include a calculation of the *third moment* about the mean for both ventilation and blood flow, and we call it Q_{skew} for blood flow and $V_{A_{skew}}$ for ventilation. It reflects asymmetry or skewing of the distribution. The formulae are very similar to Eqs. (2.17) and (2.18). The only differences are in raising the functions to the third power rather than second power. For blood flow it is:

$$MQ_3 = \Sigma\left[\dot{Q} \times \left[\log\left(Q_{\text{mean}}\right) - \log\left(\dot{V}_\text{A}/\dot{Q}\right)\right]^3\right]/\Sigma\left[\dot{Q}\right] \tag{2.19}$$

And then:

$$Q_{\text{skew}} = \text{Cube root of } MQ_3$$

For ventilation, it is:

$$MV_{\text{A3}} = \Sigma\left[\dot{V}_\text{A} \times \left[\log\left(V_{\text{A}_{\text{mean}}}\right) - \log\left(\dot{V}_\text{A}/\dot{Q}\right)\right]^3\right]/\Sigma\left[\dot{V}_\text{A}\right] \tag{2.20}$$

And then:

$$V_{\text{A}_{\text{skew}}} = \text{Cube root of } MV_{\text{A3}}$$

As before, this computation is done over the 48 compartments only, excluding shunt and deadspace. Note that the second moment, being a square root, must always be greater than or equal to zero, but that the third moment, being a cube root, can be positive, zero, or negative. The meaning of a negative skew number is the direction of skew (to the left or right of mean).

Table 2.2 indicates the common ranges of the first three moments in actual practice. Ranges for Q_{mean} and $V_{\text{A}_{\text{mean}}}$ are approximations, as are ranges for Q_{skew} and $V_{\text{A}_{\text{skew}}}$. Only LogSD,$Q$ and LogSD,V_A have been formally evaluated in terms of normal range [1]. Also, for skew, the numbers indicated could be positive or negative as just explained, although in the figure, for simplicity, the negative signs have been omitted. As far as we know, the range for skew is similar whether left- or right-skewed (i.e., whether positive or negative).

Table 2.2 Range of values of the first three moments

Moment	Value in homogeneous lung	Normal range	Moderately abnormal	Severely abnormal
Q_{mean}	0.8–1.0	0.8–1.0	0.6–0.8	<0.6
$V_{\text{A}_{\text{mean}}}$	0.8–1.0	0.8–1.0	1.0–2.0	>2.0
LogSD,Q	0	0.30–0.60	0.60–1.50	>1.50
LogSD,V_A	0	0.30–0.65	0.65–1.50	>1.50
Q_{skew}	0	0	−1 to −2; +1 to +2	<−2; >+2
$V_{\text{A}_{\text{skew}}}$	0	0	−1 to −2; +1 to +2	<−2; >+2

2.4.2 *Shunt and Deadspace*

The moments explained above cannot include shunt perfusion or deadspace ventilation because one cannot take the logarithms of 0 ($\dot{V}_A/\dot{Q}$ of shunt) or deadspace ($\dot{V}_A/\dot{Q}$ of infinity). The MIGET is however able to separately determine shunt and deadspace as should be evident from Figs. 2.5 and 2.6. This is because while we cannot use logs, the basic equation for inert gas exchange, Eq. (1.20), is perfectly able to incorporate a $\dot{V}_A/\dot{Q}$ of either zero or infinity. Thus, recall that for $\dot{V}_A/\dot{Q} = 0$, retention is 1.0 for gases of any solubility, while for $\dot{V}_A/\dot{Q}$ of infinity retention is zero for gases of any solubility.

Thus, it is inherent in the MIGET that by allowing the possibility of these two extreme $\dot{V}_A/\dot{Q}$ ratios to exist, both shunt and deadspace can be determined. This should become clearer in the section that follows dealing with the mathematics of the MIGET.

Shunt and deadspace are usually reported as fractions of total cardiac output and total ventilation (or, equivalently, of tidal volume), respectively. Of course, shunt is normally zero so that any shunt present is considered abnormal. For deadspace, the normal conducting airways of the lung create deadspace because in essence there is no steady-state exchange of gases across the walls of these airways. For gas exchange purposes, these airways are thus considered to be ventilated but not perfused. As explained above, conducting airway deadspace ventilation is normally some 30% of total ventilation (or tidal volume) because airway volume is about 150 ml and tidal volume is normally about 500 ml. Thus, 150/500 or 30% of each breath sits in the conducting airways not contributing to gas exchange.

It is worth pointing out however that expressing deadspace as a % of tidal volume is not always adequate. We often convert deadspace to an absolute volume in ml per breath (simply by multiplying fractional deadspace by tidal volume). This helps when tidal volume is higher or lower than normal because we know that deadspace in ml/lb body weight is about 1, or about 150 ml total, in an average person. Thus, when absolute deadspace in ml/breath exceeds (lean) body weight in lb, it suggests that there is more deadspace than can be accounted for by the conducting airways alone. In such cases, the conclusion is that there are some alveoli that are unperfused, thus contributing to total deadspace, something that is considered abnormal and which may occur in pulmonary embolism, in hypovolemia with low vascular pressures, and in emphysema and states of high lung inflation when zone 1 conditions of blood flow may exist in certain regions of the lung.

Just looking at % deadspace can thus be misleading: when tidal volume is low, % deadspace will be high even if absolute deadspace is normal; when tidal volume is high, % deadspace may be low yet absolute deadspace may still be abnormally high, as Fig. 2.9 shows.

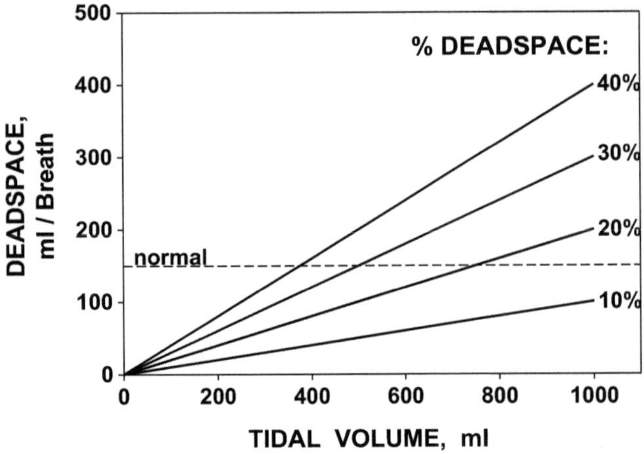

Fig. 2.9 Calculation to illustrate dependence of deadspace expressed as a percentage of tidal volume. It shows, for example, that if deadspace is 150 ml, it would be 40% of a 375 ml tidal volume, but only 20% of a 750 ml tidal volume

2.4.3 *Fractions of the Total Cardiac Output and Ventilation Within Each Decade of* $\dot{V}_A/\dot{Q}$

Simply by summing up blood flows and ventilations over all of the compartments in any one decade of the $\dot{V}_A/\dot{Q}$ distribution, we can derive a simple table of decade by decade fractions of perfusion and ventilation, and this can be useful. The MIGET software routinely calculates these values, and the output file routinely contains this table. The normal range over which MIGET can determine the $\dot{V}_A/\dot{Q}$ distribution is from a low $\dot{V}_A/\dot{Q}$ ratio of 0.005 to a high $\dot{V}_A/\dot{Q}$ ratio of 100. Thus, lower values cannot be distinguished from shunt, and higher values cannot be distinguished from deadspace. The decade table is divided therefore as follows:

Shunt ($\dot{V}_A/\dot{Q} < 0.005$)
$\dot{V}_A/\dot{Q}$ 0.005–0.01
$\dot{V}_A/\dot{Q}$ 0.01–0.1
$\dot{V}_A/\dot{Q}$ 0.1–1
$\dot{V}_A/\dot{Q}$ 1–10
$\dot{V}_A/\dot{Q}$ 10–100
Deadspace ($\dot{V}_A/\dot{Q} > 100$)

In normal subjects, the table is usually empty for the first three lines $\left(\dot{V}_A/\dot{Q} < 0.1\right)$ and for the second last line $\left(\dot{V}_A/\dot{Q}\ \ 10-100\right)$ as units with such ratios generally are not present in normal lungs. Thus, all of the blood flow and usually ~70% of the ventilation is found within the two decades of $\dot{V}_A/\dot{Q}$ 0.1 – 10. The remaining ventilation is of course the deadspace due to conducting airways.

In disease, considerable fractions of blood flow and ventilation can lie outside the normal range, and this table conveniently quantifies that for you.

There is a limitation on the use of this table that must be kept in mind. Simple translational shifts in the entire $\dot{V}_A/\dot{Q}$ distribution can cause the $\dot{V}_A/\dot{Q}$ distribution (Fig. 2.8) to cross decade boundaries. Shifts may occur whenever total ventilation and/or total cardiac output change significantly. The key is that this may happen without any change in the actual degree of mismatch (i.e., the second moments, LogSD,Q and LogSD,V_A, may not change). *Thus due to translation of the curves, the decade table values could change substantially but these changes may occur with no change in amount of mismatch.*

2.4.4 Modality and Fractional $\dot{Q}$, $\dot{V}_A$ in Each Mode

The $\dot{V}_A/\dot{Q}$ distribution that MIGET is capable of determining can have up to three distinct modes plus shunt and deadspace. The method does not however allow more than this complexity to be uncovered—the same properties of the retention curves (smooth and regular) that make their measurement feasible with a small number of gases make fine detail in the $\dot{V}_A/\dot{Q}$ curve unable to be assessed.

That said, it is possible to examine the modes individually should more than one mode appear in the $\dot{V}_A/\dot{Q}$ distribution. Mathematically, one could apply some of the above analyses (moments, lumping by decades) to each mode separately. Mostly, it is going beyond the information content of the method to rely on moments calculated for each mode. The exceptions are the first moments (Q_{mean}, $V_{A_{mean}}$) which are generally reliable. But the second and third moments should not be used for individual modes in a multimodal distribution. However, adding up all of the blood flow and ventilation in each mode (irrespective of which decades of $\dot{V}_A/\dot{Q}$ may be included) is quite robust and a very useful outcome parameter.

2.4.5 DISP, R; DISP, E; DISP, E*; DISP, R–E*

This family of four descriptors can be useful outcome variables. The abbreviation DISP stands for the word "dispersion." The letter R stands for retention, E for excretion, and E* to excretion corrected for any deadspace (which is normally present and thus should usually be corrected for).

These parameters are analogous to the second moments of the $\dot{V}_A/\dot{Q}$ curves but are based not on ventilations or blood flows of the 48 compartments but rather on the measured excretions and retentions of the 6 gases used in MIGET.

DISP, R is the root mean square (square root of the sum of the squares) of the differences between *actual* (measured) and *reference* retentions of each of the six gases. What is the reference retention for each gas? Simply the calculated retention for each gas in a homogeneous lung that has the same total non-deadspace (alveolar) ventilation and blood flow as the lung under study.

DISP, E is, symmetrically, the root mean square (square root of the sum of the squares) of the differences between actual (measured) and reference excretions of each of the six gases. What is the reference excretion for each gas? Simply the calculated excretion for each gas in a homogeneous lung that has the same total non-deadspace (alveolar) ventilation and blood flow as the lung under study. This definition means that normal conducting airway deadspace will dominate this parameter, and this diminishes its value.

DISP, E* is very similar to DISP, E. It has one difference: the measured excretions are all divided by the factor (1.0—fractional deadspace). The reference excretion is the same as above. This correction eliminates the dominating influence of anatomic deadspace on the value of DISP, E and thus yields a number reflective of $\dot{V}_A/\dot{Q}$ mismatch *per se*.

DISP, R–E* is the root mean square (square root of the sum of the squares) of the differences between actual (measured) retentions (R) and deadspace-corrected excretions (E*) of each of the six gases. This parameter has therefore also removed the dominant effect of normal deadspace on the result and is closely analogous to the more familiar alveolar-arterial P_{O_2} difference. In this case, it is actually the arterial (retention)-alveolar (excretion) difference averaged over the six gases by the root mean square calculation.

DISP, R–E is the only outcome parameter of the MIGET that, as a single variable, indicates the total inert gas exchange disturbance (excluding anatomic deadspace).* This gives it a unique place in describing the lungs by MIGET because every other parameter discussed above reflects a component of the distribution, but not the entire distribution.

Mathematically, the sum of DISP, R and DISP, E* comes close to equaling DISP, R–E*, but this is not exact because of the taking of square roots of sums.

There is an important issue to lay out in the calculation of these DISP-related parameters. One does not take the actual, fractional retentions and excretions on the simple, direct scale of 0–1 as shown, for example, in Fig. 2.6. This is because the retentions and excretions for each gas have intrinsically different variances in their measurement (this is discussed later in much more detail). Thus, we first have to normalize the retentions and excretions among the six gases by dividing each by its respective variance. This gives a variance-balanced set of six retentions and excretions that can be rationally combined in the root mean square calculation in a manner that assures that each of the six gases is properly represented in the calculation. Thus, no gas is either under- or overrepresented. More than that, dividing retention by variance means that the weighted retention for each gas has a variance numerically equal to 1.0. What this means is that the DISP indices are actually expressed in units of standard deviation.

DISP, R; DISP, E*; and DISP, R–E* would all be exactly 0 in a homogeneous lungs, just as the second moments of the $\dot{V}_A/\dot{Q}$ distribution described above. In normal lungs, the first two of the above indexes are about 1 each and the third about 2. Moderately high values of DISP, R and DISP, E* would be 3–5 and very high values 5–8. Corresponding ranges for DISP, R–E* would be 6–10 and >10, respectively.

2.5 The Relationship Between the Retention/Excretion Curves and the Blood Flow/Ventilation Curves ($\dot{V}_A/\dot{Q}$ Distribution)

The preceding pages have provided lots of information about retention and excretion on the one hand and ventilation and blood flow distributions on the other. They were briefly connected in terms of the Riley three-compartment model in Fig. 1.2 and the retention/solubility curve in Fig. 2.1. Now is the time to generalize their relationships more broadly as a bridge to the formal mathematics of the MIGET. Basically, this involves a reconsideration and expansion of Eqs. (2.5) for retention and (2.11) for excretion. Recall that these two equations described the relationship between retention and blood flow distribution (Eq. 2.5) and excretion and ventilation distribution (Eq. 2.11) in the Riley three-compartment model.

The objective of the following section is to expand those equations to the 50-$\dot{V}_A/\dot{Q}$ compartment construct used in the MIGET and to incorporate actual gases into the calculations. First, the 50 $\dot{V}_A/\dot{Q}$ ratios are specified. The concept is that we have a standard set of 50 ratios, and the question MIGET will answer is how cardiac output and total ventilation are distributed among all 50 of these compartments in any given case. Realize that "offering" MIGET a choice of all 50 compartments does not in any way require all of them to exist (i.e., have any blood flow or ventilation) in any given case.

Compartment 1 always represents shunt with a $\dot{V}_A/\dot{Q}$ ratio of 0.

Compartment 2 has the lowest $\dot{V}_A/\dot{Q}$ ratio separable from shunt in actual practice and is 0.005 as determined from simulations.

Compartment 49 has the highest $\dot{V}_A/\dot{Q}$ ratio separable from deadspace in actual practice and is 100, again determined from simulations.

Compartment 50 always represents deadspace with a $\dot{V}_A/\dot{Q}$ ratio of infinity.

Compartments 3–48 inclusive are equally spaced on a logarithmic scale between 0.005 and 100. The 48 $\dot{V}_A/\dot{Q}$ values other than 0 and infinity are easily calculated as follows:

$$D = [\text{natural LOG}\,(100/0.005)]/47 \tag{2.21}$$

Table 2.3 The six inert gases used in MIGET

Gas	Blood/gas partition coefficient
SF_6	0.005
Ethane	0.1
Cyclopropane	0.5
Isoflurane	2.0
Diethyl ether	12.0
Acetone	300.0

$$\dot{V}_A/\dot{Q}_j = 0.005 \times \exp[D \times (j-2)] \tag{2.22}$$

over the j compartments ($j = 2,3,4,...,49$). Remember we then add compartment 1 ($\dot{V}_A/\dot{Q}$ of 0, shunt) and compartment 50 ($\dot{V}_A/\dot{Q}$ of infinity, deadspace) to complete the 50 compartment $\dot{V}_A/\dot{Q}$ set.

Now turn to the inert gases. We could use any number we wish. Their defining characteristic relevant to MIGET is of course their solubility or partition coefficient. We have come to use six different gases. The choice of six, and the choice of which actual gases make up those six, reflects optimizing several factors:

(a) Having enough gases to specify experimentally the entire domain of any retention curves experienced in actual practice
(b) Having the gases each differ in solubility enough so as to roughly evenly cover the solubility range from very low to very high
(c) Having gases that can be relatively easily measured when they coexist in a given blood or gas sample
(d) Having gases that are nontoxic and which do not by their presence themselves affect the distribution of ventilation or blood flow
(e) Avoiding so many gases that measurement becomes problematic

Table 2.3 lists the present set of gases (isoflurane recently replaced enflurane which had previously replaced halothane, based on availability) and their approximate blood/gas partition coefficients in round numbers. *Just as with any variable, such as hemoglobin concentration, the solubility of each of the gases will differ from individual to individual and across species.*

Now we have the 50 compartments and the 6 gases specified. We now generalize Eqs. (2.5) for retention and (2.11) for excretion as follows:

$$R_i = \Sigma\left[\dot{Q}_j \times \lambda_i/\left(\lambda_i + \dot{V}_A/\dot{Q}_j\right)\right]/\Sigma\left[\dot{Q}_j\right] \tag{2.23}$$

$$E_i = \Sigma\left[\dot{V}_{A_j} \times \lambda_i/\left(\lambda_i + \dot{V}_A/\dot{Q}_j\right)\right]/\Sigma\left[\dot{V}_{A_j}\right] \tag{2.24}$$

We need to use matrix shorthand terminology now because otherwise the equations will become too unwieldy to write down. That means first subscripting the gas-related variables R, E, and λ by the identifier "i" and the distribution-related variables $\dot{V}_A$, $\dot{Q}$, and $\dot{V}_A/\dot{Q}$ by the identifier "j." What this means is simply that we have six gases with six R (and E and λ) values, R_1, R_2,. . ., R_6, and that to make it easy to write down we indicate this simply as "R_i, $i = 1{,}6$." For the $\dot{V}_A$, $\dot{Q}$, and $\dot{V}_A/\dot{Q}$ data, we have 50 and thus need to use another letter of the alphabet, but the idea is exactly the same: $\dot{V}_{A_1}$, $\dot{V}_{A_2}$,..., $\dot{V}_{A_{50}}$ is written simply as "$\dot{V}_{A_j}, j = 1, 50$."

The Greek symbol **Σ** indicates summation of all terms over the 50 values of j for a single value of i, which action is then repeated for each of the other five values of i.

Equation (2.23) addresses blood flow and retention; Eq. (2.24) symmetrically addresses ventilation and excretion. Equation (2.23) states in words that in the mixed arterial blood, the ratio of systemic, mixed arterial to mixed venous inert gas concentration (i.e., retention, R_i) for gas number i is given by summing over the 50 $\dot{V}_A/\dot{Q}$ compartments the fractional blood flow-weighted individual compartmental retention values calculated from λ_i and $\dot{V}_A/\dot{Q}_j$. This is simply a statement of mass conservation, extending the two-compartmental model presented earlier to 50 compartments.

Similarly, Eq. (2.24) states in words that the in the mixed expired gas, the ratio of mixed expired to mixed venous inert gas concentration (i.e., excretion, E_i) for gas number i is given by summing over the 50 $\dot{V}_A/\dot{Q}$ compartments the fractional ventilation-weighted individual compartmental excretion values calculated from λ_i and $\dot{V}_A/\dot{Q}_j$. This too is simply a statement of mass conservation. And remember that in any 1 of the 50 compartments, retention = excretion as Eqs. (2.23) and (2.24) imply. But, because Eqs. (2.23) and (2.24) employ different weightings (i.e., $\dot{Q}$ versus $\dot{V}_A$), overall retention will differ from (and be greater than) overall excretion in any lung that is not perfectly homogeneous.

The relationships between $\dot{V}_A/\dot{Q}$ distributions and their associated retention/excretion curves defined by Eqs. (2.23) and (2.24) are best appreciated visually. While any plotting software could be used, we have settled on Sigmaplot, for better or worse. We have written a Sigmaplot program, which Sigmaplot calls a "Transform" and saves in an ".xfm" file to automatically plot a chosen $\dot{V}_A/\dot{Q}$ distribution and its associated retention and excretion curves. Table 2.4 presents the entire program with quite a bit of explanatory English, and Figs. 2.10, 2.11, 2.12, 2.13, 2.14 and 2.15 exemplify six selected cases. They should be looked at to see how certain types of distributions lead to characteristic changes in the retention and excretion curves. You can use the transform program and create any desired pattern (with up to two modes, shunt and deadspace) for yourself.

What this chapter should have done for you is prepare you for the guts of MIGET—which is the mathematics of how $\dot{V}_A/\dot{Q}$ distributions can be recovered from retention and excretion curves. In particular, you need to be comfortable with Eqs. (2.23) and (2.24) and the relationships in Figs. 2.10–2.15. If you are, the next section will be digestible. If you are not yet comfortable, you may need to spend more time on the entire preceding sections.

Table 2.4 Sigmaplot program to calculate and plot $\dot{V}_A/\dot{Q}$ distributions and retention/excretion curves

This is the Sigmaplot transform program that takes as input the parameters of the $\dot{V}_A/\dot{Q}$ distribution **(all in red)** and plots both the $\dot{V}_A/\dot{Q}$ distribution and the retention-solubility curves – automatically. The name of the file is:

$\dot{V}_A/\dot{Q}$ GENERATE COMPS bimodal.xfm

Comments are in purple; input variables are in red; actual code is in red and blue and black.

When you want to run this program, you first load Sigmaplot and create a sample with 50 compartmental values of $\dot{V}_A$, $\dot{Q}$ and $\dot{V}_A/\dot{Q}$.

Select the data table, then click on the "transform" button, select the "user defined" button and then click on "open" and finally select the above .xfm title from the list that appears. The transform will appear in a window exactly as below (without English comments though).

You then can edit the data in the first 8 red data lines, then simply select "RUN" and it will all just happen—autoscaled and everything. Note that the stored version of the .xfm file may contain different values for the 8 red variables than appear below, depending on prior use of the file.

Table 2.4 (continued)

The program starts here:

$\dot{Q}_T$ =6 (Cardiac output, L/min)

SDQ1=0.3 (SDQ for main mode)

SDQ2=0.7 (SDQ for secondary mode)

QMEAN1=1 (Mean $\dot{V}_A/\dot{Q}$ ratio of $\dot{Q}$ distribution main mode)

QMEAN2=10 (Mean $\dot{V}_A/\dot{Q}$ ratio of $\dot{Q}$ distribution secondary mode)

QRATIO=0.05 (ratio of TOTAL blood flow in secondary to main mode)

V_D VT=30 (deadspace, %)

QSQT=10 (Shunt, %)

QREST=QT*(1-QSQT/100)

The following standard data in blue are the conventional data for constructing the 50-compt $\dot{V}/\dot{Q}$ curves, and also the retention curves for 11 chosen values of partition coefficient. While these data are easily changed, they are standard and usually do not need to be altered.

vqlo=0.005

vqhi=100.0

ncomp=48

slo=0.001

shi=1000.0

nsol=11

B1=1

B2=2

Table 2.4 (continued)

```
B3=3

B4=4

B5=5

B6=6

B7=7

B8=8

B9=9

B10=10

B11=11

delta=(ln(vqhi/vqlo))/(ncomp-1)

col(2)=vqlo*(exp(delta*(col(1)-1)))
```

Column 2 contains the 48 $\dot{V}_A/\dot{Q}$ ratios 0.005 to 100

```
D11=LN(QMEAN1) - LN(COL(2))

D12=LN(QMEAN2) - LN(COL(2))

REXP1=-0.5*D11*D11/(SDQ1*SDQ1)

REXP2=-0.5*D12*D12/(SDQ2*SDQ2)

COL(13)=(EXP(REXP1))

COL(14)=QRATIO*(EXP(REXP2))

COL(15)=COL(13)+COL(14)
```

Column 13 contains main mode blood flow; column 14 secondary mode blood flow and column 15 both combined.

Table 2.4 (continued)

```
SUMQ=TOTAL(COL(15))
COL(3)=QREST*COL(15)/SUMQ
COL(4)=COL(3)*COL(2)
```

Column 3 is the same as column 15 normalized to make all flows sum to the input value of cardiac output. Column 4 contains the ventilation values that must be present (you have no choice), by multiplying $\dot{Q}$ and $\dot{V}_A/\dot{Q}$ in each compartment.

```
V̇A =TOTAL(COL(4))
V̇E =V̇A/(1- VDVT/100)
CELL(16,1)=V̇A
CELL(16,2)=V̇E
```

$\dot{V}_A$ is total alveolar ventilation, $\dot{V}_E$ total minute ventilation. They will appear in column 16 ($\dot{V}_A$ in row 1 and $\dot{V}_E$ in row 2). You can also find out what these numbers are by clicking on "view" in the tool bar and selecting "statistics". The 7th row of that statistical table indicates sum of all values in each column. That for column A is $\dot{V}_A$, not $\dot{V}_E$. If the values of $\dot{V}_A$ and $\dot{V}_E$ are not to your liking, re-run the transform program selecting different values for any/all of the variables, especially QMEAN1, QMEAN2 and QRATIO.

The following code calculates the retention (column 7) and excretion (column 8) values for just the 48 compartments in the distribution selected for 11 (nsol) gases with partition coefficients equally spaced between 0.001 (slo) and 1000 (shi). Thus, no shunt or deadspace effects are included yet. The partition coefficients appear in column 6, identified as A1 through A11. B1 through B11 simply indicate the row in column 6 to pull each partition coefficient from.

Table 2.4 (continued)

```
deltas=(ln(shi/slo))/(nsol-1)

col(6)=slo*(exp(deltas*(col(5)-1)))

A1=CELL(6,B1)

col(7,B1)=total(col(3)*A1/(A1+col(2)))/QREST

col(8,B1)=total(col(4)*A1/(A1+col(2)))/VA

A2=CELL(6,B2)

col(7,B2)=total(col(3)*A2/(A2+col(2)))/QREST

col(8,B2)=total(col(4)*A2/(A2+col(2)))/VA

A3=CELL(6,B3)

col(7,B3)=total(col(3)*A3/(A3+col(2)))/QREST

col(8,B3)=total(col(4)*A3/(A3+col(2)))/VA

A4=CELL(6,B4)

col(7,B4)=total(col(3)*A4/(A4+col(2)))/QREST

col(8,B4)=total(col(4)*A4/(A4+col(2)))/VA

A5=CELL(6,B5)

col(7,B5)=total(col(3)*A5/(A5+col(2)))/QREST

col(8,B5)=total(col(4)*A5/(A5+col(2)))/VA

A6=CELL(6,B6)

col(7,B6)=total(col(3)*A6/(A6+col(2)))/QREST

col(8,B6)=total(col(4)*A6/(A6+col(2)))/VA

A7=CELL(6,B7)

col(7,B7)=total(col(3)*A7/(A7+col(2)))/QREST

col(8,B7)=total(col(4)*A7/(A7+col(2)))/VA

A8=CELL(6,B8)

col(7,B8)=total(col(3)*A8/(A8+col(2)))/QREST
```

Table 2.4 (continued)

```
col(8,B8)=total(col(4)*A8/(A8+col(2)))/VA

A9=CELL(6,B9)

col(7,B9)=total(col(3)*A9/(A9+col(2)))/QREST

col(8,B9)=total(col(4)*A9/(A9+col(2)))/VA

A10=CELL(6,B10)

col(7,B10)=total(col(3)*A10/(A10+col(2)))/QREST

col(8,B10)=total(col(4)*A10/(A10+col(2)))/VA

A11=CELL(6,B11)

col(7,B11)=total(col(3)*A11/(A11+col(2)))/QREST

col(8,B11)=total(col(4)*A11/(A11+col(2)))/VA
```

Column 9 is the column of retentions when shunt is included. Column 10 is the column of excretions when deadspace is included. Column 11 contains the retentions for the homogeneous lung having the actual overall $\dot{V}_A$ and $\dot{Q}_T$, and column 12 contains the excretions for the homogeneous lung having the actual overall $\dot{V}_A$, $\dot{Q}_T$ and deadspace.

```
COL(9)=QSQT/100 + (1-QSQT/100)*COL(7)

COL(10)=(1-VDVT/100)*COL(8)

COL(11)=COL(6)/(COL(6)+V̇A/Q̇T)

COL(12)=(1-VDVT/100)*COL(6)/(COL(6)+V̇A/Q̇T)
```

Table 2.4 (continued)

The final set of cells below contain the shunt data for automatic plotting of shunt. Deadspace has NOT been added to the graph because it is usually so large it would dwarf the rest of the distribution.

CELL(2,50)=0.0016

CELL(2,51)=0.0016

CELL(3,50)=0.01*QSQT*$\dot{Q}_T$

CELL(3,51)=0

CELL(4,50)=0

CELL(4,51)=0

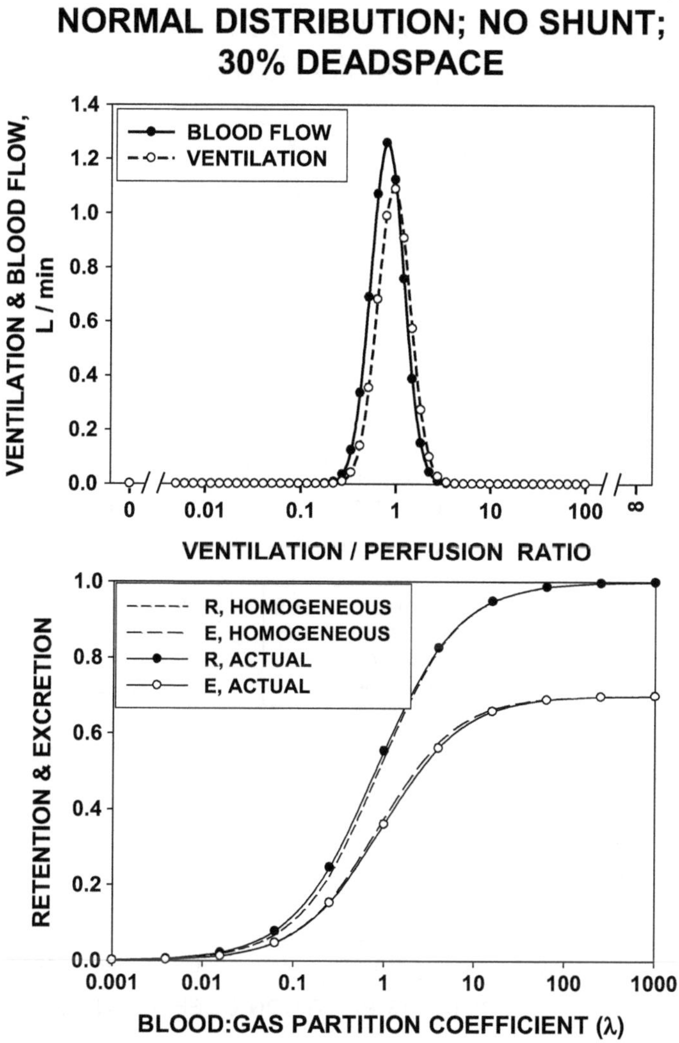

Fig. 2.10 $\dot{V}_A/\dot{Q}$ distribution (upper panel) as might be recovered using MIGET retention and excretion data (lower panel) in a young normal upright subject at rest. Note that for presentation, the actual retention curves appear to show 10 gases by the solid and open circles. Recall that MIGET uses only six as in Fig. 2.4a. Shunt is zero; deadspace is ~30%; no low or high $\dot{V}_A/\dot{Q}$ areas are present, and all $\dot{V}_A$ and $\dot{Q}$ is confined to the $\dot{V}_A/\dot{Q}$ decade of ~0.3 to ~3.0, centered on a $\dot{V}_A/\dot{Q}$ of ~1.0. Retention and excretion curves from a hypothetical homogeneous lung with the same total ventilation, cardiac output, and deadspace are shown (lower panel, dashed lines) to point out the small effects of $\dot{V}_A/\dot{Q}$ inequality on gas exchange that is seen in normal subjects

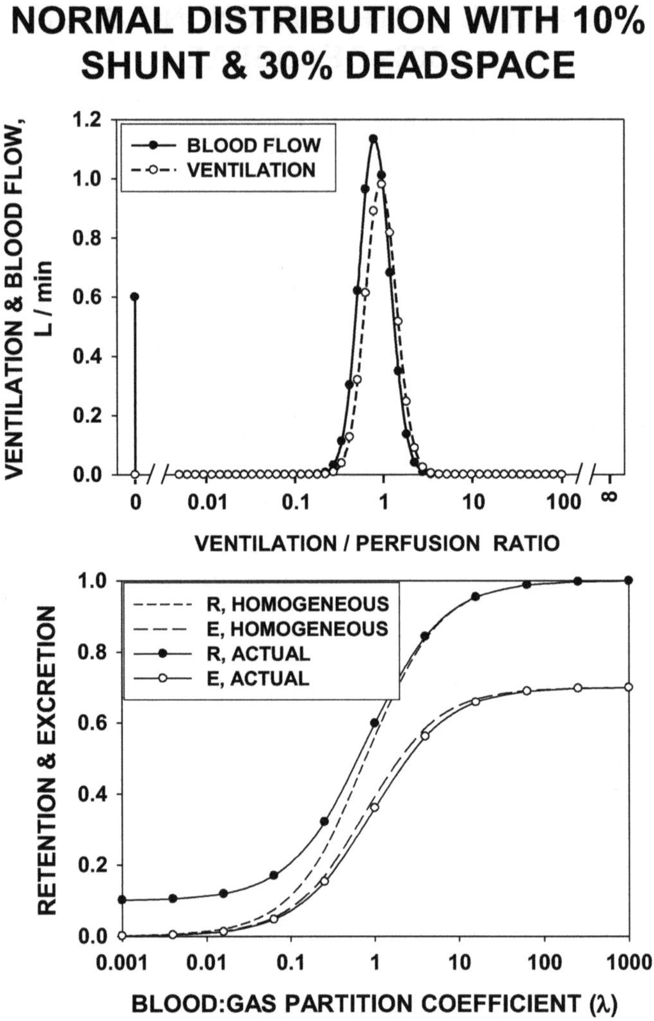

Fig. 2.11 $\dot{V}_A/\dot{Q}$ distribution (upper panel) as might be recovered using MIGET retention and excretion data (lower panel) in a subject with otherwise normal lungs but a 10% shunt. Note that the retention curve asymptotes at a retention equal to the shunt fraction (here 0.1) as solubility approaches zero

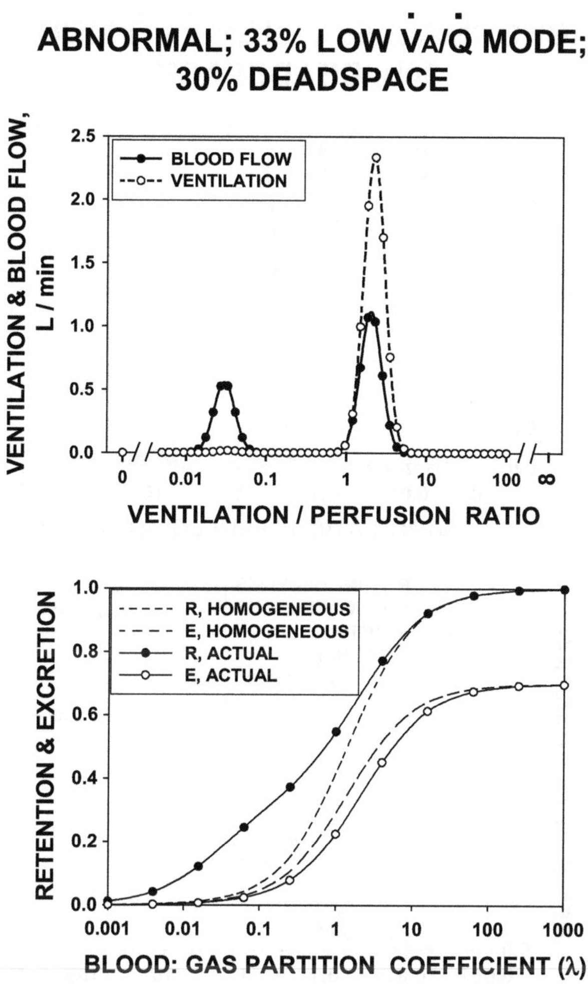

Fig. 2.12 $\dot{V}_A/\dot{Q}$ distribution (upper panel) as might be recovered using MIGET retention and excretion data (lower panel) in a subject with asthma or with chronic bronchitis, showing a distinct population of low $\dot{V}_A/\dot{Q}$ gas exchange units receiving in all 33% of the cardiac output. There is no shunt in this case

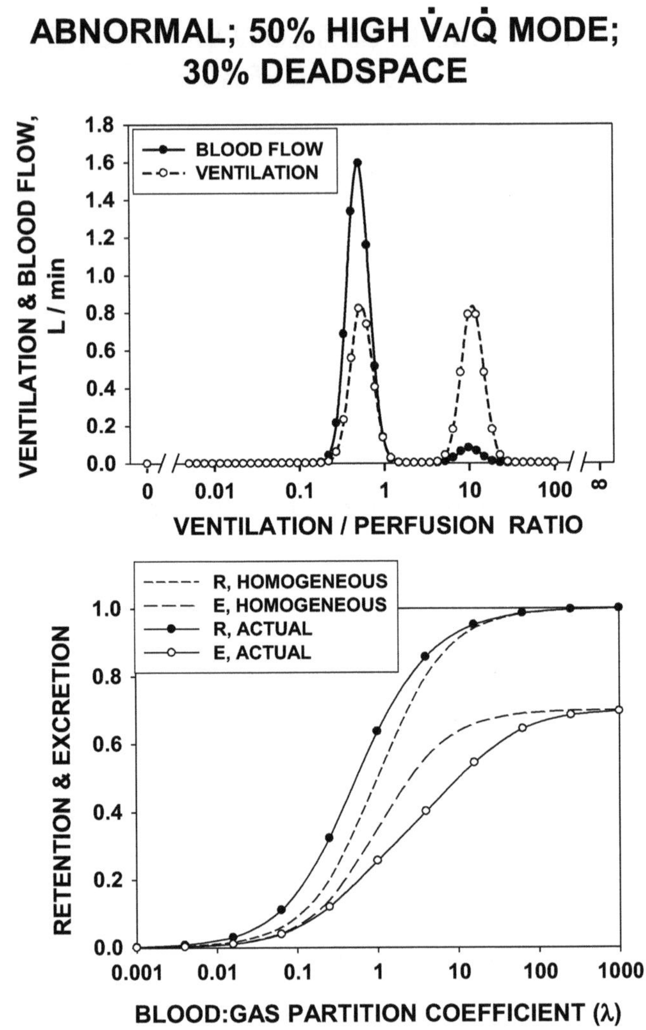

Fig. 2.13 $\dot{V}_A/\dot{Q}$ distribution (upper panel) as might be recovered using MIGET retention and excretion data (lower panel) in a subject with a pulmonary embolus or with emphysema, showing a distinct population of high $\dot{V}_A/\dot{Q}$ gas exchange units receiving in all 50% of the alveolar (non-deadspace) ventilation

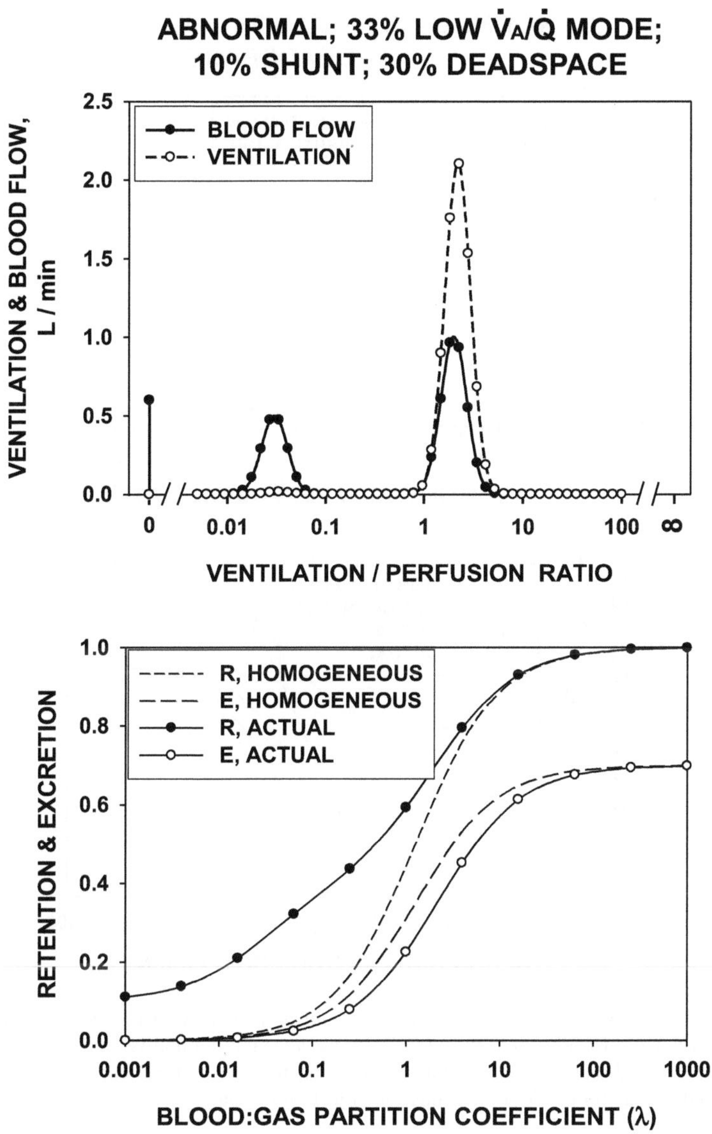

Fig. 2.14 $\dot{V}_A/\dot{Q}$ distribution (upper panel) as might be recovered using MIGET retention and excretion data (lower panel) in a subject with both shunt (10%) and a distinct population of low $\dot{V}_A/\dot{Q}$ gas exchange units receiving 33% of the cardiac output. Complex, severely abnormal, distributions like this are usually seen only in the critically ill

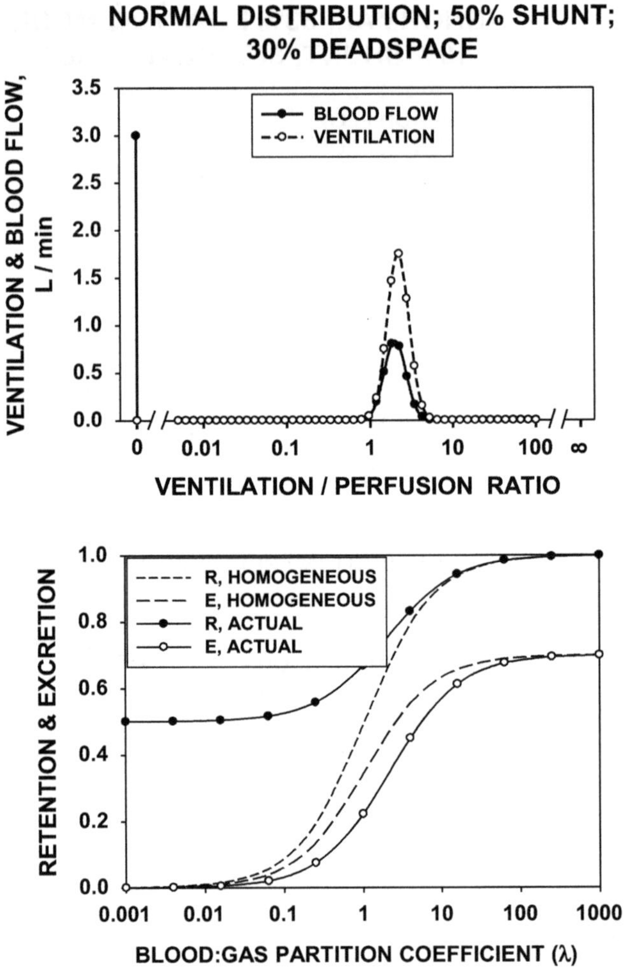

Fig. 2.15 $\dot{V}_A/\dot{Q}$ distribution (upper panel) as might be recovered using MIGET retention and excretion data (lower panel) in a subject with a massive shunt but normal lungs, as may occur in congenital heart disease or in extensive lobar pneumonia (where consolidated areas are unventilated but the remainder of the lung may function normally)

Reference

1. Wagner PD, Hedenstierna G, Bylin G, Lagerstrand L. Reproducibility of the multiple inert gas elimination technique. J Appl Physiol (1985). 1987;62:1740–6.

Chapter 3
Mathematical Underpinnings

Abstract Up until now we have largely been discussing the "forward problem": given a distribution of ventilation-perfusion ($\dot{V}_A/\dot{Q}$) ratios, what are the expected concentrations of various gases in venous and arterial blood? This chapter discusses the reverse problem, which is the fundamental basis of the MIGET: taking a set of measured retentions and excretions of six gases and their blood/gas partition coefficient (λ) to derive the underlying $\dot{V}_A/\dot{Q}$ distribution. The retention, $R_i, = \Sigma\left[\dot{Q}_j \times \lambda_i/\left(\lambda_i + \dot{V}_A/\dot{Q}_j\right)\right]/\Sigma\left[\dot{Q}_j\right]$, and excretion, $E_i, = \Sigma\left[\dot{V}_{A_j} \times \lambda_i/\left(\lambda_i + \dot{V}_A/\dot{Q}_j\right)\right]/\Sigma\left[\dot{V}_{A_j}\right]$, equations are solved using the 50 known values of $\dot{V}_A/\dot{Q}$ ratio and the 6 measured λ, retentions, and excretions. This results in 12 equations solved simultaneously for the 50 unknown values of $\dot{V}_A/\dot{Q}$ ratio. Smoothing allows this to be done with six gases and stabilizes the solution. In addition to smoothing, other constraints include that the compartmental blood flows must sum to cardiac output and that no compartmental flow may be negative. The gases used in the multiple inert gas elimination technique vary in measurement error, because of different absolute concentrations, measurement accuracy, and intrinsically different retentions/excretions. The high $\dot{V}_A/\dot{Q}$ region of the distribution is best resolved by excretion data measured in expired gas, especially of the most soluble inert gases, while the low $\dot{V}_A/\dot{Q}$ region is best resolved by retention data measured in arterial blood, especially of the low-solubility inert gases. However, all six gases provide information from both expired gas and arterial blood. Thus, using error propagation theory, weighted averages of retention and excretion data are used, depending on gas solubility. The residual sum of squares (RSS) from the fit to thc 50-compartmcnt model is used as an index of data quality. RSS follows a chi-square distribution; RSS should be 5.3 or lower in 50% of cases. It should be 10.6 or lower in 90%, and 99% of values should be <16.8.

3.1 Introduction

Equations (2.23) and (2.24) are posed in the following way: they allow you to calculate for any gas retention and excretion in a 50-compartment lung of 50 specified (i.e., known) $\dot{V}_A/\dot{Q}$ ratios, based on the known solubility (aka partition

S.R. Hopkins, P.D. Wagner, *The Multiple Inert Gas Elimination Technique (MIGET)*, Methods in Physiology, https://doi.org/10.1007/978-1-4939-7441-2_3

coefficient) of the gas and the quantitative distribution of both ventilation and blood flow to all 50 compartments.

We call that the "forward problem," and this problem is very easily solved, as Eqs. (2.23) and (2.24) and Table 2.4 lay out.

But what we really want is to be able to reverse Eqs. (2.23) and (2.24), thereby executing the "reverse problem," posed as follows:

Given a set of measured retentions and excretions of six gases of known partition coefficient, can we extract from those data the distribution of ventilation and blood flow across the 50 compartments that must exist to give rise to the particular measured retention/excretion set?

The answer is yes—within limits. The limits will be discussed much later—Sect. 3.4 deals with only that issue. For now, dismiss the idea of limits and focus just on the mathematics of the reverse problem.

3.2 Multicompartment Versus Continuous Distributions

First, let's agree on using the multicompartment model as in Eqs. (2.23) and (2.24). Sure, it is possible to use an algebraic equation with unknown coefficients and determine the values for the coefficients that produce a distribution that most closely fits the measured retentions, but the problem here is how to select a useful and rational algebraic function to use. One choice is to assume the distribution is both unimodal and lognormal, use the algebraic equation of such a curve, and best-fit the retention/excretion data by adjusting the moments of the distribution by some technique. That has been done, and when the distribution is both unimodal and lognormal, it clearly works well [2], as it should. But, when those assumptions cannot be verified, this approach is quite limiting because we now know that distributions can be multimodal and/or skewed.

Second, we could use a multicompartment model in which we chose *not* to pre-specify the 50 $\dot{V}_A/\dot{Q}$ values of the compartments but rather let them "float." To some, this seems intellectually more satisfying than pre-specifying 50 values (and thereby not allowing any others), but this is more than offset by two features of the pre-selected approach: the pre-selected approach keeps the equations (2.23 and 2.24) *linear*, such that the $\dot{V}_A$'s and $\dot{Q}$'s are the only unknowns we are trying to solve for. This is because they are only on the numerator of Eqs. (2.23) and (2.24), which keeps the problem linear. However, if $\dot{V}_A/\dot{Q}$ ratios are also unknowns, those unknowns appear in the denominator of terms in Eqs. (2.23) and (2.24). Thus there are unknowns in both the numerator ($\dot{V}_A/\dot{Q}$) and denominator ($\dot{V}_A/\dot{Q}$)—a very nasty nonlinear problem we choose not to have to face. The second good feature of the pre-selected approach is that offering a choice of 50 compartments in essence allows any values to be chosen because each pair of adjacent values is so close together in value that the intellectual concern of preselection is not actually a problem.

The mathematical problem then boils down to applying Eqs. (2.23) and (2.24) by inserting into them the 50 known values of $\dot{V}_A/\dot{Q}$ ratio, the 6 known values of partition coefficient, the 6 measured retentions, and the 6 measured excretions. This creates 12 equations: 1 for retention and 1 for excretion for each of the 6 gases. ***Then, we solve those 12 simultaneous linear equations for the 50 unknowns***. Note that while there are 50 unknown $\dot{V}_A$ values and 50 unknown $\dot{Q}$ values, there really are only 50 unknowns and not 100 because, via Eq. (2.14) above, if we know $\dot{V}_A$, we know $\dot{Q}$ (or vice versa) because their ratio is known for every one of the 50 compartments.

We know how to solve simultaneous linear equations. For example, we can illustrate with just two "gases" and two "compartments":

$$2A + B = 8 \tag{3.1}$$

$$A - B = 1 \tag{3.2}$$

The coefficients of A and B in the two equations are the equivalents of the (known) $\lambda/\left[\lambda + \dot{V}_A/\dot{Q}\right]$ terms from Eqs. (2.23) and (2.24) for $\dot{V}_A/\dot{Q}$ compartment A and $\dot{V}_A/\dot{Q}$ compartment B, while A and B are themselves the equivalents of the two (unknown) blood flows of the two compartments in those equations. The right-hand side numbers are equivalent to the measured (known) retentions. Equation (3.1) represents one gas; Eq. (3.2) represents a second gas.

It is apparent from Eqs. (3.1) and (3.2) that A uniquely equals 3, while B uniquely equals 2. There can be no other solution for A and B.

This is all that the MIGET does, using Eqs. (2.23) and (2.24), but with some special twists, now discussed.

3.3 Smoothing as a Constraint Concept

Perhaps the most fundamental puzzle requiring a special twist comes from the observation that solving Eqs. (3.1) and (3.2) can only be done meaningfully when the number of equations (=gases) equals the number of compartments (=unknown variables, $\dot{Q}$). The formulation of Eqs. (2.23) and (2.24) clearly creates a set of 12 equations [of which only 6 are actually independent as R and E can be calculated from one another by Eq. (2.8)] but specifies 50 independent $\dot{V}_A/\dot{Q}$ compartments. How can we possibly solve Eqs. (2.23) and (2.24) for all 50 compartments with only 6 independent equations?

The answer is to use smoothing. Think of it like this, as mentioned earlier: Imagine 50 individual tennis balls thrown randomly onto a tennis court. When they have stopped moving, you are asked to describe the location of every one. Can you do that by pinning down the location of only six of them? No way; the other 44 could be anywhere, and you need precisely as many measurements as there are balls (50 in this case) to solve the problem of describing where the balls lie.

However, now we will repeat this exercise with one difference: Before tossing the balls onto the court, we will connect them by a fairly short, fairly stiff rope into a single string of all 50—just like a string of pearls. Then we toss the string out onto the court. Do we now have to measure the position of all 50 balls to know pretty well where they all are? NO. If we pin down the positions of six carefully chosen ball—probably, roughly, balls 4, 12, 20, 28, 37, and 46 (chosen to be equally spaced along the length of the whole string)—the approximate positions of all the 44 others can be inferred because they have to be on the string between each pair of measured balls. If the string joining each pair of balls is sufficiently short and stiff, the 44 balls not directly measured must be within a narrow domain of space on the court. Smoothing is the same as connecting the balls by a string. Using smoothing, we can identify the positions of all 50 balls (blood flow and ventilation in the 50 $\dot{V}_A/\dot{Q}$ compartments) from six actual position measurements of well-chosen balls in the string (six inert gas retentions and excretions) ***with useful degrees of accuracy, albeit not with complete precision.***

Why is this a reasonable approach? ***First***, as reasoned much earlier, with thousands of potential gas exchange units in the lung, it is very likely that their distribution is indeed smooth. Just as the distribution of height in a school of several thousand students will be a smooth function. ***Second***, as will be discussed later, smoothing stabilizes the solution (i.e., the derived $\dot{V}_A/\dot{Q}$ distribution) in the presence of random experimental errors. Thus, with smoothing, two duplicate data sets differing only by random error will not result in differently appearing distributions, which, if it happened, would clearly be confusing. ***Third***, due to the inherently smooth nature of the retention curves also discussed much earlier, adding gases above the six we use really gains us very little information into the fine structure of the distribution—it is simply not possible with MIGET to see very fine structure in the distribution, because of the basic equations that govern the exchange of inert gases. Thus, we are not losing potential information by having smoothing—because even without smoothing, that information is simply not available. ***Fourth***, and this is subtle, the fit to the data obtained with smoothing is statistically closer to the true data (and thus a better outcome) than would be the fit obtained in the absence of smoothing. This claim depends on the interaction between random error and the laws of gas exchange (i.e., as specified by Eq. 1.20) and will be explained below in due course in the context of the "orange slice" diagram.

3.3.1 How Do We Apply Smoothing to Solve Eqs. (2.23) and (2.24)?

We apply smoothing in a remarkably simple way. But first, the basic approach must be further elucidated. We solve Eqs. (2.23) and (2.24) (modified below by smoothing) by a *least squares* method: we search for the 50 compartmental blood flow values that make the sum of squares of the differences between the six measured retentions (i, $i = 1{,}6$) and those calculated from the 50 flows (j, $j = 1{,}50$) by Eq. (2.23) (i.e., the residual sum of squares, **RSS**) ***minimal*** (summed over all six gases).

The RSS is written as follows, which comes directly from Eq. (2.23). It is simply the right side subtracted from the left side, squared, and summed over the gases.

$$\mathrm{RSS} = \Sigma_i\left[R_i - \Sigma_j\left[\dot{Q}_j \times \lambda_i/\left(\lambda_i + \dot{V}_\mathrm{A}/\dot{Q}_j\right)\right]/\Sigma\left[\dot{Q}_j\right]\right]^2 \quad (3.3)$$

This is not itself the minimal RSS. To get there, we have to differentiate RSS with respect to every one of the flows, $\dot{Q}$, and set those 50 results each to zero.

That actually gives us a set of 50 equations to solve for $\dot{Q}$. We still have 50 compartments but have not yet moved from the 50 independent tennis ball problem to one in which the balls are linked (smoothing). This approach (50 equations in 50 unknowns) does not work well as laid out—that is, without smoothing.

This least squares concept is identical to the simpler problem of conventional linear regression: finding the parameters (slope and intercept) of a single straight line that best fit a cloud of (X,Y) data points on a two-dimensional X-Y plot.

Suppose we have a set of n X and Y data points: (X_i,Y_i), $i = 1,n$.

We are looking for a straight line of the form $Y = mX + b$ where m is the slope and b the intercept. We can write the RSS, similarly as in Eq. (3.3), as follows:

$$\mathrm{RSS} = \Sigma_i[Y_i - (mX_i + b)]^2 \quad (3.4)$$

In words, RSS is the sum of squares (over all n data points) of the differences between actual Y values (Y_i) and their value on the regression line ($mX_i + b$) at the same X_i value. We differentiate Eq. (3.4) once with respect to m and once more, now with respect to b, and set these two differentials equal to zero. Each of the two differentiations gives us one equation in m, b, X_i, and Y_i. Thus, we have two simultaneous equations in two unknowns (m and b) and can uniquely solve for m and b as a result. ***What is key is to realize that differentiating and setting the result to zero gives us m and b that minimize RSS.***

$$d\mathrm{RSS}/dm = -2 \times \Sigma_i[X_i \times [Y_i - (mX_i + b)]] = 0 \quad (3.5a)$$

$$d\mathrm{RSS}/db = -2 \times \Sigma_i[Y_i - (mX_i + b)] = 0 \quad (3.6a)$$

Let us write **SX** = $\boldsymbol{\Sigma_i[X_i]}$; **SY** = $\boldsymbol{\Sigma_i[Y_i]}$; **SXY** = $\boldsymbol{\Sigma_i[X_i \times Y_i]}$; and **SXX** = $\boldsymbol{\Sigma_i[X_i \times X_i]}$. SX is the sum of all the X values and SY is the sum of all the Y values. SXY is the sum of all the ($X \times Y$) products, and SXX is the sum of all the ($X \times X$) products. These two equations simplify to:

$$\mathrm{SXY} - m \times \mathrm{SXX} - b \times \mathrm{SX} = 0 \quad (3.5b)$$

and

$$\mathrm{SY} - m \times \mathrm{SX} - n \times b = 0 \tag{3.6b}$$

These are two linear simultaneous equations in two unknowns, ***m*** and ***b***, which are then easily solved for *m* and *b*. How? Multiply all terms in Eq. (3.6b) by (−SX) and all terms in Eq. (3.5b) by *n*:

$$n \times \mathrm{SXY} - n \times m \times \mathrm{SXX} - n \times b \times \mathrm{SX} = 0 \tag{3.7}$$

$$-\mathrm{SX} \times \mathrm{SY} + m \times \mathrm{SX} \times \mathrm{SX} + n \times b \times \mathrm{SX} = 0 \tag{3.8}$$

Now add Eqs. (3.7) and (3.8) together, which eliminates the term in *b*:

$$(n \times \mathrm{SXY} - \mathrm{SX} \times \mathrm{SY}) - m \times (n \times \mathrm{SXX} - \mathrm{SX} \times \mathrm{SX}) = 0 \tag{3.9}$$

and thus:

$$m = (n \times \mathrm{SXY} - \mathrm{SX} \times \mathrm{SY})/(n \times \mathrm{SXX} - \mathrm{SX} \times \mathrm{SX}) \tag{3.10}$$

And now back-substitute this value of *m* into Eq. (3.6b):

$$\begin{aligned} b &= \mathrm{SY}/n - m \times \mathrm{SX}/n \\ &= \mathrm{SY}/n - [(n \times \mathrm{SXY} - \mathrm{SX} \times \mathrm{SY})/(n \times \mathrm{SXX} - \mathrm{SX} \times \mathrm{SX})] \times \mathrm{SX}/n \end{aligned} \tag{3.11}$$

which simplifies to:

$$b = (\mathrm{SY} \times \mathrm{SXX} - \mathrm{SX} \times \mathrm{SXY})/(n \times \mathrm{SXX} - \mathrm{SX} \times \mathrm{SX}) \tag{3.12}$$

What does best fit then mean? It means we have found the slope, ***m***, and intercept, ***b***, of the line that *minimizes the sum of the squares of the vertical (Y axis) differences between each actual data point and the line* $Y = mX + b$.

In the absence of experimental error, the minimal sum of squares RSS would be zero—all points would lie right on the regression line. Translated into the inert gas system, the six measured retentions and excretions would be able to be perfectly accurately calculated by using Eqs. (2.23) and (2.24) and the recovered $\dot{V}_\mathrm{A}$ and $\dot{Q}$ values for the 50 compartments. But of course, there are random (and systematic) errors in the MIGET just as with every other measurement ever made. Thus, the RSS will never be zero (the correlation coefficient of the *X*-*Y* plot will never be exactly 1.0). It is theoretically possible for RSS to be zero, just as it is possible to win the mega lottery, but don't hold your breath. Random error will be discussed in detail below.

Back to smoothing: ***Instead of trying to minimize only the squared differences between measured and calculated (by Eq. 2.23) retentions, a procedure that has no smoothing, we minimize a compound function that minimizes the sum of two functions each of which contains the 50 blood flows.***

The first function is exactly Eq. (3.3). ***The second function is simply the sum of squares of the 50 blood flows.*** That's really it!! Thus:

$$\mathrm{RSS} = A + B \tag{3.13}$$

where

$$A = \Sigma_i \left[R_i - \Sigma_j \left[\dot{Q}_j \times \lambda_i / \left(\lambda_i + \dot{V}_A / \dot{Q}_j\right)\right] / \Sigma\left[\dot{Q}_j\right]\right]^2 \tag{3.3}$$

$$B = Z \times \Sigma_j \left[\dot{Q}_j\right]^2 \tag{3.14}$$

This trick now includes smoothing, by virtue of including Eq. (3.14) (see below). The number "Z" is the famous smoothing coefficient. Its magnitude determines how much weight will be given to term B versus term A when the system is solved (the solution method will be explained later). The higher Z, the more importance will term B have; the smaller Z, the more importance term A will have. Minimizing A achieves the fit to the retention data; minimizing B achieves smoothing. Too much emphasis on A leads to an unstable result of limited use; too much emphasis on B produces a result in which blood flow in all 50 compartments will become, with large enough Z, exactly equal to each other at 1/50th of the total. This result occurs no matter what the retention data look like, because with a too high value of Z, term A is simply ignored.

An entire subsection on smoothing will be found later in this book.

How does the term B confer smoothing, and more importantly, why does this make the problem of determining the 50 flows tractable and stable in the presence of error, and why does it "work"?

How minimizing the sum of squares of the blood flows creates smoothing can be understood from just a 2-compartment simplification of the 50-compartment model: Suppose we have two compartments with fractional flows of $\dot{Q}_1$ and, therefore, $1 - \dot{Q}_1$. The sum of squares, SS, is:

$$\begin{aligned} \mathrm{SS} &= \dot{Q}_1 \times \dot{Q}_1 + \left(1 - \dot{Q}_1\right) \times \left(1 - \dot{Q}_1\right) \\ &= 1 - 2 \times \dot{Q}_1 + 2 \times \dot{Q}_{1^2} \end{aligned} \tag{3.15}$$

Differentiating with respect to $\dot{Q}_1$ and setting this result to zero (which identifies the lowest value of SS) gives:

$$d(\mathrm{SS})/d\left(\dot{Q}_1\right) = -2 + 4 \times \dot{Q}_1 = 0 \tag{3.16}$$

This in turn is easily solved for $\dot{Q}_1$: $\dot{Q}_1 = 0.5$

Thus, the sum of squares, SS, is minimized when $\dot{Q}_1 = 0.5$—that is, when both compartments have the same flow. This concept applies to 50 compartments as well as to 2: ***Thus, smoothing is favored when the sum of the 50 squared flows is***

minimized, simply because equalizing flow between compartments IS smoothing. The converse, unequal flows, is tantamount to jagged irregular flow allocation.

3.3.2 *The Key Issue: Solving the Equations to Derive the $\dot{V}_A/\dot{Q}$ Distribution*

Now we are ready to ask how smoothing makes the MIGET process tractable. We need to restate Eqs. (3.3), (3.13), and (3.14) in matrix algebra terms or things appear too complex. Do not let this throw you. This restatement looks like this, actually much simpler than the three equations it completely replaces:

$$\mathrm{RSS} = \left\| \mathrm{R} - \mathrm{A}.\dot{\mathrm{Q}} \right\| + \mathrm{Z} \times \left\| \dot{\mathrm{Q}} \right\|^2 \tag{3.17}$$

R is the ***vector column*** of six retentions—one column with six rows, thus with the dimension (6 × 1). A is the basic retention ***matrix***:

$$\mathrm{A} = \lambda / \left[\lambda + \dot{V}_A/\dot{Q} \right] \tag{3.18}$$

A is simply the matrix form designation of Eq. (1.20) generalized for many (here 6) gases and many (here 50) $\dot{V}_A/\dot{Q}$ compartments. The A matrix therefore has 6 rows and 50 columns and thus dimension (6 × 50). $\dot{\mathrm{Q}}$ is the ***vector row*** of 50 blood flows, with 50 columns but 1 row, with dimension (1 × 50). The matrix product $\mathrm{A}.\dot{\mathrm{Q}}$ in Eq. (3.17) has the dimension of (6 × 50) by (50 × 1), that is, (6 × 1), the same as the dimension of R (as it should).

Z is again a single number, called a scalar, as opposed to a vector (i.e., *Z* is not a column or row or matrix) and is the numerical value of the smoothing coefficient.

Differentiating Eq. (3.17) with respect to $\dot{Q}$ works essentially as for ordinary equations, and setting the result equal to zero will minimize RSS just as for the linear regression example worked above. This operation gives:

$$d\mathrm{RSS}/d\dot{Q} = -2 \times \mathrm{A_T} \times \left(\mathrm{R} - \mathrm{A}.\dot{\mathrm{Q}} \right) + 2 \times \mathrm{Z} \times \dot{\mathrm{Q}} = 0 \tag{3.19}$$

where $\mathrm{A_T}$ is the transpose of the A matrix (i.e., 50 × 6 form of A with 50 rows and 6 columns). In other words, when the original A matrix has its rows now become its columns and at the same time its columns become its rows, we say it has been transposed. A simple example using a 3 × 3 matrix filled with the numbers 1 through 9 is shown below:

If a matrix

$$\mathrm{A} = \begin{matrix} 1 & 2 & 3 \\ 4 & 5 & 6 \\ 7 & 8 & 9 \end{matrix}$$

Its transpose,

$$A_T = \begin{matrix} 1 & 4 & 7 \\ 2 & 5 & 8 \\ 3 & 6 & 9 \end{matrix}$$

Thus, from Eq. (3.19),

$$A_T \times \left(R - A.\dot{Q}\right) = Z \times \dot{Q} \tag{3.20}$$

Now a sleight of hand:
If the residual vector, which we call "r," is defined as:

$$r = R - A.\dot{Q} \tag{3.21}$$

r is the difference between the data, **R**, and the least squares best fit to the data, $\mathbf{A}.\dot{\mathbf{Q}}$. It is, like **R** itself, a single column vector with six rows, with dimension (6 × 1).

Substituting Eq. (3.21) into Eq. (3.20) gives:

$$A_T \times r = Z \times \dot{Q}$$

Or, turning it around,

$$\dot{Q} = A_T \times r/Z \tag{3.22}$$

Now substituting Eq. (3.22) into Eq. (3.21) yields:

$$r = R - A \times A_T \times r/Z$$

Or, after collecting the two terms in **r**:

$$(I + A \times A_T/Z) \times r = R \tag{3.23}$$

Here **I** is the matrix equivalent of the number 1—that is, a 6 × 6 matrix with all the diagonal entries equal to 1 and all other entries equal to zero. $\mathbf{A} \times \mathbf{A_T}$ is the matrix product of **A** with its transpose, thus with dimension (6 × 50) × (50 × 6) or (6 × 6).

Hence, both **I** and $\mathbf{A} \times \mathbf{A_T}$ have the same (6 × 6) dimension, and so must their sum, $\mathbf{I} + \mathbf{A}.\mathbf{A_T}$. Now, **r** has dimension (6 × 1), and so the final left-hand side product is a column vector with dimension (6 × 1), the same as the data column containing the six retentions, **R**!!

What have we done in moving from Eq. (3.20) to Eq. (3.23)?

We have gotten rid of the 50 unknowns $\dot{Q}$ and replaced them with just 6 unknowns, r. These r values are the residuals—the differences for each of the

six gases—between measured and least squares best-fit retentions. In addition, we have reduced the large, rectangular ("underdetermined") equation matrix (6 equations in 50 unknowns) to a small, square ("exactly determined") matrix of six equations in six unknowns. These are six simultaneous, linear equations. The six unknowns are the residuals (r) for the six gases—that is, the six differences between each measured gas retention and the best fit to the corresponding measured value.

Please now refer back to Eqs. (3.1) and (3.2). They are two simultaneous equations in two unknowns. Imagine Eqs. (3.1) and (3.2) simply expanded from two to six unknowns and from two to six equations. They can be solved for A, B, . . ., F by high school algebra-level methods of progressive elimination of variables followed by back substitution, as shown above for Eqs. (3.1) and (3.2).

Now that we have determined the 6 values of the residuals, r, by solving the 6 Eq. (3.23) system, we can compute the 50 values of $\dot{Q}$ directly and uniquely from Eq. (3.22).

That is MIGET. Well, almost.

3.3.3 *Additional Constraints on the Equation System*

3.3.3.1 The Blood Flows Must Sum to Cardiac Output

There remain some very important yet somewhat subtle additions to this scheme before it will actually work physiologically, and these are now discussed. The first of these is quite simple:

The sum of all the compartmental blood flows must equal the cardiac output. We actually simplify the computer program by using fractional blood flows, meaning that the sum of the factional flows must equal 1.0. Summing to 1.0 is simply commonsense and reflects nothing more than mass conservation of flow.

Note that this requirement was NOT guaranteed by solving Eq. (3.23) and calculating $\dot{Q}$ values from Eq. (3.22) as formulated from the six equations. We can encourage this to happen, however, rather easily. We do this by adding to the system one more equation in the 50 blood flows, so we now have a 7 × 50 A matrix and 7 data points. The seventh equation is simply:

$$\Sigma\dot{Q} = 1.0 \tag{3.24}$$

The values of A in this seventh row are thus all 1.0. Thus, when this seven-equation system is solved, all seven residuals, **r**, are now minimized. The seventh is the difference between 1.0 and the sum of all fractional blood flows. Note that Eq. (3.24), being only one of seven equations, will ***help*** the flows sum to one, but ***the effect may not be absolute***—it factors into the minimization of residuals as one of seven influences. The sum of the blood flows will be very close to 1.0 but may not precisely equal 1.0.

3.3.3.2 No Compartment May Have Blood Flow Less Than Zero

This is yet one more very important constraint to add to the system to make it work physiologically—the non-negativity constraint. The concept is again simple, but the implementation is a bit complex. The concept is that in the real lung, the minimum value of blood flow in any region is zero. There cannot be negative blood flows in any compartment.

But, if you have followed the preceding paragraphs concerning the equation system and its solution (from about Eq. (3.17) on down), you may have picked up on the fact that the seven-equation system defined to this point has no inbuilt or inherent limitation on the ***sign*** (i.e., $\pm$) of any of the seven residuals. Indeed, we ***expect*** that in the set of seven residuals, roughly half will be negative, and roughly half will be positive. This is exactly analogous to the linear regression on a two-dimensional *X-Y* plane discussed previously: about as many data points will lie above the best-fit regression line as below it. In fact, if this were not the case, it would not be the best possible fit line!!

The point is that some of the seven residuals computed from Eq. (3.23) must be negative. Now turn to Eq. (3.22) which computes the 50 flows from the seven residuals. $\dot{Q}$ depends on three things in that equation: A_T, r, and *Z*. All 50 $\times$ 7 cells of A_T contain numbers that must be positive. In addition, *Z* is positive. But r must have some negative numbers in it. Thus, the product of Eq. (3.22), $A_T.r/Z$, that is, the set of 50 $\dot{Q}$ values, may well have some negative numbers if all we do is solve Eq. (3.23) and substitute into Eq. (3.22) as currently formulated.

We have just stated that negative values of $\dot{Q}$ have no meaning and cannot be allowed. ***So, the question becomes how to ensure that the outcome of this process contains no negative values of*** $\dot{Q}$.

Here is the stepwise (iterative) logic of how we achieve this:

Step 1 *Solve Eqs. (3.22) and (3.23) including all 50* $\dot{V}_A/\dot{Q}$ *compartments in the A matrix*. This will in almost every case return a set of 50 $\dot{Q}$ values in which some are negative.

Step 2 Throw out of the A matrix all those compartments whose $\dot{Q}$ values from step 1 were negative. This results in a smaller A matrix of course—fewer $\dot{V}_A/\dot{Q}$ compartments.

Step 3 Again solve Eqs. (3.22) and (3.23) from the beginning but with this smaller A matrix. Look and see if this time any compartments came back with negative flows. If so, repeat Step 2, followed by this step 3, as many times as necessary ***until no more compartments appear in the solution with negative flow. Is it possible that all 50 compartments will end up with negative flow in this iterative process? NO, as the flows must sum to +1.0. But in theory, this could result in just 1 of 50 compartments remaining, containing all of the blood flow.***

Step 4 Now look at each of the compartments that was *not* used after steps 1–3 (not used because they wanted to have blood flow that was negative). Ask whether forcing ***increased*** blood flow (from zero, the least allowable) in any of these compartments will ***reduce*** or ***increase*** the residual sum of squares. Either outcome is possible. Basically, find the compartment, if any, that would have the ***biggest*** RSS-reducing effect by increasing its blood flow from zero, and ***throw it back into the available A matrix.*** Then go back to Step 3—that is, solve Eqs. (3.22) and (3.23) once more.

Step 5 Now repeat the Step 2/Step 3 exercise of throwing out any compartments that this time around have negative flow until the result comes back with no negative flow compartments.

Step 6 Now execute Steps 4 and 5 once again.

Eventually this iterative looping will end with a definitive, unique result, as follows:

1. For all $\dot{V}_A/\dot{Q}$ compartments having positive flows, either increasing or decreasing their flows from the values found will only increase the RSS.
2. For all $\dot{V}_A/\dot{Q}$ compartments that were tossed out because they wanted negative flows, increasing their flows from zero to a tiny amount above zero will increase the RSS.

Thus, we will have found a result in which 50 or fewer compartments show positive flow, the remainder are allocated zero flow, and ANY changes to the flow distribution that still maintain the non-negativity requirement would worsen the RSS.

Now that we have the set of compartmental $\dot{Q}$ values, we can trivially compute the corresponding set of $\dot{V}_A$ values from Eq. (2.14):

$$\dot{V}_A = \dot{Q} \times \dot{V}_A/\dot{Q} \tag{2.14}$$

That, truly, is MIGET. ***Well, almost.***

We have used only the measured retentions and have ignored the simultaneously measured excretions and not even considered the consequences of experimental errors.

3.4 Error Propagation, Using Both *R* and *E*, and Weighting

As this subsection title implies, we need to do something about experimental error and include so-far-unused data, the excretions. These two aspects are closely connected and are thus discussed together. What, why, and how?

What: Properly weight each of the seven equations in the system so that each influences the result appropriately—not too little nor too much. Add information from the excretions to improve the accuracy of the recovered distribution.

Why: Because by their nature each of the six gases will be subject to intrinsically different levels of measurement error. This is because of different absolute concentrations, because of different accuracies in their measurement, and because of their intrinsically different degrees of retention versus elimination. That means weighting each equation to allow for differences in error and also reducing variance by combining the excretion data with those of retention.

How: First, we need to remember that in the laboratory we are measuring both retention R and excretion E directly, but, as Eq. (2.8) showed, these are NOT independent variables—they are necessarily related by the need to conserve mass. Here again is Eq. (2.8):

$$\dot{V}_{\mathrm{E}} \times E = \lambda \times \dot{Q} \times [1 - R] \tag{2.8}$$

This might imply that it is redundant to measure both R and E because one can be computed from the other. That implication, while correct in a perfect, error-free world, is not correct in the real world because of the existence of experimental error. R and E offer two different but complementary views of the same picture, much like a posteroanterior and lateral chest radiograph looking at the same chest but from different, complementary perspectives. As with any measurement, having two estimates and averaging them must reduce total variance and improve the result.

It turns out that due to experimental error, the high $\dot{V}_{\mathrm{A}}/\dot{Q}$ end of the $\dot{V}_{\mathrm{A}}/\dot{Q}$ spectrum ($\dot{V}_{\mathrm{A}}/\dot{Q}$ greater than 1) is best resolved in MIGET from the values of E, while the low $\dot{V}_{\mathrm{A}}/\dot{Q}$ end ($\dot{V}_{\mathrm{A}}/\dot{Q}$ less than 1) is best resolved by R. But that does not mean we use only E for high $\dot{V}_{\mathrm{A}}/\dot{Q}$ regions and only R for low $\dot{V}_{\mathrm{A}}/\dot{Q}$ regions—we use a weighted average of information from both R and E for every gas, with the relative weight of R falling and that of E rising as we progress from low to high gas solubility.

Thus, it makes sense to use both R and E. We do this using Eq. (2.8) to compute R from E and averaging this set of "R from E" values with the directly measured R values:

$$\mathrm{R, from\,E} = 1 - \left(\dot{V}_{\mathrm{E}} \times E\right)/\left(\dot{Q} \times \lambda\right) \tag{3.25}$$

To make the following easier, let us define **ALPHA** as $\dot{V}_{\mathrm{E}}/\left(\dot{Q} \times \lambda\right)$, so that:

$$\mathrm{R, from\,E} = 1 - \mathrm{ALPHA} \times E \tag{3.26}$$

And then:

$$\mathrm{R, average} = (\mathrm{R, direct} + \mathrm{R, from\,E})/2$$

This average, as written above, can be seen to be an equally weighted average of the two estimates (R,direct and R,from E). That is the concept, but in reality we do not equally weight them. Unfortunately, we cannot in general simply average the above two R estimates equally as the above form of the equation implies. Using error propagation theory (laid out below for our case), we find that the least variance in R,average will *not* be seen when the R,direct and R,from E estimates are equally weighted. Here is how we solve this problem and come up with the right weighting for each of the six gases.

If we use the letter $\boldsymbol{t}$ as the fractional weight for R,direct, then the fractional weight of R,from E must simply be $\mathbf{1-\boldsymbol{t}}$. The above equation is then properly expressed as follows:

$$\text{R, average} = t \times \text{R, direct} + (1 - t) \times \text{R, from E} \tag{3.27}$$

In general, t is not 0.5 as stated. Moreover, it is not the same for each gas but varies systematically from close to 1 for low-solubility gases to close to 0 for high-solubility gases. How so? ***We use differential calculus once again to identify for each gas the value of t that minimizes the variance of the estimate R,average.***

Putting Eq. (3.26) into Eq. (3.27), we have:

$$\text{R, average} = t \times R + (1 - t) \times (1 - \text{ALPHA} \times E) \tag{3.28}$$

In Eq. (3.28), both R and E will contribute variance to **R,average.** The amount of variance contributed by R and by E varies according to the gas as explained above. Conventional error propagation theory can be used to compute this and allow the value of $\boldsymbol{t}$ that produces the least amount of propagated error to be determined.

We now need to find the value of $\boldsymbol{t}$ (different for each of the gases) that will minimize the variance in R,average. We do that by expressing variance in R, average as a function of the variances of its components on the right side of Eq. (3.28). This results in an equation that contains t^2 and which, when differentiated with respect to $\boldsymbol{t}$ and set equal to zero, identifies the value of $\boldsymbol{t}$ that minimizes variance in R,average. See Evans and Wagner 1977 [1] for further details; this is a very involved piece of algebra and calculus.

At the end of the day, $\boldsymbol{t}$ for minimum variance in **R,average** is, differently for each gas, given by:

$$t = (Y + Z)/(X + Y + 2.0 \times Z) \tag{3.29}$$

where, for each gas:

$$X = R \times R(\text{VAR}_\text{a} + \text{VAR}_\text{v})$$
$$Y = \text{ALPHA} \times \text{ALPHA} \times E \times E \times (\text{VAR}_\text{e} + \text{VAR}_\lambda + \text{VAR}_\text{v})$$
$$Z = \text{ALPHA} \times E \times R \times \text{VAR}_\text{v}$$

Here, $\boldsymbol{R}$ is directly measured retention; $\mathbf{VAR_a}$ is the variance in the arterial gas concentration; $\mathbf{VAR_v}$ is the variance in the mixed venous gas concentration; **ALPHA** is the ratio of total ventilation to the product of total blood flow and partition coefficient $\dot{V}_\text{E}/(\dot{Q} \times \lambda)$; $\boldsymbol{E}$ is directly measured excretion; $\mathbf{VAR_e}$ is the variance in mixed expired gas concentration; and **VARλ** is the variance in the partition coefficient.

From this, the minimum variance value of retention (**RMV**) is, as above, given by:

$$\mathrm{RMV} = t \times R + (1 - t) \times (1.0 - \mathrm{ALPHA} \times E)$$

While the corresponding value of excretion (**EMV**) is given by:

$$\mathrm{EMV} = (1.0 - \mathrm{RMV})/\mathrm{ALPHA}$$

The variance of **RMV** is given by:

$$\mathrm{VARIANCE} = Y - 2.0 \times t \times (Y + Z) + t \times t \times (X + Z + 2.0 \times Z) \qquad (3.30)$$

And the weight **WT** given to **RMV** is the inverse of the square root of the variance:

$$\mathrm{WT} = 1.0/\mathrm{SQUARE\,ROOT}(\mathrm{VARIANCE})$$

Remember, these calculations are performed separately for each gas.

In this equation set, R, E, and λ have already been defined and are the measured data for the inert gas in question in the particular sample being analyzed. One just inserts the measured values of R, E, λ for the actual data set being analyzed, plus the previously measured, and hence known, variances for the arterial, expired, and venous inert gas measurements, and $\boldsymbol{t}$ can be computed.

You will be relieved to know that $\boldsymbol{t}$ is automatically calculated internally in the MIGET software. All that needs to be worried about are the variance values, and you should use appropriate numbers for the particular laboratory doing the measurements. The MIGET software allows for entry of your own values of variance (it asks for the coefficient of variation as a fraction).

Where are we going with all of this stuff? The answer, fortunately, is now fairly simple. We now have the variance values calculated for each of the six gases by the above process, Eq. (3.30). Then, to balance the six gas equations (i.e., to give each gas its proper influence in the solution procedure), we divide the entire equation for each gas (i.e., both R and every cell of the row of the A matrix for each gas) by the ***square root of its own variance***. This changes NOTHING within any one row—both sides of the equation $R = \mathrm{A} \times \dot{Q}$ above are divided by this number, which does not change the equality. However, each row (each gas equation) has been altered in absolute magnitude and altered differently for each gas.

Why divide the equation for each gas by the square root of its own variance? The square root of variance is standard deviation. Think of a variable ***R*** with standard deviation **SD**. Now divide BOTH ***R*** and **SD** by **SD**! We now have (***R*/SD**) as a new, ***weighted*** variable, with a standard deviation of 1!!

That is what we have done. ***Every one of the six equations has been modified by dividing by its own SD, so as to have its retention have the same SD of 1.0. That is, the 6 equations have been balanced so that based on their individual errors, each gas contributes appropriately to the final result, that is, the set of 50 compartmental flows.***

Finally, we assign a weight to the seventh equation (the equation that enforces the summation of all flows to 1.0). It is chosen empirically, big enough to have some teeth, not so big as to dominate the process. That said, the seventh equation does very little work when the data for the six gases are of good quality, simply because the lung, and not just our software, also obeys mass conservation principles. In other words, with good-quality data, we will achieve summation to one without having to really enforce it. The weight for this seventh equation is also hardwired into the software and does not need further consideration.

3.4.1 More on Weighting

We have just talked a lot about weighting the seven equations to equalize their influence on the outcome according to how much experimental error they each are subject to. In a very similar way, we need to weight each of the 50 compartments as well. This is because the basic algorithm fits the retention data (R,average) to the blood flow distribution, after which we calculate the obligatory corresponding ventilation distribution (see Eq. 2.14 above). But when $\dot{V}_A/\dot{Q}$ ratio is very high, small errors in $\dot{Q}$ propagate to large errors in $\dot{V}_A$, and we need to consider this. We do so by placing a weight on each $\dot{V}_A/\dot{Q}$ compartment that is basically determined by its own $\dot{V}_A/\dot{Q}$ ratio so that the likelihood of large errors of this type is mitigated. Again, this is all done internally within the program.

3.5 The Residual Sum of Squares

Remember that the solution to the seven-equation system proceeds by first calculating the seven residuals, **r**, as in Eq. (3.23). Using just the six residuals related to the six gases (i.e., setting aside the residual of the seventh equation pushing the sum of blood flows to be 1.0, because we are interested in how well the inert gas retentions are being matched by the fitting process), we compute the sum of squares of the six gas residuals. This number is what we call the residual sum of squares (RSS).

Since we have weighted the equations so that they each have unit variance, the RSS has direct meaning we can use. As the sum of the (squared) individual residuals, RSS represents the overall ability to fit a $\dot{V}_A/\dot{Q}$ distribution to the retention data. The actual value of the RSS is in units of variance. An RSS of, say, 6.0 means that the average for each gas is 1.0 (as there are six gases). That means that the least squares best-fit retentions on this specific occasion differ from the measured retentions on average by 1 SD for each gas.

The cumulative chi-square distribution with six degrees of freedom can be used to determine the probability of seeing a given RSS. That table says that 50% of RSS values should not exceed 5.3. It says that 90% of RSS should be less than 10.6, that 95% should be less than 12.6, and that 99% should be less than 16.8.

Figure 3.1 shows the whole cumulative chi-square distribution for six gases. It is important to understand what this approach is saying and thus how to use it properly.

It says that suppose you do a study in which 100 separate MIGET distributions are obtained, perhaps several each from a group of subjects. You will get from that 100 different RSS values. You can and should compare your cumulative frequency distribution of the 100 RSS values to Fig. 3.1. That will tell you whether your experimental errors are larger or smaller than one would expect on a random basis with the supposed amount of error you have entered into the MIGET program during data entry. That is a way to do quality control—to know if your methods are good enough or not.

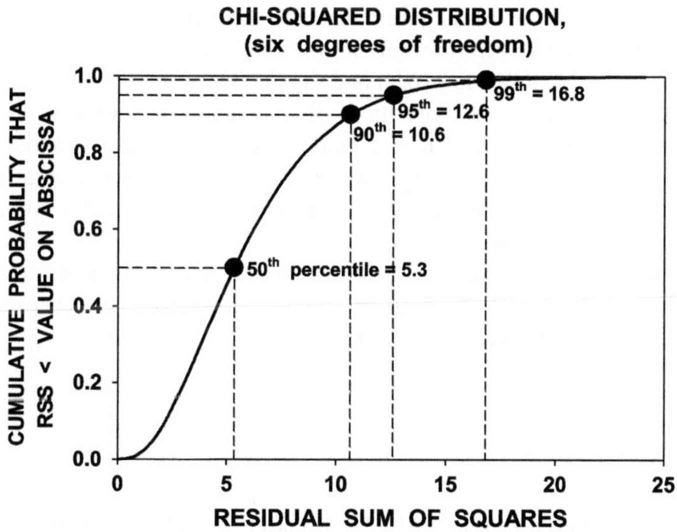

Fig. 3.1 The cumulative chi-square distribution for six degrees of freedom. This curve shows the expected probability of seeing indicated residual sums of squares (RSS) when fitting random error-containing retentions from the six gases to the $\dot{V}_A/\dot{Q}$ distribution. The numbers are based on retentions normalized to their variance; thus, by definition, each retention has variance = 1.0. The way to use this relationship is to collect all RSS values given in the MIGET computer output from all subjects/conditions in a given study and plot the actual cumulative frequency of measured RSS against the line in the figure. RSS should be 5.3 or lower in 50% of cases. It should be 10.6 or lower in 90% of cases and so on. This exercise evaluates data quality

What you should not usually try and do is assess an ***individual*** MIGET distribution by its RSS when RSS is in the expected range (less than about 12, 95th percentile). In any given case, because errors are random, RSS may be anything from near zero to perhaps 12, with a reasonable probability that would not cause you to question the input data or worry that RSS is too high and reflects a problem. However, if RSS is higher than about 12, then the probability that this simply reflects random error becomes more remote. This is a signal for you to go back and check the input data for accuracy on that data set. This is especially true if other data sets on the same day produced much lower RSS so that the case in question is an outlier. A decimal point error, something in the wrong units, etc., can all cause RSS mayhem, and often looking at the raw data will allow you to find and correct the error and resubmit the data to the MIGET software.

One more quality control-related use of the residuals is by examining the residual for each gas. If any one gas has a residual of more than 2 (or less than −2), it means that gas value is more than 2 sd away from expected. This has a probability of happening by chance only 5% of the time (you know that ±2 sd encompasses about 95% of values). If this is an isolated value, look for a blood or gas sampling error or a data entry error. If this is a systematic error seen in a whole set of many MIGET samples on a given day, look for something systematic—perhaps the solubility measurement was in error (this would affect all MIGET data from a single subject), or the measurement method was contaminated, or perhaps the absolute concentration of the particular gas is for some reason very low, which increases the relative variance for that gas and causes a disparity between the value of variance you told MIGET to expect and the real variance, which for this gas was much greater. On rare occasions, the high molecular weight gases (enflurane/isoflurane in particular) may show a residual in which the measured retention exceeds the best-fit value on a systematic basis. If technical errors have been excluded, this outcome is compatible with a role for molecular weight in the elimination of the inert gases.

This completes discussion of how a set of retentions and excretions is transformed into the $\dot{V}_A/\dot{Q}$ distribution. What follows now is discussion of what we have really gotten as a result of doing this.

References

1. Evans JW, Wagner PD. Limits on VA/Q distributions from analysis of experimental inert gas elimination. *J Appl Physiol Respir Environ Exerc Physiol*. 1977;42:889–98.
2. Stewart WE, Mastenbrook SM Jr. Graphical analysis of multiple inert gas elimination data. *J Appl Physiol Respir Environ Exerc Physiol*. 1983;55:32–6.

Chapter 4
Strengths and Limitations from a Mathematical Perspective

Abstract This chapter goes into further detail about the underlying mathematics of the multiple inert gas elimination technique (MIGET), discussing the strengths and weaknesses. One limitation is because of the mass conservation equations that govern MIGET; smooth retention curves are present when there is fine irregularity between adjacent compartments. Another reason that fine detail is limited is experimental error: Smoothing stabilizes the distribution to random errors, but this limits the ability to discern high-frequency biological irregularity in the distributions. MIGET can separate regions of ventilation-perfusion ($\dot{V}_A/\dot{Q}$) ratio $\geq$0.005 from shunt and $\dot{V}_A/\dot{Q}$ regions $\leq$100 from deadspace. When shunt and regions with $\dot{V}_A/\dot{Q} = 0.005$ coexist, precise separation is difficult, but the total perfusion to the two regions is robust. The same is true for separating $\dot{V}_A/\dot{Q}$ regions close to 100 from deadspace. Minimal detectable mismatch is limited by smoothing: When the log standard deviation (LogSD) is $\geq$0.3 smoothing does not affect the recovered distribution. Although removing smoothing would allow distributions with LogSD ~ 0.1 to be measured, the results become variable because of random experimental error; thus, MIGET cannot tell trivial inequality from a homogeneous lung. MIGET can identify unimodal distributions when symmetrical or skewed and multimodal distributions up to three modes. The chapter concludes with a discussion of the "orange slice" diagram, the inert gas equivalent of the O_2–CO_2 diagram of Rahn and Fenn, and a graphical analysis based on the retention equation. It serves as the basis to discuss how random experimental errors affect MIGET.

4.1 Fine Structure of the Distribution (Including Shunt Versus Very Low $\dot{V}_A/\dot{Q}$ regions; Deadspace Versus Very High $\dot{V}_A/\dot{Q}$ regions)

It has already been mentioned that fine structure of the $\dot{V}_A/\dot{Q}$ distribution is inherently unable to be discerned. In part because of the very nature of the mass conservation equations that govern MIGET: They produce very regular, smooth retention curves even when there is fine irregularity between adjacent compartments. The additional reason is experimental error. Smoothing (i.e., the value of Z) does a great job of stabilizing the distribution to random errors but at the same time

S.R. Hopkins, P.D. Wagner, *The Multiple Inert Gas Elimination Technique (MIGET)*, Methods in Physiology, https://doi.org/10.1007/978-1-4939-7441-2_4

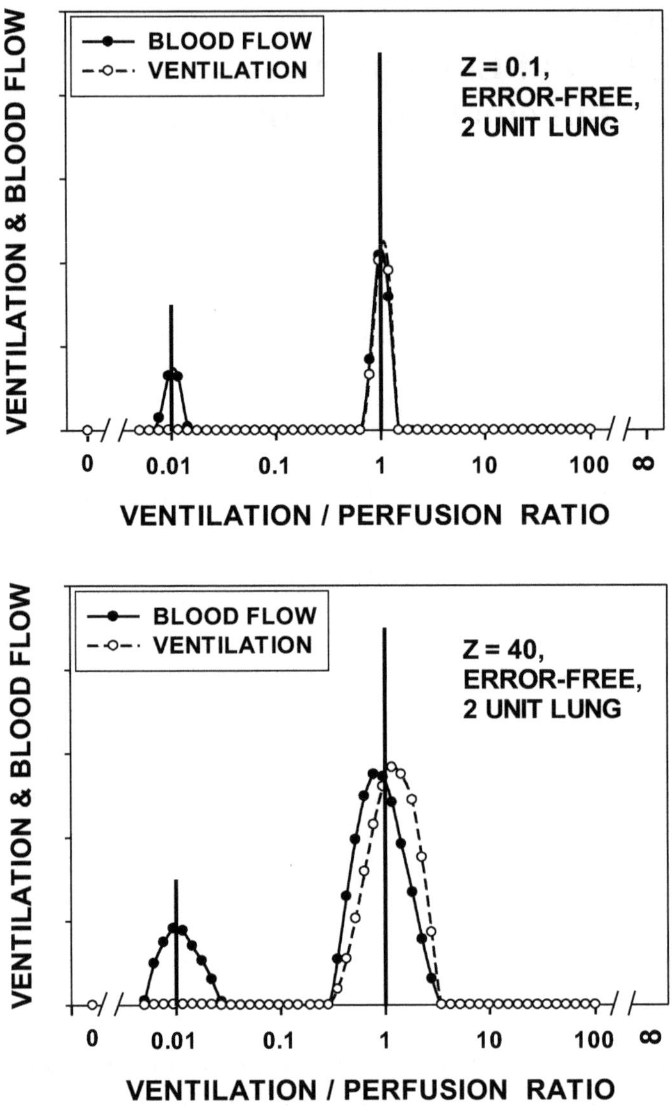

Fig. 4.1 Recovery of a bimodal $\dot{V}_A/\dot{Q}$ distribution from the associated retention values using theoretical, error-free data to illustrate the effects of imposing smoothing in the algorithm. What is seen is that the presence of two populations, their location along the $\dot{V}_A/\dot{Q}$ axis, and the relative blood flow totals attributed to each population are identifiable with smoothing, but the shape and width of each population is beyond resolution

limits the ability to discern high-frequency biological irregularity in the distributions.

To make this section a bit more quantitative, Fig. 4.1 shows, using *error-free* data, fits by the MIGET software to retention data from a particular two-compartment (i.e., bimodal) lung (one compartment of normal and one of reduced $\dot{V}_A/\dot{Q}$ ratio). In this case the $\dot{V}_A/\dot{Q}$ difference between the two compartments is two decades, and the blood flows are 75% (to the unit with normal $\dot{V}_A/\dot{Q}$ ratio = 1) and 25% (to the unit with low $\dot{V}_A/\dot{Q}$ ratio = 0.01) in each. Both panels show these flows in the two compartments as vertical straight lines at the above two $\dot{V}_A/\dot{Q}$ ratios. Recovered $\dot{V}_A/\dot{Q}$ distributions are shown, with minimal smoothing in the upper panel (Z = 0.1) and with usual smoothing (Z = 40) in the lower panel. The scales for the recovered distributions are different from that used for the two given compartments to make for a clearer comparison. It is evident that MIGET is quite able to recover the position and separation of the two modes, and also quantify the total split of blood flow (25% vs. 75%), but cannot tell if each mode is more or less narrow. We wish it could discern the width of each mode, but there is just not enough information in the data for this. That said, the most important characteristics (modality, position, total flow) are well identified.

Figure 4.2 shows in the same format how MIGET behaves when the two compartments are separated in $\dot{V}_A/\dot{Q}$ ratio by only one decade, and it is clear that while for Z = 0.1 it works well, the system has a tough time when the normal value of Z = 40 is used, when the modes are closer together. However, it still indicates more than one mode, the approximate positions along the $\dot{V}_A/\dot{Q}$ axis of each, and the total flow distribution (based on summed flows to the left and right of the nadir between the two partially joined modes).

Figure 4.3 shows the outcome when the two compartments are separated by only half a decade. Again, if we use a low value of Z = 0.1, the recovery is quite good, but we lose the ability to identify two modes when Z = 40 is used. Remember that the greater Z, the more smoothing is enforced.

Figures 4.1–4.3 may lead you to ask why then use a Z = 40 in actual practice when the results are obviously better when Z = 0.1? The answer is experimental error. Figures 4.1–4.3 were created purposefully using computed inert gas data free of experimental error. Realistic levels of random experimental error will not significantly affect the results presented in the lower panels of the figures (when Z = 40) but will create unstable results when Z = 0.1, meaning that duplicate samples processed using Z = 0.1 will yield visually different outcomes for each particular error-containing data set. On the other hand, using Z = 40 sacrifices the potential for fine detail recovery but affords stability in the outcome. But the better way to look at it is that because of experimental error, fine detail is unable to be resolved anyway, Z choice aside. A final comment however is this: One could stabilize the outcome to some extent and have more detail by using a low value of Z if one were willing to make multiple duplicate measurements and average the results. Given the previously provided arguments that actual $\dot{V}_A/\dot{Q}$ distributions should be relatively smooth because they are based on perhaps 100,000 acinar gas exchange units, the theoretical advantage of a low Z and multiple duplicates is not likely to be actually realized.

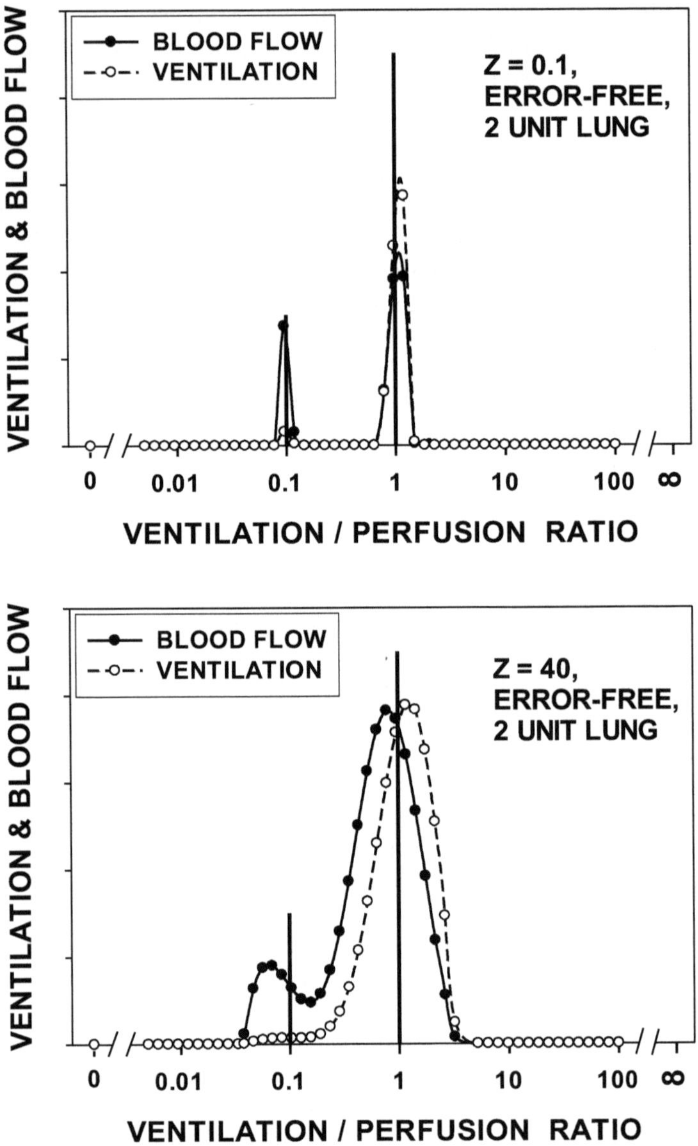

Fig. 4.2 Recovery of another bimodal $\dot{V}_A/\dot{Q}$ distribution from the associated retention values using theoretical, error-free data to illustrate the effects of imposing smoothing in the algorithm. Here the two populations were separated by only one decade in $\dot{V}_A/\dot{Q}$ ratio. What is seen is that the presence of two populations, their approximate location along the $\dot{V}_A/\dot{Q}$ axis, and the relative blood flow totals attributed to each population are still identifiable with smoothing, but the shape and width of each population is beyond resolution, and they begin to merge

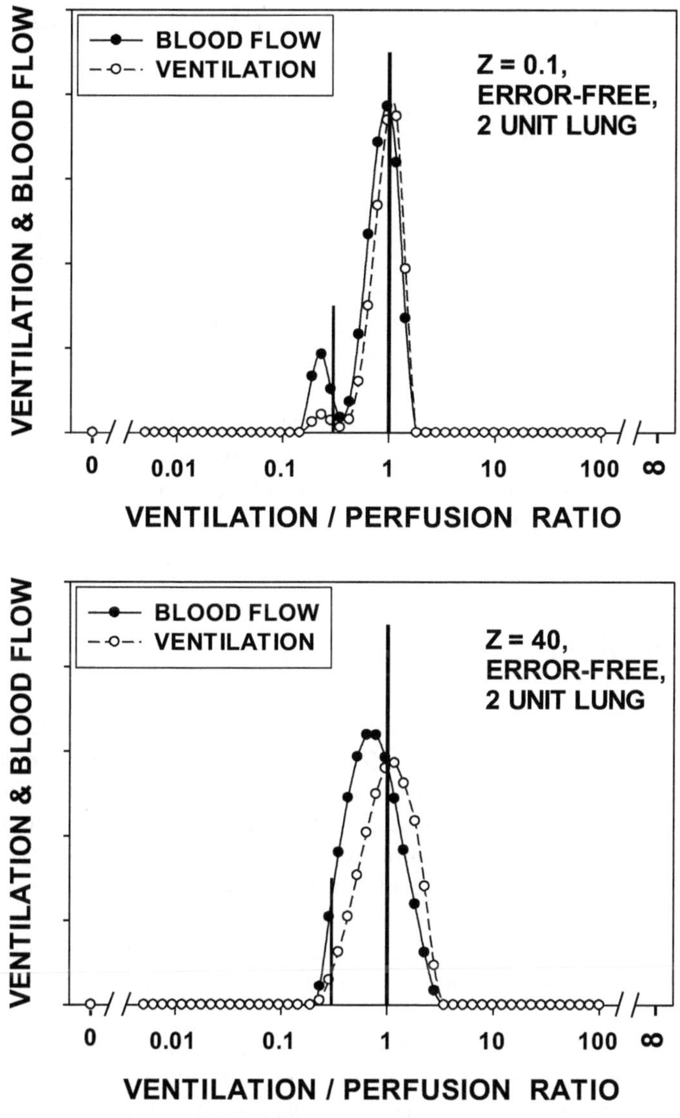

Fig. 4.3 Recovery of yet another bimodal $\dot{V}_A/\dot{Q}$ distribution from the associated retention values using theoretical, error-free data to illustrate the effects of imposing smoothing in the algorithm. Here the two populations were separated by only one half decade in $\dot{V}_A/\dot{Q}$ ratio. What is seen is that the presence of two populations is no longer identifiable. However, the overall dispersion of the distribution (i.e., moments) remains robust, as would the arterial P_{O_2} predicted to result from the $\dot{V}_A/\dot{Q}$ inequality

4.1.1 Lowest $\dot{V}_A/\dot{Q}$ and Shunt; Highest $\dot{V}_A/\dot{Q}$ and Deadspace

Calculating retention sets from lungs having a range of $\dot{V}_A/\dot{Q}$ ratios using the actual gases employed in MIGET and recovering the responsible distributions by the MIGET software has shown that using the usual six gases, even in the presence of usual random experimental error (which in our hands has always been a coefficient of variation of 3% except SF_6 at 5%), we can reliably separate regions of $\dot{V}_A/\dot{Q}$ as low as 0.005 from shunt ($\dot{V}_A/\dot{Q}$ of zero). Similarly, we can separate regions of $\dot{V}_A/\dot{Q}$ as high as 100 from deadspace ($\dot{V}_A/\dot{Q}$ of infinity). We therefore cannot identify and resolve regions of $\dot{V}_A/\dot{Q}$ lower than 0.005 as different than shunt; we similarly cannot identify and resolve regions of $\dot{V}_A/\dot{Q}$ higher than 100 as different than deadspace.

When shunt and regions with $\dot{V}_A/\dot{Q}$ right at 0.005 actually coexist, their precise separation is difficult, but the total perfusion associated with the two types of regions is robust. The same is true for regions of $\dot{V}_A/\dot{Q}$ near 100 in the additional presence of true deadspace.

4.1.2 Minimal Detectable Mismatch

Figure 4.4 shows how MIGET performs with error-free data derived from hypothetical distributions having LogSD, Q values equal to 0 (i.e., homogeneous lung), 0.1, 0.2, 0.3, 0.4, 0.5, 1.0, 1.5, and 1.8. This range encompasses the whole spectrum of observed values from little or no $\dot{V}_A/\dot{Q}$ inequality to that seen in critically ill patients with acute lung injury. Normal subjects usually have LogSD, Q in the range of 0.3–0.5 with the 95% upper confidence limit at 0.6, moderate inequality is a LogSD, Q of about 1, moderately severe 1.5, and severe is anything above 1.5. In each case in Fig. 4.4, the result reflects the use of essentially no smoothing ($Z = 0.1$, closed circles) and the conventionally used level of smoothing ($Z = 40$) in open circles.

What this figure tells us is that when LogSD, Q is 0.3 or greater, smoothing ($Z = 40$) does not prevent accurate recovery of distribution parameters. Removing essentially all smoothing ($Z = 0.1$) would in theory allow distributions with LogSD as low as 0.1 to be measured, but as stated above, this would have the disadvantage of rendering the results unstable in the presence of experimental error.

We simply accept that with usual smoothing, we are limited to identifying inequality associated with LogSD values of 0.3 or greater. Fortunately, in 40 years of using MIGET, we do not find LogSD values this low at all often. Furthermore, it is well established from 60 years of pulmonary ventilation and perfusion imaging by radioactive tracers, CT, and MRI that there are identifiable degrees of inequality due to both gravitational and nongravitational factors that sum to provide just about this degree of inequality [4].

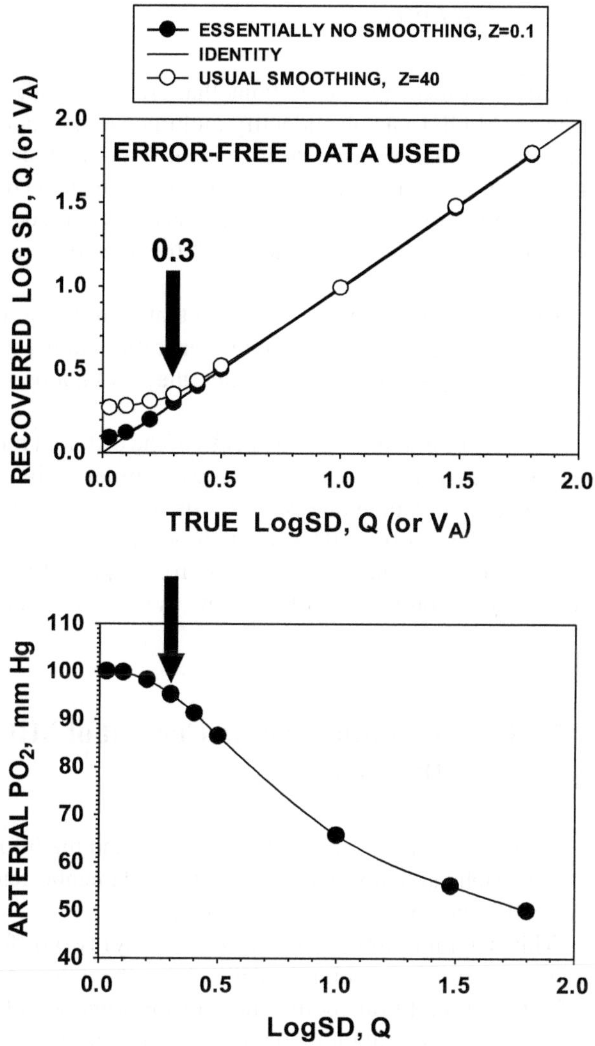

Fig. 4.4 Effect of imposed smoothing on the ability to accurately determine low values of dispersion (LogSD, Q) (*upper panel*). *Lower panel* plots the associated arterial P_{O_2} as a function of LogSD, Q. The outcome is that $\dot{V}_A/\dot{Q}$ distributions with actual LogSD, Q of 0.3 or greater are recoverable in spite of imposed smoothing and that this equates to the alveolar-arterial P_{O_2} difference exceeding 5 mm Hg. The message is that trivial degrees of heterogeneity cannot be resolved, but since real lungs contain at least the amount of heterogeneity associated with LogSD, Q of about 0.3, this is not considered a serious limitation

The lower panel of Fig. 4.4 is helpful in biologically calibrating the values of LogSD, Q. This panel shows the arterial P_{O_2} that would be expected at any given value of LogSD, Q. *The bottom line here is that when arterial* $\mathbf{P_{O_2}}$ *is high in relation to mean alveolar* $\mathbf{P_{O_2}}$ *(i.e., when inequality is trivial), MIGET cannot tell trivial inequality from a perfectly homogeneous lung. That is a limitation we live with in exchange for results that are stable to the effects of random error. The greater the amount of inequality, the better can the software correctly identify the mismatch. When LogSD, $Q = 0.3$ or more (i.e., when arterial* $\mathbf{P_{O_2}}$ *is about 95 mm Hg or less), we are in good shape.*

4.1.3 Modality

MIGET can identify distributions that are unimodal, either when symmetrical or skewed. MIGET can also identify multimodal distributions, as shown in Figs. 4.1–4.3. The best MIGET can do is to identify three modes. With six gases, this is the theoretical limit since in essence each mode is defined principally by its values of total ventilation and blood flow. Thus, each mode consumes two degrees of freedom. As a result, 3 (modes) $\times$ 2 (degrees of freedom/mode) $=$ 6 data points needed.

However, it is important to realize that identifying three modes depends not just on having six gases but also on the separation (along the $\dot{V}_A/\dot{Q}$ ratio axis) of those modes: If they are too close together, they will not be separable, as shown above in Figs. 4.1–4.3.

The rule of thumb is that the peaks of the different modes need to be separated by one decade of $\dot{V}_A/\dot{Q}$ ratio (or more) in order to be resolved by MIGET. When separated by less than one decade, it is still likely that a partially merged bimodal/trimodal distribution will be able to be seen, suggesting the presence of two or more different modes, but only statistics from repeated measurements will be able to determine the probability of modality in such circumstances.

4.2 Limits on Information Content of MIGET: The Orange Slice Diagram

There is a very useful way to better understand the information content of MIGET. It is a graphical analysis based on the fundamental retention equation, Eq. (1.20). We call it the "orange slice diagram" [3].

This diagram plots the retentions of (any) two different inert gases on an X–Y graph, where each point is defined by the retention values of the two gases (of different partition coefficients) for the ***same*** $\dot{V}_A/\dot{Q}$ ratio. If you have followed things so far, you might recognize that this plot is nothing more than the inert gas equivalent of the venerable O_2–CO_2 diagram of Rahn and Fenn [1].

Figure 4.5 shows in the top panel the famous O_2–CO_2 diagram and in the lower panel the inert gas equivalent-based on Eq. (1.20)—for two gases with partition coefficients of 1 (X axis) and 10 (Y axis). The O_2–CO_2 diagram plots the possible pairs of O_2 and CO_2 values for a given $\dot{V}_A/\dot{Q}$ ratio, over the continuous $\dot{V}_A/\dot{Q}$ range from zero to infinity. The figure shows the P_{O_2} and P_{CO_2} values for eight specific $\dot{V}_A/\dot{Q}$ ratios by filled circles and the intervening points by the solid curved line.

The inert gas equivalent has the same general shape, but the orientation is different because O_2 is taken up and CO_2 eliminated, while both inert gases are depicted as being eliminated, as is the case for MIGET.

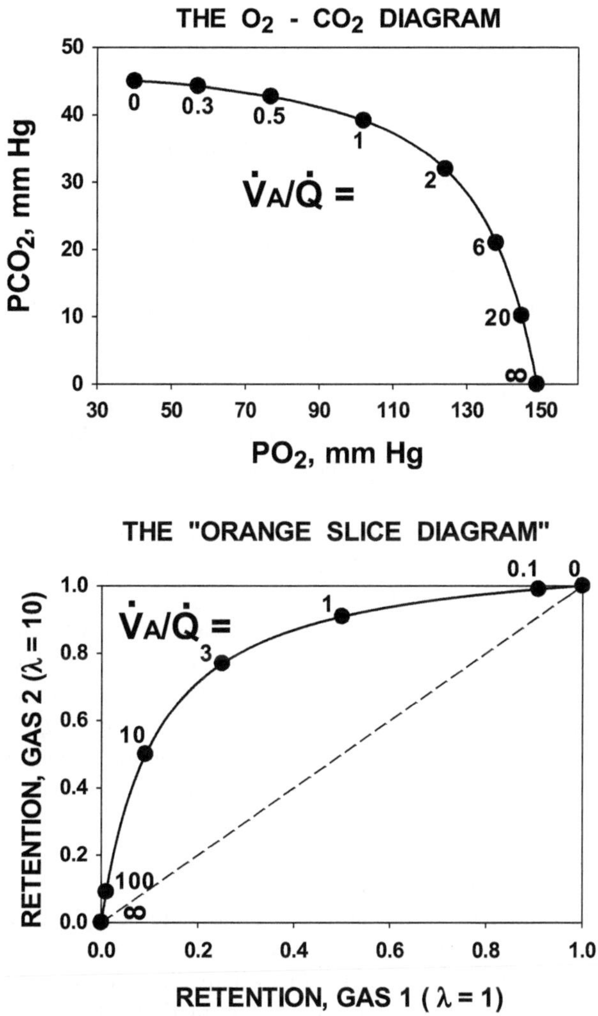

Fig. 4.5 The O_2/CO_2 diagram of Rahn and Fenn (*top panel*) and the corresponding diagram for two inert gases of different partition coefficient both being eliminated (*lower panel*). In each panel, the *solid curved line* shows the location of individual lung units with $\dot{V}_A/\dot{Q}$ ratios as indicated. The end points correspond to shunt and deadspace as shown, and the *dashed line* between these two extreme points encloses an area that looks like an *orange slice*. That enclosed area defines the entire feasible space of retention pairs for the two gases in lungs with all possible patterns of $\dot{V}_A/\dot{Q}$ inequality: That is, no retention pairs can lie outside the enclosed region

The formula for the curved line is easily stated based on Eq. (1.20):

$$Y = 10/\left(10 + \dot{V}_A/\dot{Q}\right) \quad \text{and} \quad X = 1/\left(1 + \dot{V}_A/\dot{Q}\right)$$

when $\dot{V}_A/\dot{Q}$ is eliminated between these two equations, we get:

$$Y = 10 \times X/[1 + 9 \times X] \tag{4.1}$$

This particular equation is for two gases of stated partition coefficient. The graph can of course be constructed for any pair of gases, and the particular shape will reflect the two partition coefficients chosen. Whatever the choice of partition coefficients, Eq. (4.1) is hyperbolic, and must be smooth (no derivatives can be zero). The curve must rise continuously from bottom left to top right as $\dot{V}_A/\dot{Q}$ ratio falls. There can be no irregularities in the curve.

A dashed straight line joins the two ends of the inert gas $\dot{V}_A/\dot{Q}$ line in Fig. 4.5. Its significance will be pointed out below, but this is where the similarity to an orange slice arises: *The area enclosed by the curved line and the dashed straight line looks like an orange slice. This enclosed area has special significance, as discussed below in detail.*

The curved $\dot{V}_A/\dot{Q}$ line in Fig. 4.5 shows the values of retention for the two gases, $\lambda = 1$ and 10, and each point on the line corresponds to a specific, different $\dot{V}_A/\dot{Q}$ ratio indicated on the figure. Retention is zero when $\dot{V}_A/\dot{Q}$ is infinitely high (the gas is completely excreted from the blood) and is 1.0 when $\dot{V}_A/\dot{Q}$ is zero (the gas is completely retained in the blood).

Another way to say it is, if these two gases were exchanged by a perfectly homogeneous lung, their retention values would have to lie somewhere along the curved line. Precisely where would be determined by the specific $\dot{V}_A/\dot{Q}$ ratio of that homogeneous lung.

Now, the minute the lung develops $\dot{V}_A/\dot{Q}$ *inequality, the retention values of the two gases in the mixed arterial blood coming from the lung cannot lie on the curved line any more*. Remember, inequality is defined as the situation in which the lung contains gas exchange units of at least two different $\dot{V}_A/\dot{Q}$ ratios. Suppose that we have inequality, and, for simplicity, have just two different units, with $\dot{V}_A/\dot{Q}$ ratios of 0.1 and 10, respectively. *Each of the two units* would show retentions lying on the curved line at points dictated by Eq. (1.20) for $\dot{V}_A/\dot{Q}$ ratios of 0.1 and 10, as shown in Fig. 4.6. However, the *endcapillary blood from these two units* comes together and *mixes* to form the systemic arterial blood. This mixed blood has a retention value that:

(A) *Cannot* lie on the curved $\dot{V}_A/\dot{Q}$ line.
(B) *Must* lie somewhere along the solid straight line in Fig. 4.6 joining the two points (each on the curved line) that indicate the retentions for the two gases in each of the two units. Just where along that straight line the mixed arterial point will lie depends simply on the fractional distribution of blood flow between the two compartments.

This second point leads to a more general conclusion: The area bounded by the curved line and straight dashed line, shown in Figs. 4.5 and 4.6, identifies what is the only feasible domain of arterial retentions. The arterial blood simply cannot have values for retentions of the two gases that lie outside of this "orange slice."

Figure 4.7 identifies the location of the mixed arterial blood retention point (indicated by the magenta dot) for the two gases we have been considering in this particular two-compartment construct. The magenta dot *must* lie *inside* the orange slice and must line somewhere on the solid straight line connecting the retention points of the two $\dot{V}_A/\dot{Q}$ compartments. This is because it is a simple linear combination (also called weighted average) of the retention values from the two

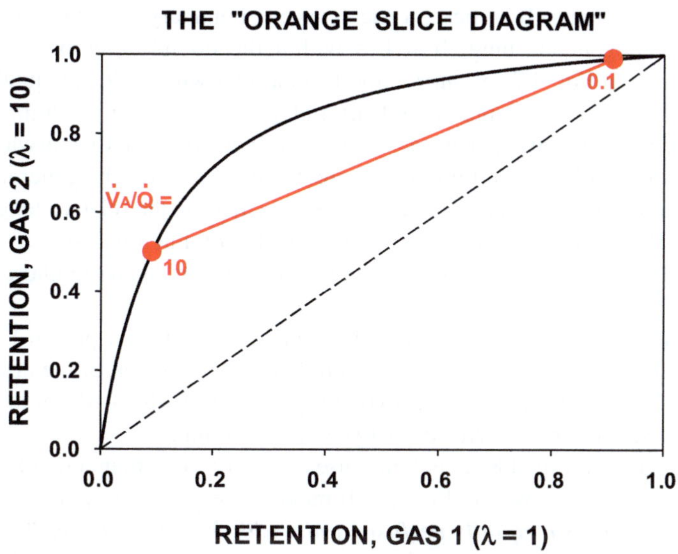

Fig. 4.6 *Orange slice, Part II.* Illustration of a two-compartment lung ($\dot{V}_A/\dot{Q}$ ratios of 0.1 and 10) for two hypothetical gases (λ of 1 and 10). Mass conservation principles demand that the systemic arterial blood coming from mixing of the pulmonary venous blood from each compartment must have retentions for the two gases that lie somewhere on the *straight red line* joining the positions of the two compartments on the *curved line*. Just where on that *line* is determined by the fractional blood flow values of the two $\dot{V}_A/\dot{Q}$ compartments

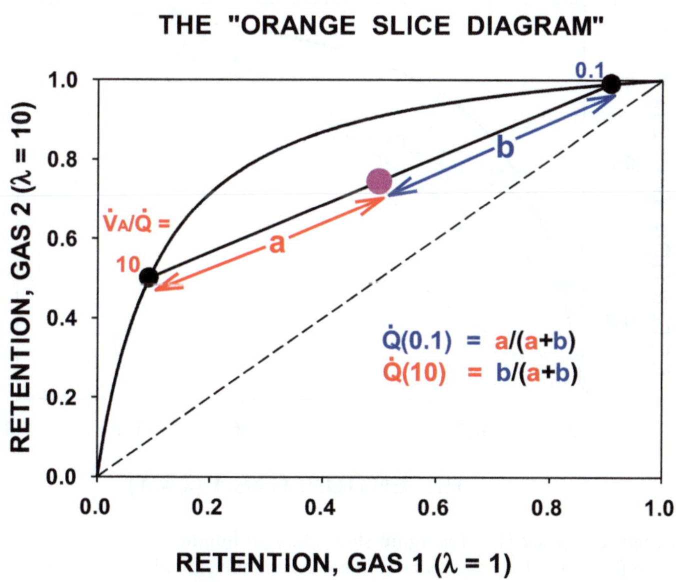

Fig. 4.7 *Orange slice, Part III.* Suppose that the actual retention values for the two gases lie at the *magenta* spot indicated. Mass conservation allows us to compute the precise and only fractional blood flow values associated with each compartment that can account for the *magenta* spot by the length ratios shown on the figure: $a/(a + b)$ and $b/(a + b)$ as indicated

$\dot{V}_A/\dot{Q}$ compartments. If it lies on the (curved) boundary, it *must* come from a perfectly homogeneous lung. If it lies within the bounded region (as does the magenta point in Fig. 4.7), it *must* come from a lung with $\dot{V}_A/\dot{Q}$ inequality.

In Fig. 4.7, that magenta mixed arterial point is shown assuming the two compartments are equally perfused. The point is therefore exactly halfway between the two individual $\dot{V}_A/\dot{Q}$ unit points. Figure 4.7 also indicates the general rule: The distribution of flow between the two units is indicated by the relative lengths of the segments. Thus, fractional blood flow in the low $\dot{V}_A/\dot{Q}$ unit must be $a/(a + b)$, while that in the high $\dot{V}_A/\dot{Q}$ compartment must be $b/(a + b)$. In this particular case, a and b are equal.

It should be evident that this principle can be applied to any pair of $\dot{V}_A/\dot{Q}$ units and any distribution of blood flow between them: The position of the arterial point along the straight line joining the two compartments is dictated uniquely by the fractional distribution of perfusion between the two units.

Now proceed to Fig. 4.8. This figure uses the very same magenta mixed arterial retention point as was used in Fig. 4.7. It makes a very, very important point:

A lung having two equally perfused units with $\dot{V}_A/\dot{Q}$ values of 0.1 and 10—which were the actual choices in the present example—perfectly explains the data point (purple points and line), but it is by no means the only two-compartment lung that would have the exact same arterial retention for the two gases.

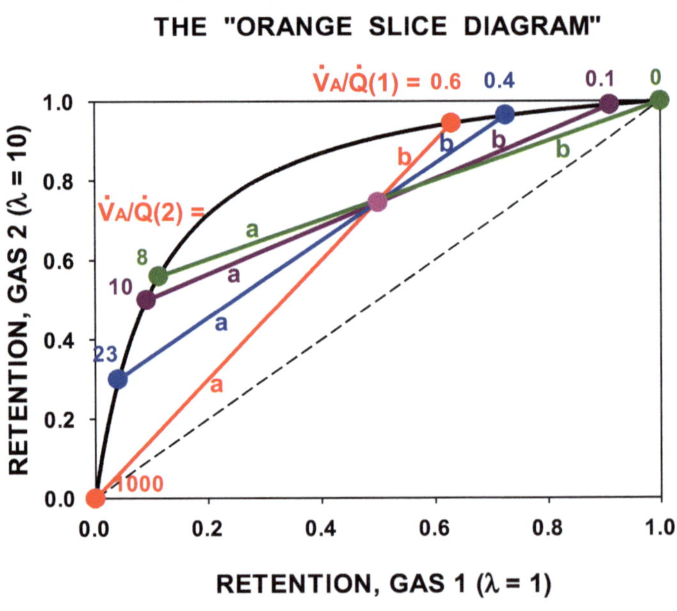

Fig. 4.8 *Orange slice, Part IV*. This figure shows that an infinite number of compartment pairs could equally well explain the *magenta* retention values in the case of Fig. 4.7. However, there are severe constraints on the possibilities: Each such compartment pair must lie at points on the *curved line* such that the *straight line* joining each pair passes through the *magenta* spot. This limits and defines uniquely the $\dot{V}_A/\dot{Q}$ ratio of the second compartment given the $\dot{V}_A/\dot{Q}$ ratio of the first compartment. Moreover, the fractional flows for each pair are precisely defined as in Fig. 4.7 by the segmental length ratios $a/(a + b)$ and $b/(a + b)$ in each case

Figure 4.8 actually demonstrates that an infinite number of different two compartment lungs could explain the same arterial point for the two gases, although it shows only three additional such lungs to the original mentioned above. These three lungs are shown in orange, blue and green. All that is required is that the two units have $\dot{V}_A/\dot{Q}$ ratios on the curved line such that the straight line joining them passes through the actual arterial point. The pairs of $\dot{V}_A/\dot{Q}$ ratios that can explain the arterial point are infinite in number, but are constrained. Thus, if one unit has a $\dot{V}_A/\dot{Q}$ of 0, the other must have a $\dot{V}_A/\dot{Q}$ of 8 (green line) and no other value than 8 is possible. What is more, the fractional distribution of flow between these two units, given by a/(a + b) and b/(a + b) is unique, and must be different from the fractions that must exist for all other feasible pairs of units.

Figure 4.8 is the graphical equivalent of what we call linear programming (LP). LP is a technique that identifies by linear algebraic methods the *maximal possible* value of a constrained function, in this case the maximal possible value of perfusion in a unit of any $\dot{V}_A/\dot{Q}$ ratio. The fractional perfusions shown geometrically by the segment ratios are the graphical equivalent of what a linear programming analysis of the problem would show. Thus, in this figure, the fraction $a/(a+b)$ for the *green* unit pair indicates *maximal possible shunt* ($\dot{V}_A/\dot{Q}$ of zero) *perfusion* in the lung giving rise to the magenta arterial point. Similarly, the fraction $a/(a+b)$ for the *purple* unit pair indicates maximal possible perfusion in a unit with $\dot{V}_A/\dot{Q} = 0.1$ in the lung giving rise to the magenta arterial point—and so on for the other color-coded unit pairs. As you might now appreciate, the four pairs of units indicated in Fig. 4.8 are but four examples out of an infinite number of pairs formed by slowly rotating the straight line through the magenta point from the point at top right ($\dot{V}_A/\dot{Q} = 0$) to bottom left ($\dot{V}_A/\dot{Q} =$ infinity).

Note by extension of these arguments, a two-compartment lung whose $\dot{V}_A/\dot{Q}$ ratios were, say, 0.6 and 8 cannot possibly explain the magenta arterial point in Fig. 4.8, because NO combination of segment lengths can give rise to a point that lies at the actual, magenta arterial position.

What have we learned from this exercise?

1. Given a measured pair of retentions for two gases of different partition coefficient, we can identify an *infinite number of two-compartment lungs* that will perfectly explain the observed retentions.
2. Despite this, the number of combinations of two units is limited by the requirement that each unit *lies on the curved* $\dot{V}_A/\dot{Q}$ *line and must be located at the ends of a straight line passing through the arterial point.* This in turn means that for ANY selected $\dot{V}_A/\dot{Q}$ unit, there is only ONE other $\dot{V}_A/\dot{Q}$ unit that could combine with it to explain the data point, and moreover, the fractional distribution of flow between the two units is unique.
3. Finally, in this two-gas, two $\dot{V}_A/\dot{Q}$ compartment, for example, the fractional flows provided by the segment analysis are the *maximal possible flows* that could account for the data.

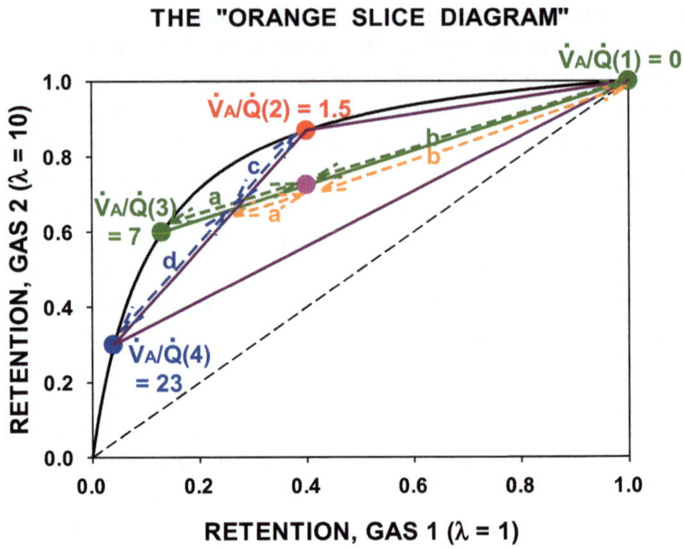

Fig. 4.9 *Orange slice, Part V*. This figure shows that there are $\dot{V}_A/\dot{Q}$ distributions of more than two compartments that can also explain the data. In this case, with three compartments (#1, 2, 4), the *magenta* data point lies inside the *triangle* bounded by these three $\dot{V}_A/\dot{Q}$ compartment points. Indeed, three compartments chosen *not* to include the *magenta* point would *not* be able to explain the data. While there is an infinite number of three-compartment choices to explain the data, all such combinations must form a *triangle* enclosing the data, thus limiting the distribution possibilities quantitatively. Blood flow through each of the three compartments is again uniquely set by the segmental length ratios as explained further in the text

This last point needs a little further elucidation, and this comes from recognizing that an arterial point (e.g., of Fig. 4.8) can also be explained by lungs with more than two compartments. Let's look at a *three*-compartment possibility (to explain the same magenta point as in Fig. 4.8) in Fig. 4.9. This figure starts with the same magenta arterial point, lying well inside the "orange slice." Next, four individual $\dot{V}_A/\dot{Q}$ ratio units are shown along the curved $\dot{V}_A/\dot{Q}$ line: $\dot{V}_A/\dot{Q} = 0$ (green), 1.5 (red), 7 (green), and 23 (blue). The two green units were selected specifically because the green straight line joining them does in fact pass through the magenta arterial point. So, from the preceding analysis, maximal possible shunt ($\dot{V}_A/\dot{Q} = 0$) flow must be given by the segment ratio in *green* on Fig. 4.9: $a/(a + b)$.

Now look at the *three* purple straight lines connecting $\dot{V}_A/\dot{Q}$ ratio units 0, 1.5, and 23. *They form a triangle and, critically, the magenta arterial point lies within the triangle. This fact makes these three chosen compartments able to explain the magenta arterial point. In contrast, if the triangle joining any three* $\dot{V}_A/\dot{Q}$ *units does NOT enclose the measured arterial point, that three unit lung cannot explain the measured arterial point by any combination of flows.*

What must the fractional flows to the three compartments be? First, shunt flow must be the segment ratio in orange: $a/(a + b)$. Second, total flow to the other two compartments must be given by the complementary segment ratio in orange: $b/(a + b)$. This must in turn be divided among the blue and red units as follows: The blue unit perfusion is $c/(c + d)$ times $b/(a + b)$, and the red unit perfusion is $d/(c + d)$ times $b/(a + b)$.

To get back to why the *two* compartment fit gives the highest fractional flows, just compare the green to the orange segment ratios, $a/(a + b)$. For both colors, b is exactly the same, but it is very clear that a (orange) is less than a (green). Thus, the indicated shunt flow is lower in the three-compartment lung than in the two-compartment lung. That in the two-compartment lung gives the maximal possible shunt flow.

This was illustrated using shunt as the target unit, but the concept is applicable to *any* $\dot{V}_A/\dot{Q}$ unit in the exact same way. Moreover, the extension from two to three compartments as satisfactory models accounting for a given data point can be continued to any greater number of compartments. *The clear requirement for a multicompartment model to be able to explain a particular arterial retention set is that the arterial point in question must lie within the polygon formed by straight lines joining all compartments on the orange slice diagram, in a manner similar to that just described for three compartments.*

This discussion of the two-dimensional (two gases, two or more $\dot{V}_A/\dot{Q}$ units) "orange slice" diagram is very useful and conceptually extremely powerful and indicates what MIGET can and cannot do in a very visual manner. However, while the concept expressed in the diagram can be expanded up to six gases, we have not figured out how to construct a six-dimensional diagram to show you how all these notions actually play out when all six gases are used. Therefore, the diagram should be used to understand basic concepts. Do not assume anything about the quantitative outcome scaled up to six gases from what you have just seen in Figs. 4.7–4.9.

Standard computerized linear programming can however expand the concept up to six (or in fact any number of) gases, and we have analyzed the outcome of doing this some time ago [2]. That paper shows examples of the *envelope* of maximal possible flows to all 50 compartments for several sets of six-gas retentions. What is absolutely critical to understand with this linear programming approach is that when maximal possible unit flows are indicated across all 50 $\dot{V}_A/\dot{Q}$ values as such an envelope, these are one-at-a-time maximal flows—of course, flows cannot be maximal in all 50 units simultaneously, or total flow would be an order of magnitude greater than cardiac output!!

In summary, the "orange slice" diagram serves a very useful purpose in showing the principles underlying the constraints on the information content of MIGET.

But we are not done with it yet—it serves an additional perhaps even more useful purpose in providing a platform to begin considering how random experimental errors can affect the MIGET. Until now, we have not dealt at all with random experimental error in exploring the information content of MIGET. All that we have done in the context of error up to this point has been to note that error propagation theory allows us to minimize data variance by weighted averaging of retentions and excretions and at the same time allows us to weight the six gases relative to one another to properly apportion their contributions to the equation matrix used in MIGET. The following section continues with the orange slice concept and brings in the effects of random experimental error.

4.3 Effects of Random Error

The orange slice diagram continues to be extremely valuable in understanding consequences of random error and also in understanding how we approach the problem to minimize its effects.

The discussion begins with Fig. 4.10. As before, the $\dot{V}_A/\dot{Q}$ line is drawn for two gases (now in **magenta**), and the line joining its end points is also shown (dashed line, same color). Recall that this enclosed "orange slice" constitutes what we have called the "feasible region." That is, it is the only area of the entire X–Y plane where actual, *error-free* retention data can exist as combinations of flows to any set of $\dot{V}_A/\dot{Q}$ compartments. This statement comes from the fact that any arterial point is the linear combination of points on the curved $\dot{V}_A/\dot{Q}$ line, weighted by compartmental flows that can only be greater than zero. In Fig. 4.10, the large **red** dot is a particular true arterial point for the two gases. True here means uncontaminated by experimental error.

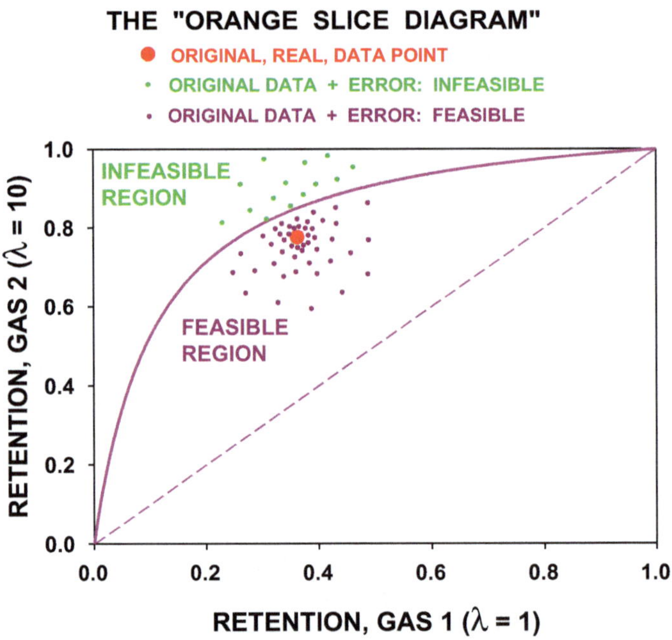

Fig. 4.10 *Orange slice, Part VI*. Consideration of experimental error. Suppose the *solid red dot* marks the actual retention point for the two gases illustrated. It is highly unlikely that any real measurement will realize retention values that exactly equal the actual values, simply due to random measurement error. Retaining the notion of feasible and infeasible regions on the diagram (see Fig. 4.5), random error added to the actual data point (for both gases) will produce a cloud of points that map the probable locations of measured (error containing) retention pairs. That probability cloud is a normal distribution in two dimensions, with more points close to the actual data point than further away. Moreover, some points (those in *purple*) will still lie inside the feasible space, while the other (in *green*) are now positioned outside the feasible space

However, in the real world, random error is present. We think of random error then as moving the true point in some direction. *If truly random, any direction on the X–Y plane is equally possible.* Random error being a normally distributed function, it is more likely that the error-perturbed data point will lie close to the true point than distant from it. So, a cloud of about 60 points has been added on Fig. 4.10 as small dots, more close to than distant from the true point in accordance with the normal distribution of random error. Ignoring for now their color differences, these points form a probabilistic cloud that is statistically symmetrical in all directions and which indicates a sampling of where the error has moved the true data point to. This cloud is called a Monte Carlo simulation of error, created by adding randomly chosen independent, random errors in retention for the two gases simultaneously to the two retentions marking true data point.

The size (diameter) of the cloud will reflect the inherent variance in the measurement of retention, which is information needed when doing a Monte Carlo simulation. In the particular example of Fig. 4.10, a very large variance is assumed, simply to make the illustration visually clearer. In reality, the coefficient of variation (CV) is no more than 3%. On the scale of Fig. 4.10, with a 3% CV, the cloud of 60 points would coalesce into a single blob overlying the true red data point, making the concept hard to discuss.

The key observation to be made in Fig. 4.10 shows that some error-perturbed estimates of the true point remain *inside* the feasible region (**magenta dots**), while some lie OUTSIDE (**green dots**).

The following discussion should be read on the explicit assumption that *for now only* we will *not* employ any smoothing when we talk about using MIGET software to come up with the $\dot{V}_A/\dot{Q}$ distribution that explains the measured data point. We will bring smoothing back into the story in a few paragraphs, but for now—no smoothing.

4.3.1 Magenta Points

Magenta points represent retention sets lying *inside* the orange slice (i.e., in the feasible area); every magenta point can be perfectly modeled (=fitted) by a multicompartmental lung. Submitting such retention sets to the MIGET software would result in a perfect fit to the data—the residual sum of squares would be zero for every magenta data set analyzed. The result would be a $\dot{V}_A/\dot{Q}$ distribution that was perfectly compatible with the error-perturbed data set in question and different for every magenta set. Note that this would *not* be the same distribution that was responsible for the true data point, because the true data point (large red dot) lies some distance away from each of these actual, error-containing, measured values (small magenta dots). You can see (using the logic and analysis of Fig. 4.8) that if we had multiple data sets that differed only by random error, the distributions recovered by MIGET (no smoothing) would perfectly fit each and every

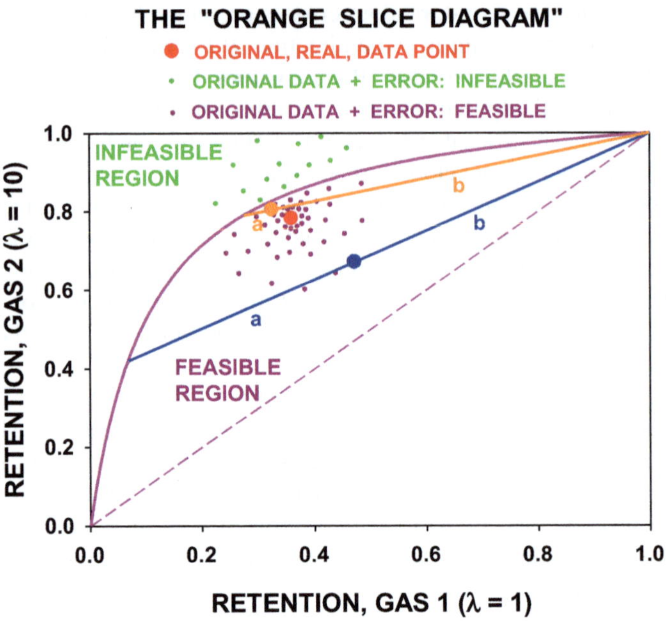

Fig. 4.11 *Orange slice, Part VII*. This figure focuses on two specific error-containing examples from the cloud of possibilities, both lying *within* the feasible space, marked in *orange* and *blue*. Suppose we choose a two-compartment model to fit the data in each case, and in each specify one of those as being shunt. Then, using the logic underlying Figs. 4.7 and 4.8, shunt fraction would be $a/(a + b)$ in each case, but it is obvious that the actual shunt fraction values will be very different comparing *orange* and *blue points*. Random error in this particular example has a major impact on ability to identify shunt (or blood flow in any $\dot{V}_A/\dot{Q}$ compartment)

magenta error-perturbed retention set in Fig. 4.10 but that the compartmental flow fractions could vary all over the place depending onto where random error had moved the true data point.

This is shown in Fig. 4.11, using the geometric analysis of Fig. 4.8 for just two sample error-perturbed points and calculating two-compartment fits to the data. First take the **orange** point, just above and to the left of the true **magenta** point. Shunt is given by the **orange segment ratio, a/(a+b).** This comes to 6.6% (of the cardiac output) in the particular example. Now examine the **blue** point. The same **blue segment ratio, a/(a+b),** is very different and gives a shunt of 43.3% of the cardiac output. Yet both outcomes reflect analysis of the same true retention data set, simply affected differently by the randomness of experimental error.

Please remember that the degree of random error in Fig. 4.11 has been greatly exaggerated upward from real levels, just to make the points clearer. Thus, the real MIGET technology is not so weak as to exhibit such uncertainty in estimates of the distribution. Also note that if there were zero random error, the least squares approach would find a distribution that perfectly fitted the true data point, red in this case, and would not show such variance in parameter estimates.

4.3.2 Green Points

Green points can and do exist in reality (precisely because true points within the feasible region have been exteriorized from the orange slice by the addition of random error). However, they cannot represent a real lung, again because they lie outside of the feasible region. How can we apply the principles laid out above to come up with the $\dot{V}_A/\dot{Q}$ distribution when technically no distribution can fit the data? Submitting such data sets to MIGET actually works just fine. The outcome is a nonzero residual sum of squares, just as for linear regression when actual data never all lie exactly on the regression line. But what will the $\dot{V}_A/\dot{Q}$ distribution look like when technically no distribution can be found to fit the data?

Figure 4.12 shows the principle that applies to such exterior points when a least squares fitting approach is used. Suppose we take one exterior point as an example—the blue point in this figure. Least squares fitting by definition of the concept will find the feasible $\dot{V}_A/\dot{Q}$ distribution whose retentions for the two gases lie closest to the exterior blue point but which are in the feasible space including lying on its boundary. *In other words, the least squares solution is that retention*

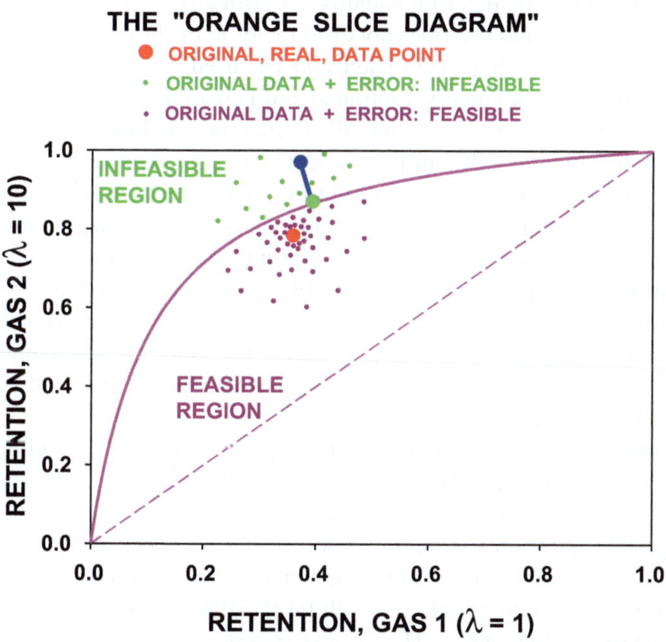

Fig. 4.12 *Orange slice, Part VIII.* This figure focuses on one specific error-containing example (in *blue*) from the cloud of possibilities lying *outside* the feasible space. No compartment model can possibly fit such data, based on the segment analysis concept of Figs. 4.6 and 4.7. Yet, with six gases, data lying outside the feasible region are the rule, not the exception. The figure shows that the best fit to the blue data point (i.e., the shortest distance to the feasible region) will be the *green point* on the *curved line*—which marks a one-compartment lung of a particular $\dot{V}_A/\dot{Q}$ ratio. Yet, the actual retentions (*red point*) lie well within the feasible space and must therefore come from a lung with heterogeneity. We have a conundrum. See text for the way forward!

point for which the ***blue straight line*** *in Fig. 4.12 is the shortest*. The residual sum of squares is in fact the squared length of the blue line. Simple geometry dictates that closest point (and thus lowest sum of squares) must in fact lie precisely on the feasible region's boundary and at a point (indicated by the large **green** point in the figure) identified simply by dropping a perpendicular line from the blue exterior point to the boundary line. Any other blue line (to any other point on the curved $\dot{V}_A/\dot{Q}$ line or to any interior point within the feasible region) must be longer (i.e., give a larger sum of squares).

Clearly, this exercise shows that when least squares fitting is done, the solution (large **green** point in Fig. 4.12) must lie at a point that cannot reproduce either the exterior (**blue**) measured data point or the true interior (**red**) data point.

More subtle, the large green dot must represent a homogeneous lung—because the curved $\dot{V}_A/\dot{Q}$ line in the figure is, as you know, the line showing the position of all possible individual, *single* $\dot{V}_A/\dot{Q}$ ratios. Yet, the true magenta point lies inside the feasible region and must represent a multicompartment lung (or else it would have been positioned precisely on the curved $\dot{V}_A/\dot{Q}$ line bounding the region). Therefore, for exterior data points, least squares fitting *without smoothing* will always find an *extremal* solution—a homogeneous lung. It is extremal because a lung cannot have less than zero inequality. Statistically, extremal solutions are by definition improbable and that does not even factor in the biological issue that when 500 million alveoli are both ventilated and perfused, it is simply impossible to imagine that they are all equally ventilated and perfused. No class of even 20 students, all of the same age, will have all its members exhibit identical height or weight. Imagine a class of 500 million students and the same question.

Looking at the true data point in red in Fig. 4.12, the least squares approach—remember, the above refers to least squares fitting *without* smoothing—gives a clearly unsatisfactory solution (large green point) when random error exists sufficient to throw the data outside the feasible space defined by the orange slice (measured data at large blue point). It also gives an unsatisfactory solution even when the error-perturbed data remain within the orange slice (small magenta points) because, while such data points are perfectly fitted, they are distant from the true red data point.

Inferred but not yet shown is that the appropriate degree of smoothing will lead to a solution where the fitted data point is statistically virtually superimposed on the actual (true) red data point within the orange slice.

An important question is whether, when scaled up to the six-gas MIGET with actual levels of experimental error (coefficient of variation about 3–5%), most measured (error-containing) data sets will lie INSIDE the six-dimensional feasible space or OUTSIDE that space. The answer is simple: *>99% of measured sets lie outside the feasible space*. How do we know that? Simply by applying MIGET to innumerable data sets and noting that the residual sum of squares is greater than zero more than 99% of the time. This can be established theoretically by starting with a feasible region retention set, randomly adding realistic levels of experimental error, and applying MIGET to the error-perturbed retentions. More than 99% of such Monte Carlo simulations return a greater than zero sum of squares (and this is without using smoothing). It can also be established experimentally by looking at the many, many data sets accumulated over the past 40+ years and noting the frequency with which the sum of squares was actually zero—vanishingly small.

What this means is that the feasible region of Figs. 4.5–4.12 is very, very small in six dimensions such that actual random errors exteriorize the data in the vast majority of cases.

4.4 Smoothing and Random Errors

So now we return to the value of smoothing in the least squares approach.

Earlier, smoothing was said to be based and justified, on three principal foundations: (a) that the real distribution is highly likely to be smooth due to the many thousands of gas exchange units present (so we lose little or no discrimination by enforcing a degree of smoothing), (b) that the smoothing confers stability in the recovered distribution in the inevitable presence of random error, and (c) that the inherent smoothness of the retention curve itself does not allow very fine detail to be discerned in the first place.

First, the preceding analysis has shown that simple least squares fitting (without smoothing) will result in potentially large differences in recovered $\dot{V}_A/\dot{Q}$ distribution parameters from duplicate data sets that differ only in random error. Adding the smoothing constraint has been shown to produce data fits and associated distributions that do not differ much as a consequence of random error. Thus, smoothing greatly mitigates this inherent instability caused by random error. In fact, the numerical amount of smoothing (the value of Z in Eq. 3.14) is titrated to the amount of random error such that when a Monte Carlo simulation of error-perturbed retention sets is performed, the outcome is shown to be acceptably stable. We choose the lowest value of Z (least amount of smoothing) that stabilizes the results. Any lower value would cause significant variations between duplicate samples; any greater value of Z would cause loss of resolution among $\dot{V}_A/\dot{Q}$ ratios.

A Z value of 40 has been found over many years to be that optimal value.

The orange slice analysis can be used to understand another, fourth, rationale for using smoothing: Adding the appropriate level of smoothing fits the measured data, whether they lie within or outside the feasible region, with a retention set that on average (almost) overlies the true (magenta) retention point! This can be shown again by Monte Carlo simulation with repeated application of MIGET to randomly perturbed retention sets. This is a very strong attribute of the approach.

The "price" paid for this is a slightly higher residual sum of squares (see Fig. 4.12). A straight line from the blue, exterior, error-containing actual data point to the red, true, data point is longer than the blue line from the same blue point to the nearest point on the boundary of the feasible region. Remember that the length of these lines (squared) is in fact the geometric equivalent of the residual sum of squares. But this price seems irrelevant: Smoothing gets you stability in the face of random error and the statistically best estimate of the true data at the same time.

The only thing that smoothing really loses us, given that random error is unavoidable, is the ability to recover distributions whose LogSD values are less than 0.3, as illustrated previously.

Believe it or not, we are essentially done with the mathematical theory of MIGET. We can now turn to more interesting biological and methodological issues. Two major classes of such issues need to be discussed:

First, the physiological (not mathematical) information content of MIGET and, secondly, the practical issues of running MIGET.

References

1. Rahn H, Fenn WO. Graphical analysis of the respiratory gas exchange. American Physiological Society: Bethseda, MD; 1955.
2. Ratner ER, Wagner PD. Resolution of the multiple inert gas method for estimating VA/Q maldistribution. Respir Physiol. 1982;49:293–313.
3. Wagner PD. A general approach to the evaluation of ventilation-perfusion ratios in normal and abnormal lungs. The Physiologist. 1977;20:18–25.
4. West JB. Ventilation/blood flow and gas exchange. Oxford: Blackwell; 1972.

Chapter 5
Physiological Information Content of MIGET

Abstract Metrics derived from the multiple inert gas elimination technique (MIGET) can be chosen, depending on the research question at hand. The overall appearance of the ventilation-perfusion ($\dot{V}_A/\dot{Q}$) distribution is informative. Normal subjects have narrow, unimodal, symmetrical distributions with $\dot{V}_A/\dot{Q}$ ratios of 0.3–3, exclusive of deadspace. In patients with lung disease, the distributions may be broad or multimodal. Fractional blood flow and ventilation can be partitioned into V_A/Q ratio domains to quantify contributions to each: deadspace ($\dot{V}_A/\dot{Q}$ ratio >100), high $\dot{V}_A/\dot{Q}$ (3.0–100), normal $\dot{V}_A/\dot{Q}$ (0.3–3.0), low $\dot{V}_A/\dot{Q}$ (0.3–0.005), and shunt <0.005. Central moments of the $\dot{V}_A$ and the $\dot{Q}$ distribution are useful parameters, especially the second moments, $\text{LogSD}, \dot{V}_A$ and LogSD, Q for the ventilation and blood flow distributions, respectively. Dispersion parameters—known as Disp,R; Disp,E; Disp,R-E; and Disp,R-E*—similar to alveolar-arterial differences for O_2 and CO_2, can be derived directly from the retentions/excretions. Since inert gases are much less vulnerable to diffusion limitation across the blood/gas barrier than oxygen, this can be exploited and the expected arterial P_{O_2} calculated from the observed $\dot{V}_A/\dot{Q}$ distribution. When diffusion limitation occurs, the actual arterial P_{O_2} is less than expected. This difference is also used to compute a whole-lung O_2-diffusing capacity. Finally, MIGET data can be used to infer information about incomplete gas-phase diffusive mixing or different lung structures such as in birds.

5.1 The $\dot{V}_A/\dot{Q}$ Distribution and Its Parameters

This section is about what information can be derived from MIGET. One of the positive features of MIGET is that many different outcome parameters can be obtained and they can be chosen to optimize answering the investigator's particular question.

This topic has been partially addressed above in discussing limits on MIGET but will be covered again here from a more practical standpoint. There are many different ways to look at the $\dot{V}_A/\dot{Q}$ distribution, and each offers something a little different. Which outcomes to use in a given study will be dictated by the biological question(s) you are trying to answer. Choosing which basically comes down to

S.R. Hopkins, P.D. Wagner, *The Multiple Inert Gas Elimination Technique (MIGET)*, Methods in Physiology, https://doi.org/10.1007/978-1-4939-7441-2_5

common sense. Some are qualitative, some quantitative (and thus amenable to statistical treatment). Mostly, more than one outcome will be useful.

5.1.1 Visual Appearance of the $\dot{V}_A/\dot{Q}$ Distribution

Visual appearance of the $\dot{V}_A/\dot{Q}$ distribution itself is the first and most obvious outcome to examine, as shown in several prior figures, e.g., Figs. 2.10–2.15. Recall that this consists of two mathematically connected frequency distributions—that of ventilation (ordinate) against $\dot{V}_A/\dot{Q}$ ratio (abscissa) and that of blood flow (ordinate) against $\dot{V}_A/\dot{Q}$ ratio (abscissa). The mathematical connection is simple:

$$\dot{V}_A = \dot{Q} \times \dot{V}_A/\dot{Q} \tag{2.14}$$

So that if you have one frequency plot (e.g., $\dot{Q}$ versus $\dot{V}_A/\dot{Q}$), you automatically have the other from this relationship.

Here is the critically important concept to understand in interpreting these two curves: When **low** $\dot{V}_A/\dot{Q}$ regions are present, there must be by definition **more blood flow than ventilation** in those regions. **It is therefore the plot of $\dot{Q}$ versus $\dot{V}_A/\dot{Q}$ that will be visually more abnormal than that of $\dot{V}_A$ versus $\dot{V}_A/\dot{Q}$**. Yet think about the actual biological conditions in the lung in those low $\dot{V}_A/\dot{Q}$ regions: How do low $\dot{V}_A/\dot{Q}$ regions come to exist? There can be only two possibilities—increased blood flow or reduced ventilation. The former is impossible mathematically in significant amounts when the low $\dot{V}_A/\dot{Q}$ ratios are, well, low (less than 0.1, say). This is simply because to have normal ventilation yet a $\dot{V}_A/\dot{Q}$ ratio of, say, 0.01 from increased perfusion mandates a 100-fold increase in blood flow, and this is simply unreasonable. On the other hand, it is easy to imagine a $\dot{V}_A/\dot{Q}$ ratio of 0.01 arising from a 99% reduction in ventilation from airway obstruction, while blood flow remains normal. What's the message?

Abnormalities characterized by <u>low</u> $\dot{V}_A/\dot{Q}$ regions affect the blood flow part of the $\dot{V}_A/\dot{Q}$ distribution much more than the ventilation component but are actually more likely to reflect ventilation, not blood flow, disturbances in the lung.

Exactly the same can be said for interpreting high $\dot{V}_A/\dot{Q}$ regions, in reverse:

Abnormalities characterized by <u>high</u> $\dot{V}_A/\dot{Q}$ regions affect the ventilation part of the $\dot{V}_A/\dot{Q}$ distribution much more than the blood flow component but actually reflect blood flow, not ventilation, disturbances in the lung.

That this happens in life is shown by the many studies showing low $\dot{V}_A/\dot{Q}$ regions as the hallmark of obstructive airway diseases (asthma, chronic bronchitis) and high $\dot{V}_A/\dot{Q}$ regions as the hallmark of obstructive pulmonary vascular diseases (pulmonary embolus).

The normal range of $\dot{V}_A/\dot{Q}$ ratios in healthy young human subjects is from about 0.3 to 3 [6]. Essentially all normal subjects should exhibit narrow, unimodal, symmetrical distributions confined to this $\dot{V}_A/\dot{Q}$ range (except, of course, for having a sizable fractional deadspace of 20–40% due to the conducting airways being in series with the gas-exchanging alveoli).

5.1.2 Modality of the Distribution

Modality of the distribution will be apparent on inspection, that is, whether the distribution consists of a single mode or of two or even three separate modes. Recall that no more than three modes can be discerned by MIGET. The distribution in health has been shown many times to be narrow, logarithmically normal, and unimodal, but in disease, modes that are separate from the normal mode and depict either or both low and high $\dot{V}_A/\dot{Q}$ regions are common. However, remember that not all patients or study subjects will exhibit multiple modes. Some disease situations result in broad, unimodal distributions that may or may not be lognormal (i.e., they may be skewed leftward or rightward or have "shoulders" connected to the main mode). When the distribution is separated into distinct modes (as opposed to a broad, single mode), it suggests that the different modes represent populations of alveoli subject to distinctly different, essentially "all or nothing" influences, e.g., airways open or (partly) closed. A broad unimodal distribution on the other hand suggests a more continuously varying influence, e.g., airways subject to a variable degree of constriction.

5.1.3 Summing Up of Fractional Ventilation and Blood Flow in Chosen Domains of the $\dot{V}_A/\dot{Q}$ Spectrum

One choice is to add up fractional blood flow (and/or ventilation) within the normal $\dot{V}_A/\dot{Q}$ range of 0.3–3.0 and do the same for the low $\dot{V}_A/\dot{Q}$ range below 0.3 (down to the lower limit, 0.005) and again for the high $\dot{V}_A/\dot{Q}$ range above 3.0 (up to the upper limit, 100). You can lump in shunt (perfusion at $\dot{V}_A/\dot{Q} = 0$) with low $\dot{V}_A/\dot{Q}$ blood flow if you wish or keep them separate. Similarly you can lump in deadspace (ventilation at $\dot{V}_A/\dot{Q} =$ infinite) with high $\dot{V}_A/\dot{Q}$ ventilation if you wish or keep them separate. You now have quantitative indices of fractional distribution of $\dot{V}_A$ and $\dot{Q}$ among potentially five domains—shunt, low $\dot{V}_A/\dot{Q}$, normal $\dot{V}_A/\dot{Q}$, high $\dot{V}_A/\dot{Q}$, and deadspace—to treat statistically.

Another choice is to sum fractional $\dot{V}_A$ and $\dot{Q}$ in each separate mode.

And, of course, identifying shunt ($\dot{V}_A/\dot{Q} = 0$ and up to the 0.005 lower limit of detectability) and deadspace ($\dot{V}_A/\dot{Q} =$ infinite and down to the 100 upper limit of detectability) is a form of expressing outcomes of MIGET as activity in particular $\dot{V}_A/\dot{Q}$ domains. This is done routinely by the MIGET software.

Yet another choice is to (arbitrarily) sum fractional $\dot{V}_A$ and $\dot{Q}$ within each decade of $\dot{V}_A/\dot{Q}$ (i.e., all flow in compartments with $\dot{V}_A/\dot{Q}$ less than 0.01 (down to the lower limit of 0.005, thus making for only part of a decade); similarly for all compartments with $\dot{V}_A/\dot{Q}$ between 0.01 and 0.1; 0.1 and 1; 1 and 10; and 10 and 100). This choice is in fact tabulated automatically by the MIGET software to save you having to do this by hand.

The summing of flow within chosen $\dot{V}_A/\dot{Q}$ ranges can be very useful if the primary object of the study is to ask whether a particular region of the $\dot{V}_A/\dot{Q}$ spectrum is affected by an intervention, because this is the most direct way of approaching the question. For example, in asthma, whether a drug reduces low $\dot{V}_A/\dot{Q}$ regions may be best tested in this way, so that influences on other parts of the $\dot{V}_A/\dot{Q}$ distribution do not distort the result for the particular region of interest.

However, there is one inescapable limitation of this kind of quantitation of the $\dot{V}_A/\dot{Q}$ distribution: changes in total $\dot{V}_A$ or $\dot{Q}$ within a specified $\dot{V}_A/\dot{Q}$ range may be due to changes in the *amount of* $\dot{V}_A/\dot{Q}$ *inequality*, or they may be due to a simple *shift of the* $\dot{V}_A/\dot{Q}$ *distribution leftward or rightward* (without change in amount of inequality) should the intervention change total ventilation or blood flow. This limitation can almost always be lifted by simultaneously examining other outcome parameters of the $\dot{V}_A/\dot{Q}$ distribution—visual appearance as above and indicators of amount of inequality (see below).

5.1.4 Central Moments

Central moments of the $\dot{V}_A$ and (separately) the $\dot{Q}$ distributions are commonly used as outcome parameters. We routinely calculate the first three moments, all on a logarithmic rather than linear $\dot{V}_A/\dot{Q}$ ratio scale. Higher-order moments could easily be computed but lack physiological significance and computational robustness. What do the moments mean?

The ***first*** moment is the mean $\dot{V}_A/\dot{Q}$ ratio of the $\dot{V}_A$ and $\dot{Q}$ distributions—that is, their position along the $\dot{V}_A/\dot{Q}$ axis. The first moment is affected not only by the total ventilation and total blood flow but also (and independently) by the degree of $\dot{V}_A/\dot{Q}$ inequality. The first moment of the $\dot{V}_A$ distribution must always be higher than that of the $\dot{Q}$ distribution. When low $\dot{V}_A/\dot{Q}$ regions develop *but total ventilation and blood flow remain constant*, the first moment of the $\dot{Q}$ distribution will fall while that of $\dot{V}_A$ must rise. Symmetrically, when high $\dot{V}_A/\dot{Q}$ regions develop (constant total ventilation and blood flow), the first moment of the $\dot{V}_A$ distribution will rise while that of $\dot{Q}$ must fall. However, suppose total $\dot{V}_A$ rises, total $\dot{Q}$ stays constant, and there is no change in amount of $\dot{V}_A/\dot{Q}$ inequality. Then, the first moment of both $\dot{V}_A$ and $\dot{Q}$ distributions will rise, and the opposite will occur if total ventilation were to fall, nothing else changing. While this should all be intuitively clear, it points out the need to carefully look at first moments of *both* $\dot{V}_A$ and $\dot{Q}$ before trying to interpret changes from an intervention.

The formulae to compute the first moments for $\dot{V}_A$ and $\dot{Q}$, respectively, are:

$$V_A, \text{mean} = \text{Antilog of}: \quad \Sigma\left[\dot{V}_A \times \text{Log}(\dot{V}_A/\dot{Q})\right] \tag{5.1}$$

$$Q, \text{mean} = \text{Antilog of}: \quad \Sigma\left[\dot{Q} \times \text{Log}(\dot{V}_A/\dot{Q})\right] \tag{5.2}$$

These equations sum the log of each of the 48 $\dot{V}_A/\dot{Q}$ ratios from lowest $\dot{V}_A/\dot{Q}$ (0.005) to highest $\dot{V}_A/\dot{Q}$ (100), weighted by the actual ***fractional*** $\dot{V}_A$ (Eq. 5.1) or ***fractional*** $\dot{Q}$ (Eq. 5.2) at that $\dot{V}_A/\dot{Q}$ ratio as it exists in the particular distribution under consideration. The result is the ventilation- or perfusion-weighted average log $\dot{V}_A/\dot{Q}$ ratio of the distribution. We then take the antilog of this to express the result as the mean $\dot{V}_A/\dot{Q}$ ratio itself. Note that if the distributions are lognormal (i.e., symmetrical), the resulting mean $\dot{V}_A/\dot{Q}$ ratios will be precisely the same as the $\dot{V}_A/\dot{Q}$ ratio associated with the peak of the distribution. Note also that shunt and deadspace must be excluded from this computation because their $\dot{V}_A/\dot{Q}$ ratios, namely, 0 and infinity, do not have a defined log value that can be plugged into the above equations. Thus, neither shunt nor deadspace indicated by MIGET will influence these two moments.

The ***second*** moments reflect dispersion about the above-computed mean $\dot{V}_A/\dot{Q}$ ratios and again are computed separately for the $\dot{V}_A$ and $\dot{Q}$ distributions. A homogeneous lung would exhibit second moments of zero, the lowest possible value (negative values are not feasible). As inequality increases, so too will the values of the second moments. The highest value ever seen is 2.5. Hence, the range encountered in actual practice runs from 0 to 2.5. However, recall from prior discussion that because of our choice to include smoothing, we cannot identify a value for this parameter if it is actually less than 0.3. The second moment is almost always in the range of 0.3–0.5 in healthy young lungs. More formally, the 95% upper confidence limit is 0.6 for the blood flow moment and 0.65 for ventilation, indicating that values above these limits would be expected in normal humans no more than 5% of the time. When these moments reach about 1.0, we consider inequality to be moderate; when 1.5, inequality is severe; and when 2.0 or greater, inequality is very severe. The lower panel of Fig. 4.4 provides a clear calibration of these numbers in terms of the arterial P_{O_2} that would accompany these degrees of inequality (under conditions of breathing air at rest at sea level).

We have come to call the second moments LogSD, $\dot{V}_A$ and LogSD, Q for the ventilation and blood flow distributions, respectively. The Log part again reflects our choice to use second moments calculated on a log scale. The SD abbreviation reflects "standard deviation" which is strictly a correct use of this term only when the distribution is completely symmetrical (i.e., lognormal). It then has the same meaning as standard deviation in ordinary statistics—defining the region of the $\dot{V}_A/\dot{Q}$ spectrum in which about 68% of the ventilation or blood flow lies. When the distribution is unimodal but asymmetrical, or when the distribution is multimodal, we need to say that LogSD reflects overall dispersion (i.e., inequality) but should not be interpreted as a true standard deviation.

LogSD, Q and LogSD, $\dot{V}_A$ have become perhaps the most useful and robust parameters of inequality. Their strength lies in their robustness, which in turn stems from their distilling all information from the six gases (except for shunt and deadspace as already explained) into just two parameters. Their weakness is that they cannot portray the complexities of the distributions: two quite differently appearing distributions could have the same LogSD values. They are used when the major question is to ask whether an intervention changes the overall amount of $\dot{V}_A/\dot{Q}$ inequality.

The formulae for calculating LogSD, $\dot{V}_A$ and LogSD, Q are:

$$\text{Variance}, \dot{V}_A = \Sigma V_{A_j} \times \left[\text{mean}\,\text{Log}(\dot{V}_A/\dot{Q}) - \text{Log}(\dot{V}_A/\dot{Q})_j\right]^2 \qquad (5.3)$$

$$\text{Variance}, \dot{Q} = \Sigma Q_j \times \left[\text{mean}\,\text{Log}(\dot{V}_A/\dot{Q}) - \text{Log}(\dot{V}_A/\dot{Q})_j\right]^2 \qquad (5.4)$$

Here, the subscript j refers to summation over the 48 $\dot{V}_A/\dot{Q}$ ratios from 0.005 ($j=1$) to 100 ($j=48$) inclusive (thus excluding shunt and deadspace as before for the first moments).

In Eq. (5.3), **mean Log**$(\dot{V}_A/\dot{Q})$ is the log of $\dot{V}_A$, mean as computed in Eq. (5.1) (i.e., the Eq. 5.1 result before taking antilogs). In Eq. (5.4), **mean Log**$(\dot{V}_A/\dot{Q})$ is the log of $\dot{Q}$, mean as computed in Eq. (5.2) (i.e., the Eq. 5.2 result before taking antilogs).

Equation (5.3) expresses the variance of the ventilation distribution as the sum of the squared differences between the mean $\dot{V}_A/\dot{Q}$ ratio and the $\dot{V}_A/\dot{Q}$ ratio of each compartment, weighted by that compartment's fractional ventilation, $\dot{V}_A$. It is a conceptually conventional expression of variance of a distribution of numbers about the mean of that number set and is essentially the same formula as might be applied to any data set to determine its variance. Equation (5.4) is the perfusion-weighted expression of the variance and therefore yields the variance of the blood flow distribution. However, we customarily prefer to speak in terms of "standard deviation" which requires taking the square root of the variance.

Thus:

$$\text{Log SD}, V_A = \text{Square root of Variance}, \dot{V}_A \qquad (5.5)$$

$$\text{Log SD}, Q = \text{Square root of variance}, \dot{Q} \qquad (5.6)$$

Interpreting LogSD, V_A and LogSD, Q has essentially been addressed above. These parameters have been the mainstay of statistically testing for differences in inequality from normal or changes in inequality after interventions. Remember they are broad parameters reflecting all 48 of the compartments that are both ventilated and perfused—those of low, normal, and high $\dot{V}_A/\dot{Q}$ ratio. Again, they do *not* reflect shunt or deadspace, since, due to the values of the associated $\dot{V}_A/\dot{Q}$ ratios, one cannot take their logarithms as required in the above equations. Differences in shunt and deadspace must be tested statistically separately. Note that, when low (but not high) $\dot{V}_A/\dot{Q}$ regions are present, LogSD, Q will be the more elevated of the two parameters and that, when high (but not low) $\dot{V}_A/\dot{Q}$ regions are present, LogSD, V_A will be the more elevated.

The ***third*** moment expresses the degree of asymmetry (skew) in a distribution. A symmetrical distribution will have zero skew. A positive or negative value of the third moment can occur and indicate asymmetry in either the low or high $\dot{V}_A/\dot{Q}$ direction. Obviously, bimodal and trimodal distributions will be reflected by non-zero values of the third moment in either the blood flow or the ventilation curve or both. We tend not to use this moment very often, but it is routinely computed in the MIGET program should you wish to make use of it.

The formulae for the third moments are almost identical to those of the second moments, with the same definitions. The only (and obviously key) difference is raising terms to the third power rather than second power, as follows:

$$\text{Skew}, V_A = \Sigma \dot{V}_{A_j} \times \left[\text{mean}\,\text{Log}(\dot{V}_A/\dot{Q}) - \text{Log}(\dot{V}_A/\dot{Q})_j\right]^3 \tag{5.7}$$

$$\text{Skew}, Q = \Sigma \dot{Q}_j \times \left[\text{mean}\,\text{Log}(\dot{V}_A/\dot{Q}) - \text{Log}(\dot{V}_A/\dot{Q})_j\right]^3 \tag{5.8}$$

Customarily, we have never taken the cube root of this parameter, leaving it simply as in the above equations.

5.1.5 *Direct Dispersion Indices*

The final set of commonly used outcome parameters of the $\dot{V}_A/\dot{Q}$ distribution itself is what is termed the "Disp" parameters—**Disp,R; Disp,E; Disp,R-E; and Disp,R-E*.**

What are these? They are best thought of as similar to alveolar-arterial differences for O_2 and CO_2, corresponding parameters of inequality when one is limited to having O_2 and CO_2 as gases to measure. The abbreviation "Disp" simply indicates these as a measure of dispersion. R refers to inert gas retention (i.e., arterial to mixed venous inert gas partial pressure ratios), E to inert gas excretion (i.e., mixed expired to mixed venous inert gas partial pressure ratios), and R-E to their difference.

We have six gases but elect to compute a single value for each of these four parameters. To do this for any one of the four Disp parameters, we compute the *square root of the sum of squares of the particular Disp parameter for each gas, summed over the six gases.* Thus, we now need to define what we mean by Disp for a single gas.

Disp,R, for a single gas, is simply the difference between the measured value of *R* and that of the homogeneous lung having the same total ventilation and blood flow as the actual lung under study (computed from Eq. (1.20) knowing the total $\dot{V}_A$ and $\dot{Q}$ and the blood/gas partition coefficient of that particular gas). Note however that to bring all six gases together as mentioned above, the only fair way to do this is to apply the weighting numbers that provide for equal variance for each gas. These were presented in detail earlier in this manual. Thus, we do not use fractional retentions but rather weighted retentions to calculate Disp,R over all six gases.

Symmetrically, Disp,E, for a single gas, is simply the difference between the measured value of E and that of the homogeneous lung having the same total ventilation, blood flow, and deadspace as the actual lung under study (again computed from Eq. (1.20) knowing the total $\dot{V}_A$ and $\dot{Q}$ and the blood/gas partition coefficient of that particular gas and using the MIGET-derived value of deadspace). Again, to bring all six gases together as mentioned above, we apply the weighting numbers that provide for equal variance for each gas. Thus, again we do not use fractional excretions directly but rather weighted excretions to calculate Disp,E over all six gases.

Disp,R-E (again for a single gas) is just the difference between measured retention and excretion for that gas, again after applying the weighting number. This parameter is generally dominated by the deadspace because that is the single largest contributor to differences between R and E. However, deadspace is a normal part of gas exchange, and it is helpful to eliminate its contribution to Disp,R-E. This is easy to do by dividing measured values of E for each gas by (1 − fractional deadspace). This correction would, for each gas, yield E values that exactly equaled R values if there were no $\dot{V}_A/\dot{Q}$ inequality. This corrected version of Disp,R-E, termed Disp,R-E*, would therefore be zero in the absence of inequality, such that values above zero would indicate positive R-E differences (analog of alveolar-arterial P_{O_2} differences) that must indicate $\dot{V}_A/\dot{Q}$ inequality.

The Disp set of parameters have the value of being derived directly from the measured retentions and excretions and do not reflect the derived distributions of ventilation and blood flow, and this appeals to some users. Again, they represent the average (root mean square) alveolar-arterial difference of the six gases and are expressed in units of standard deviation of measurement error. Thus, a Disp,R of exactly 1.0 means a root mean square alveolar-arterial difference of 1 SD of the value of R.

Disp,R reflects dispersion of the blood flow distribution, Disp,E dispersion of the ventilation distribution, and Disp,R-E and Disp,R-E* dispersion of the combined ventilation and blood flow distributions.

Thus, Disp,R correlates with LogSD, Q and Disp,E correlates with LogSD$\dot{V}_A$. However, there is no LogSD equivalent of Disp,R-E. Furthermore, shunt and deadspace *are* included in these Disp indices as they come right from the measured retentions and excretions.

Whether this last difference [inclusion of shunt and deadspace (using Disp) and exclusion (using LogSD)] is a benefit or a liability in the use of either LogSD or Disp parameters is not worth arguing. The best way to view these differences is by choosing those parameters that best reflect the physiology you are trying to elucidate with your MIGET experiment.

5.2 Diffusion Limitation of O_2 and CO_2 Exchange

The inert gases are about an order of magnitude less vulnerable to diffusion limitation in their transfer across the blood/gas barrier than are O_2/CO_2. What is meant by this statement? It means that when diffusive exchange of gases between alveolar gas and capillary blood takes place, it takes about ten times longer for the blood and gas partial pressures of O_2/CO_2 to equalize than is the case for the inert gases. This is independent of direction of diffusion (i.e., whether from alveolar gas to capillary blood (uptake) or from capillary blood to alveolar gas (elimination)). This is shown in Fig. 5.1. Why is that?

The process of diffusion across the blood/gas barrier depends on several variables: (A) the surface area and thickness of the barrier and the time available in the capillary for exchange to occur, which are all determined by cardiopulmonary structure and function, and (B) properties of the gases being exchanged: their physical solubilities in the Hb-free barrier, their effective solubilities in the blood (i.e., the actual blood solubility for inert gases and the average slope of the dissociation curve for gases like O_2, CO_2, and CO), and their molecular weights (which play a minor role). In any given lung, all of the factors in (A) are of course the same for all gases being exchanged, and so what separates the rates of diffusive exchange of the various gases are the solubilities of the gases (a) in the Hb-free barrier and (b) in the blood.

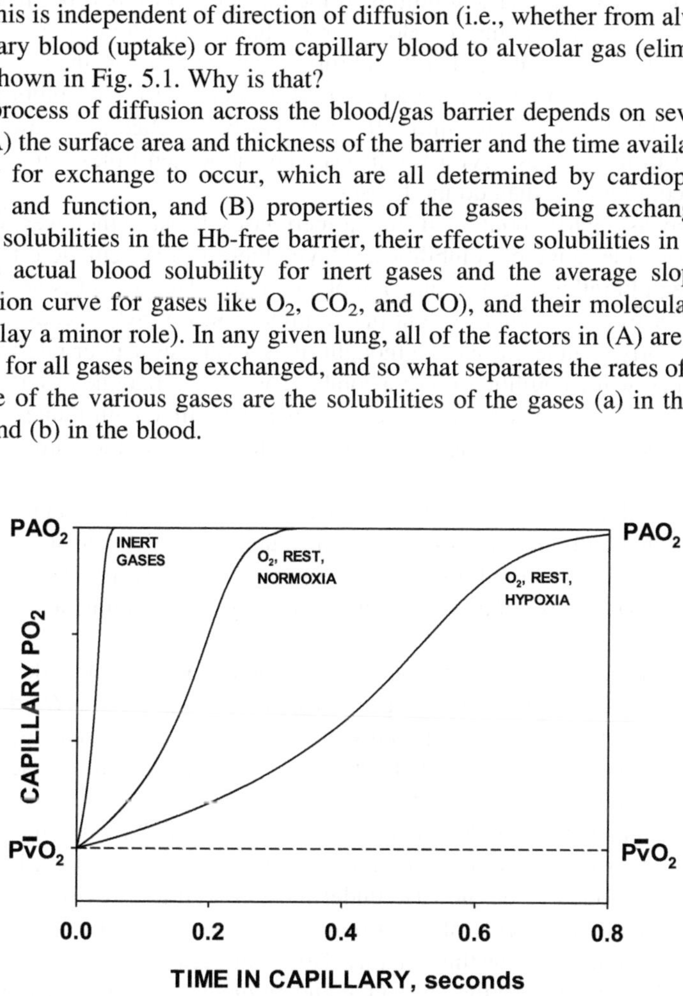

Fig. 5.1 Schematic illustration of rate of rise of P_{O_2} along the pulmonary capillary at rest in normoxia and in hypoxia and, for comparison, corresponding time course for an inert gas in the same lung. In this case, the inert gas is being taken up by the lung rather than being eliminated, but the time course would have the same half time either way. The key point is that inert gases equilibrate an order of magnitude faster than does O_2. Thus by the time a subject showed diffusion limitation for an inert gas, they would be dead from hypoxia due to insufficient O_2 uptake

In essence, the barrier solubility dictates the rate of diffusion across the barrier, while the blood solubility dictates how long it takes to load the gas into the blood for that rate of diffusion across the barrier. Think of it as similar to filling a bucket with water. The analogy here is the rate of pouring water into the bucket represents the rate of diffusion across the barrier, while the size of the bucket represents the effective solubility of the gas in blood.

Here then is the bottom line:

Inert gases have similar solubilities in the barrier as in blood because Hb has little influence on their solubility. However, O_2 has a low solubility in the Hb-free barrier and about a 30-fold higher "solubility" in blood—that is, the average slope of the Hb dissociation curve is about 30-fold greater than is the solubility in the barrier. For O_2, this translates into a large bucket being filled at a slow pour rate (compared to inert gases). As a result, O_2 takes about ten times longer to reach diffusion equilibration than do inert gases.

You may ask: what about the different inert gases, because they were chosen to span a very wide range of solubilities? Maybe there is a systematic difference in diffusion equilibration rate as a function of solubility? The answer is no (assuming that differences in molecular weight play a minor role). Why? Because equilibration of a poorly soluble gas can be thought of as a very slow pour rate into a very small bucket, while equilibration of a soluble gas can be imagined as a rapid pour rate into a large bucket: all inert gases take the same time equilibrate (subject to having the same ratio of barrier to blood solubility and the same molecular weights).

Why was all this explained in the first place? This is to understand that inert gas exchange is basically invulnerable to, and thus unaffected by, changes in the lungs that can cause O_2 diffusion limitation. Inert gases are clearly affected by the other three causes of hypoxemia: $\dot{V}_A/\dot{Q}$ inequality, hypoventilation, and shunt. See Fig. 2.5.

We take advantage of this fortunate phenomenon in MIGET in work reported by Hammond and Hempleman [1]. We can calculate from the $\dot{V}_A/\dot{Q}$ distribution (including shunt) recovered in any individual case what the arterial P_{O_2} should be based on just the distribution and shunt found. This calculation also accounts for any hypoventilation that may exist.

This calculated arterial P_{O_2}, also termed "predicted P_{O_2}," should equal the arterial P_{O_2} that is actually measured at the same time as the MIGET was used. This indeed is found to be the case under most conditions, obviously within the limits of experimental error of measurement of both inert gases and O_2. Thus, normal subjects at rest show this equality. Note that because of random experimental error, it is very unlikely that measured and predicted arterial P_{O_2} will be exactly the same in any single case, so that the above conclusion can be drawn only after examining a fair number of individual examples, plotting measured versus predicted P_{O_2}, and then determining the slope and intercept of the regression line, which should be 1 and 0, respectively.

There are several conditions during which O_2 exchange across the blood/gas barrier becomes diffusion limited. In normal subjects, this sometimes occurs during intense exercise at sea level, and the more so, the fitter the athlete (presumably due to higher cardiac output and thus lower capillary red cell transit time). At altitude, diffusion limitation occurs (not at rest but during exercise) in all normal subjects, fit or not, and even at moderately light exercise levels. The pathological condition in which diffusion limitation is most common is diffuse interstitial fibrosis, where in severe cases, diffusion limitation is observed at rest. More commonly, it occurs only during exercise and is the major factor in the common fall in arterial P_{O_2} that is seen during exercise in these patients. It may occur in patients with cirrhosis, where the mechanism is thought to be diffusion limitation—not for O_2 crossing the alveolar blood/gas barrier but rather for O_2 exchange within very dilated pulmonary blood vessels. These findings are discussed more in the final chapter encompassing the MIGET literature.

When diffusion limitation occurs, the actual (i.e., measured) arterial P_{O_2} falls to values lower than that predicted by MIGET, because, as explained above, inert gases are invulnerable to the problem. The difference between the predicted and measured P_{O_2} then reflects the effects of diffusion limitation over and above any consequence of $\dot{V}_A/\dot{Q}$ mismatch, shunt, and hypoventilation, and this difference is used within the MIGET software to compute a whole-lung O_2-diffusing capacity consistent with the difference.

This computation makes one explicit assumption: The ratio of diffusing capacity to blood flow in all lung units is the same. We make this assumption not because we think it is necessarily likely, but because we have no data to make any more complex assumption about how diffusing capacity is distributed in relation to blood flow. Thus, we are making the simplest assumption, and the resultant value for the diffusing capacity is therefore the lowest possible whole-lung value. That is because any maldistribution of diffusing capacity with respect to blood flow would further drop arterial P_{O_2} (compared to uniform distribution) such that the measured P_{O_2} could be maintained only by a higher total lung diffusing capacity.

While this aspect of MIGET—being able to identify and quantify diffusion limitation from the difference between measured and predicted arterial P_{O_2}—is well-established and generally well-accepted, it is necessary to mention some caveats:

First, post-pulmonary shunts also would cause predicted arterial P_{O_2} to exceed the measured value. Post-pulmonary shunts include bronchial and Thebesian venous drainage into pulmonary veins and left ventricular cavity, respectively. Why do these emulate alveolar diffusion limitation in their differential effect on inert gases versus O_2? Because after pulmonary capillary blood has exited the lungs and reached the pulmonary veins and left ventricle, for inert gases there is no further possibility of their exchange. Thus, their blood concentrations in these locations stay fixed at the values in existence when the blood exited the lungs. However, the bronchial and Thebesian veins contain O_2-depleted blood because they are part of the circulation supplying the airways and heart muscle, respectively. As a result, as they empty their O_2-depleted blood into the pulmonary veins or left ventricular cavity, they will cause a drop in blood P_{O_2} [comparing P_{O_2} at the end of the pulmonary capillaries with P_{O_2} in the systemic arteries from which MIGET samples are taken (radial, brachial, femoral)].

Interestingly, we have never observed a systematic difference with resting measured lower than predicted arterial P_{O_2}, suggesting that the contribution of post-pulmonary shunts to arterial desaturation is negligible and smaller than our ability to discern differences between measured and predicted P_{O_2}.

The other caveat is that even when breathing room air, but especially breathing gas with elevated O_2 concentrations, the random errors in both measured and predicted P_{O_2} start to become large as a result of the very Shallow slope of the Hb dissociation curve. It is our practice to never make anything of the predicted/measured P_{O_2} difference at F_{IO_2} above that of room air and to be conservative about the difference even when breathing room air (unless measured P_{O_2} is substantially subnormal, say, less than 80 mm Hg on average). In other words, when measured and predicted P_{O_2} are both above 80 mm Hg, random error is more likely the cause of differences between them. However, with enough data points, even under these conditions, one can ask whether the measured and predicted P_{O_2} values are similar or different. In practice, if a study is focused on identifying diffusion limitation, we prefer to assess measured/predicted P_{O_2} differences in hypoxia (for normal subjects).

5.3 Molecular Weight Dependence of Inert Gas Exchange

In the preceding, molecular weight differences among the inert gases were discounted as a negligible factor affecting their diffusive exchange across the blood/gas barrier. This was because, even though there are substantial molecular weight differences (e.g., ethane at 30, SF_6 at 146, and halogenated anesthetics such as enflurane and isoflurane at close to 200), those differences result in minor differences in the time to diffusion equilibration, which even for the heaviest gases in the group is complete within the first 10% of the pulmonary capillary.

It is however possible that diffusive mixing of inert gases between alveolar gas and each inspired gas breath taken can be incomplete. But please be clear—here we are talking about diffusion within the gas phase along the airways, not diffusion between alveolar gas and capillary blood. If this gas-phase mixing is incomplete, the heaviest gases would have a harder time being eliminated than would the lighter gases (in the very same lung). This would lead to a systematically higher retention of the halogenated gases (isoflurane/enflurane and SF_6) and a systematically lower retention of the other gases when the least squares MIGET algorithm is applied to the data set. This will result in two recognizable findings in the MIGET computer program output. First, the residual sum of squares will be higher than expected (because the MIGET equation system does not incorporate any effects of molecular weight), and second, the residuals for the heavy gases will be systematically positive, while those of the light gases will be systematically negative. We have observed such behavior in dogs subjected to partial pneumonectomy [4] and attribute this to the compensatory alveolar hyperinflation of the remaining lung tissue that creates longer diffusion distances within the terminal airways and thus requires more time for equilibration. It has also been reported in the horse [3]. But even then, this has never been of sufficient magnitude to interfere with the previously discussed measured/predicted P_{O_2} difference analysis of alveolar-capillary O_2 diffusion. However some estimates in rodents suggest that it may be important in this species [2].

5.4 MIGET and the Role of Extrapulmonary Factors in Gas Exchange

This is a very big and clinically important topic. It reflects the fact that for any given set of intrapulmonary conditions (i.e., $\dot{V}_A/\dot{Q}$ distribution, shunt, diffusion limitation), arterial P_{O_2} is sensitive to several extrapulmonary physiological variables.

These contributing variables include F_{IO_2}, total alveolar ventilation, cardiac output, $\dot{V}_{O_2}/\dot{V}_{CO_2}$, Hb/Hct/$P_{50}$, temperature, and acid/base state.

The arterial P_{O_2} that is predicted from any given $\dot{V}_A/\dot{Q}$ distribution and shunt (see above discussion of diffusion limitation for what the term predicted means) is therefore sensitive to the values of, and changes in, each of the above variables.

The MIGET software contains the code necessary to predict arterial P_{O_2} from the $\dot{V}_A/\dot{Q}$ distribution (and shunt) as mentioned many times. Explicit inputs required for this prediction are all of the above extrapulmonary variables. What this means is that it is possible to isolate the quantitative contribution of any one of these extrapulmonary factors to arterial P_{O_2} and, furthermore, to model the effects of changes in any of their values (one at a time or any/all in combination) on arterial P_{O_2}.

Thus, one can ask, for example, for any measured $\dot{V}_A/\dot{Q}$ distribution, how arterial P_{O_2} would be predicted to respond to any given change in F_{IO_2}, or any change in Hb concentration, or temperature, and so on. This is done by re-running the MIGET software, changing the values of the extrapulmonary variable(s) as desired. Over the years, this aspect of MIGET has proven to be one of its most useful and enduring.

5.5 MIGET and Mammalian Versus Avian Gas Exchange

A unique opportunity presented itself many years ago when we chose to apply MIGET to measure the $\dot{V}_A/\dot{Q}$ distribution in birds [5]. Bird lungs have quite a different structure (parabronchi with parallel air and blood capillaries like a car radiator and unidirectional alveolar gas flow) than do mammalian lungs with their alveolar structure and tidal gas flow. The result is that gas exchange in the bird lung happens by a crosscurrent process. To cut a long story short, this allows for more efficient gas exchange and the potential for a higher mixed arterial P_{O_2} than mixed expired P_{O_2}, and this is discussed further in the last chapter.

When we applied MIGET, we were not sure how the algorithms would perform because of the fundamentally different gas exchange process but went ahead anyway (discussed in the MIGET literature chapter). We found massive residual sums of squares (RSS) between the measured arterial retentions and the closest least squares best-fit values. At first we thought we must have large experimental errors, which are the usual cause of high RSS. After excluding this, we realized that Eq. (1.20), correct for mammalian alveolar gas exchange, was simply the wrong equation for crosscurrent exchange.

Equations (5.9) and (5.10) are the avian equivalents to Eq. (1.20) for retention, R, and excretion, E, respectively, in a single homogeneous gas exchange unit of bird lung [5].

If we set $T = \exp\left(-\lambda/\left(\dot{V}/\dot{Q}\right)\right)$:

$$R = 1 - \left[\left(\dot{V}/\dot{Q}\right)/\lambda\right] \times [1 - T] \tag{5.9}$$

and:

$$E = 1 - T \tag{5.10}$$

Figure 5.2 shows plots of these equations which you can compare to those of the mammalian curves in Fig. 2.6. The biggest difference is the sudden flattening out of the excretion curve. When the MIGET software was rewritten to replace the mammalian retention/excretion Eq. (1.20) with those for the bird, Eqs. (5.9) and (5.10), and applied to the very same data that gave massive RSS as indicated above, the RSS normalized, and nice, narrow $\dot{V}_A/\dot{Q}$ distributions were recovered that matched predicted to measured arterial P_{O_2} well.

The lesson learned from this exercise is that MIGET has yet another attribute—it has the potential, realized here in the avian lung studies, of recognizing when the gas exchange process obeys rules different than for the standard mammalian lung. The standard algorithms simply failed and suggested the need for developing the "bird version" of MIGET.

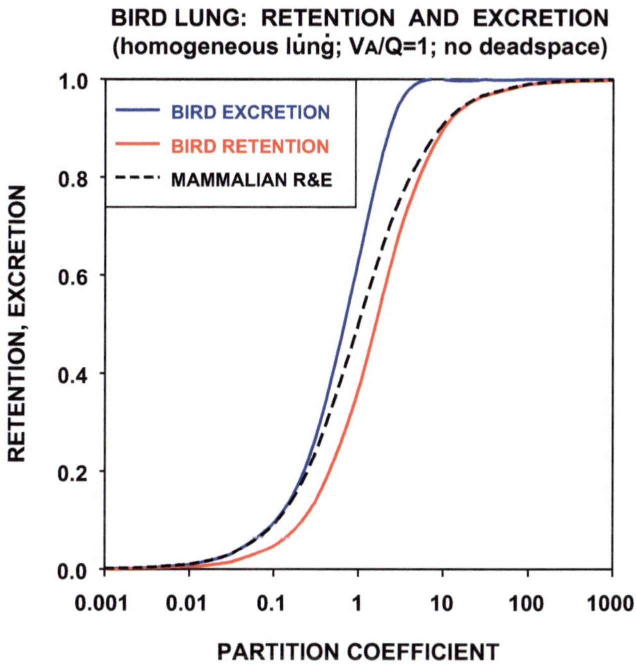

Fig. 5.2 Retention (red) and excretion (blue) curves in the *homogeneous* avian lung compared to a mammalian lung with the same total ventilation and blood flow. Because of the crosscurrent anatomy underlying gas exchange, excretion in the avian lung actually exceeds retention

References

1. Hammond MD, Hempleman SC. Oxygen diffusing capacity estimates derived from measured VA/Q distributions in man. Respir Physiol. 1987;69:129–47.
2. Hlastala MP, Scheid P, Piiper J. Interpretation of inert gas retention and excretion in the presence of stratified inhomogeneity. Respir Physiol. 1981;46:247–59.
3. Hopkins SR, Bayly WM, Slocombe RF, Wagner H, Wagner PD. Effect of prolonged heavy exercise on pulmonary gas exchange in horses. J Appl Physiol (1985). 1998;84:1723–30.
4. Hsia CC, Herazo LF, Ramanathan M, Johnson RL Jr, Wagner PD. Cardiopulmonary adaptations to pneumonectomy in dogs. II. VA/Q relationships and microvascular recruitment. J Appl Physiol (1985). 1993;74:1299–309.
5. Powell FL, Wagner PD. Measurement of continuous distributions of ventilation-perfusion in non-alveolar lungs. Respir Physiol. 1982;48:219–32.
6. Wagner PD, Laravuso RB, Uhl RR, West JB. Continuous distributions of ventilation-perfusion ratios in normal subjects breathing air and 100 per cent O2. J Clin Invest. 1974;54:54–68.

Chapter 6
MIGET: Practical Aspects

Abstract The preceding pages have laid out in detail the theoretical basis of the multiple inert gas elimination technique (MIGET). However, the actual execution of MIGET is also important to present. Two important issues arise in executing MIGET: (a) knowing how it is done, step by step, and (b) becoming familiar with, and able to execute, the important details with expertise and accuracy that make the difference between good and bad results. To calculate a set of retentions and excretions for each of the six gases, the partial pressures of the six gases in three locations—systemic arterial blood, pulmonary arterial blood, and mixed expired gas—must be measured or estimated. By using the Fick principle and measuring cardiac output, systemic arterial and mixed expired partial pressures, and ventilation, the pulmonary arterial catheter can be omitted, and the pulmonary arterial partial pressure for each gas can be calculated. This chapter discusses in detail the hands-on steps of conducting a MIGET study including making up a solution of the six gases for infusion, subject preparation, sample collection, sample analysis by gas chromatography, measurement of the blood/gas partition coefficient (λ), data entry into the MIGET software, and evaluating the results.

6.1 Different Ways of Doing MIGET

The measurement of gases in the blood by gas chromatography employing the techniques that form the basis of MIGET was first reported in 1974 [35]. Before dealing with the issues related to running a MIGET experiment in detail, in detail, some general discussion of *different ways of running MIGET* is necessary.

The objective is to end up with a set of retentions and excretions for each of the six gases (plus the ancillary data; see 6.6). Remember that retention is for each gas the ratio of its systemic to pulmonary arterial partial pressure, while excretion is the ratio of its mixed expired to pulmonary arterial partial pressure. Thus, one needs to end up with partial pressures of the six gases in three locations—systemic arterial blood, pulmonary arterial blood, and mixed expired gas.

The direct approach is obviously to sample both systemic and pulmonary arterial blood and also mixed expired gas. However, because pulmonary arterial blood sampling requires a pulmonary arterial catheter, it is rarely done in man, although it is much easier in most experimental animals. We can easily get around that

S.R. Hopkins, P.D. Wagner, *The Multiple Inert Gas Elimination Technique (MIGET)*, Methods in Physiology, https://doi.org/10.1007/978-1-4939-7441-2_6

limitation by using the Fick principle and measuring cardiac output as well as the systemic arterial and mixed expired partial pressures and ventilation. Then, from Eq. (2.6), which reflects the Fick principle, the pulmonary arterial partial pressure is easily calculated for each gas. Cardiac output can be measured without a pulmonary artery catheter by dye dilution, by echocardiography, by impedance cardiography, and by acetylene uptake or other methods, depending on circumstances. Fortunately, it turns out that having a precise value for cardiac output is not critical in that the dispersion (second moment) of the $\dot{V}_A/\dot{Q}$ distribution is very *insensitive* to errors in cardiac output [36]. However, the first moment (mean, i.e., the position of the distribution on the $\dot{V}_A/\dot{Q}$ axis) will have the same percentage error as does cardiac output. However, that is also usually not problematic given the log scale of that $\dot{V}_A/\dot{Q}$ axis.

Another modification of the sample collection paradigm substitutes peripheral venous sampling for systemic arterial sampling. The concept underlying this is straightforward: the inert gases are not metabolized in tissues, so once well distributed in the body tissues, the peripheral venous levels should be similar to those in the arteries supplying the tissues. We evaluated this formally many years ago [36] and found this to work quite well. We used veins on the back of the hand to substitute for the radial artery and found that after 90 min of inert gas infusion, heated-hand venous levels were reliably at 95% of the arterial values, allowing their use with a 5% correction factor as a substitute for direct arterial blood. Since with direct arterial sampling one usually needs about 20–30 min of infusion before sampling, this modification avoids arterial catheterization if one can afford to wait an extra 60 min for equilibration in the tissues. Unfortunately, therefore, this approach does not allow for studies in which $\dot{V}_A/\dot{Q}$ changes might occur rapidly, as in exercise or with drug therapies, etc. It is best suited to studies in which patients are required to have basic MIGET studies at frequent intervals—such as daily or weekly—without interventions that might change the $\dot{V}_A/\dot{Q}$ distribution rapidly. Multiple hand vein catheterizations are feasible, but multiple arterial catheterizations are not.

In summary, there are three procedural options in using MIGET. In each case, mixed expired gas must be collected and analyzed for the inert gases, and minute ventilation must be measured:

1. Direct sampling of systemic and pulmonary arterial blood
2. Direct sampling of arterial blood with cardiac output measurement
3. Sampling of peripheral venous blood with cardiac output measurement

Let's now go through a list of 11 steps that identify the key elements of the method:

Running MIGET requires, in order:

1. Making up a solution of the six gases in sterile saline (or 5% dextrose)
2. Preparing the subject
3. Infusing the inert gas solution intravenously
4. Collecting blood and expired gas samples simultaneously

5. Measuring ancillary variables such as ventilation, cardiac output, temperature, blood gases, [Hb], acid/base status, etc.
6. Analyzing the samples for inert gas concentrations by gas chromatography
7. Measuring blood/gas partition coefficients of the gases again using the GC
8. Measuring the (random) error of the method
9. Entering the necessary inert and blood/gas data and ancillary variables onto a paper or electronic spreadsheet form
10. Entering the data from the spreadsheet into the MIGET software—either in the original DOS format or now using a Windows GUI environment—and then running the MIGET software
11. Looking at and analyzing the results—for data errors and physiological meaning

6.2 Making Up a Solution of the Six Gases in Sterile Saline (or 5% Dextrose)

One starts with a one liter plastic bag of sterile normal saline (or 5% dextrose) and a commercially prepared tank (obviously with its own regulator assembly) containing three of the six inert gases, SF_6 (20%), ethane (70%), and cyclopropane (10%). These are the three least soluble of the six gases used. The tank cannot be at more than about 300 psi or the cyclopropane will liquefy. A clean but not necessarily sterile plastic tube is used to connect the low-pressure port of the regulator to the saline bag as follows. A sterile ~18 gauge spinal needle is inserted into the bag port. This needle has a sterile three-way plastic stopcock attached to its hub, and attached to the stopcock is a sterile 0.22 μm Millipore filter. The filter is then connected to the clean but not sterile plastic tubing leading to the tank low-pressure outlet. The saline bags always have air in them, and the air is bled out through the needle/stopcock before we start (see Fig. 6.1). Thus the sequence is tank, regulator, tube, filter, stopcock, needle, and saline bag.

The tank is opened, the tubing flushed into the room with a little tank gas, to wash out the air in the line, and under low pressure, about 50–100 ml of tank gas is bled into the saline bag until it is fairly, but not ridiculously, taut. The system is disconnected between the stopcock and the filter, keeping the parts sterile, and the saline bag is vigorously shaken to the point of foam, taking about 20 s. The bag is left on a bench for the foam bubbles to coalesce (1–2 min), and the gas is bled out of the bag as initially done with the air originally present in the bag.

Next, the gas tank is connected up once more to the same saline bag, and another 50–100 ml "tank gas" is slowly bled into the saline bag until taut, and the process of shaking is repeated.

After this second shaking, all gas bubbles are bled out of the saline bag into the room through the spinal needle, staying sterile. Then, using sterile technique, 8 ml pure (analytical grade) acetone liquid, drawn up into a sterile 10 ml syringe, is injected into the saline bag, and 1 ml of liquid diethyl ether is similarly injected into

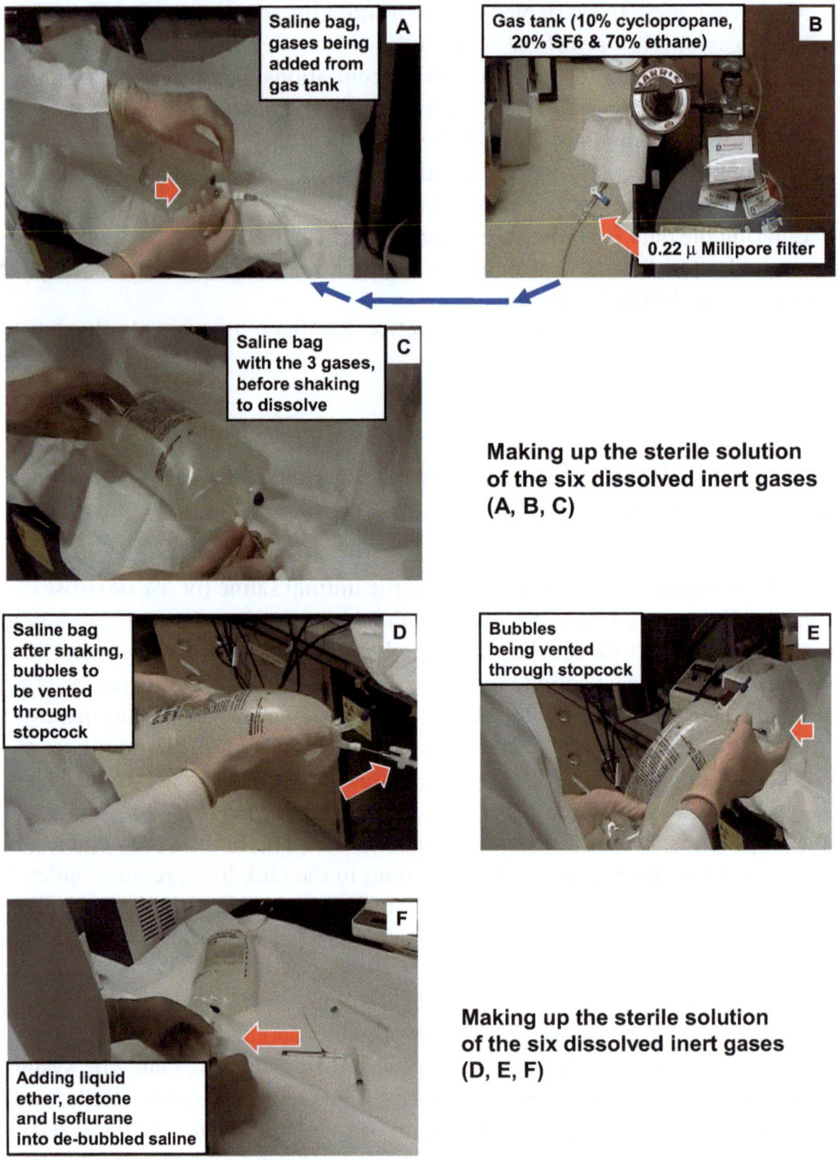

Fig. 6.1 Making up the sterile solution of the six dissolved inert gases. Panels (**a**) and (**b**) show the process of filling the plastic saline bag with gas from the tank of SF_6, ethane, and cyclopropane. A plastic tube from the tank regulator connects to the saline bag by means of a stopcock and spinal needle inserted into the access port of the saline bag. Panel (**c**) shows the spinal needle and the three inert gases expanding the saline bag. The bag is now shaken, and Panels (**d**) and (**e**) show the spinal needle with stopcock and the process of expelling all the undissolved gas from the saline bag through the stopcock. Panel (**f**) shows the remaining three gases (ether, acetone, isoflurane) being directly injected from a syringe (since they remain as liquids in their storage bottles) one at a time into the saline bag. The entire procedure is performed with sterile components, while the gas from the tank passes through a 0.22 μ Millipore filter as a precaution

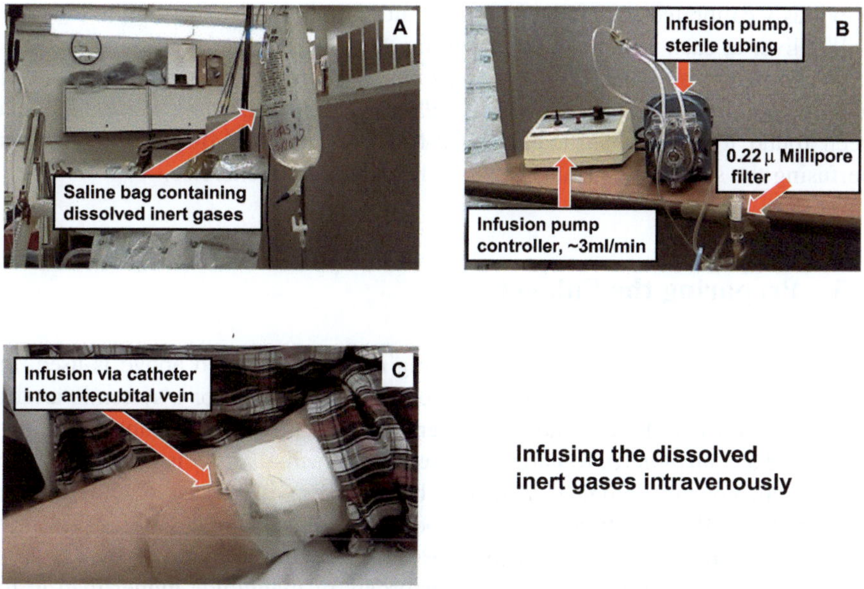

Fig. 6.2 Infusing the dissolved inert gases intravenously. Panel (**a**) shows the inert solution, constituted as depicted in Fig. 6.1, hanging from a stand, and connected via sterile tubing to a roller pump (panel **b**) for constant infusion at the desired rate of about 3 ml/min into the antecubital vein of the subject (Panel **c**). Note another 0.22 m Millipore filter as a precaution (Panel **b**)

the bag. Finally, about 3–4 ml of liquid enflurane (or isoflurane) is similarly injected into the saline bag. The bag is then gently rotated and turned end over end to disperse the three liquids throughout the saline, but the bag is *not* shaken. Some residual enflurane/isoflurane will remain (as this gas is fairly water insoluble) and sink to the bottom of the bag. That's OK.

The one liter saline bag, now containing all six gases, is done. It is then labeled as MIGET infusion and hung vertically on a regular IV pole. We leave the spinal needle (plus attached stopcock) in the bag, protecting the stopcock ports with caps for sterility. In fact, the spinal needle needs to be pushed up about 1–2 in. into the bag, and the needle tip needs to be kept away from the bag sides (to avoid the needle puncturing the bag!! The needle will of course be facing upward (stopcock down, needle point up) as the ports for the saline bag are on the lower end of the bag. Pushing the needle 1–2 in. up into the saline ensures that any excess undissolved enflurane or isoflurane cannot run into the subject inadvertently and anesthetize him or her, because that undissolved liquid falls to the bottom of the bag due to its density.

The infusion bag is next connected to a roller pump using sterile tubing, and further sterile tubing leads to the subject, where it will be connected to a peripheral venous catheter (Fig. 6.2). Interposed between the catheter and the tubing will be another, fresh, sterile 0.22 μm Millipore filter for added sterility and as insurance against gas bubble transmission into the subject. The roller pump will have been previously calibrated so that the pump operator knows how many ml/min the pump will deliver at each numerical pump setting.

When ready, *the infusion rate is set to numerically equal about 1/3000 of the minute ventilation (i.e., infusion rate, ml/min = 1/3 ventilation, l/min).* This has been found to provide sufficient signal-to-noise ratio in MIGET without over-perfusing the subject with saline, both at rest and during exercise.

6.3 Preparing the Subject

Depending on which of the three operating modes mentioned above is used, the subject will be prepared accordingly. The usual setup calls for a systemic arterial catheter and a peripheral venous catheter. Usually, the nondominant hand radial artery and a *contralateral* forearm vein are used. Brachial artery is fine—the choice is medical, not physiological, and may have to do with access or other factors unrelated to MIGET. What is very important, even essential, is that these two catheters be on opposite arms. This is because the concentration of the six gases in the infusate, and thus forearm vein, is orders of magnitude higher than in the artery, and we must avoid any chance of direct diffusion of any of the inert gases from vein to artery, for hopefully obvious reasons: such a mistake would raise the arterial concentrations in a manner unrelated to pulmonary gas exchange and destroy the experiment completely. If the peripheral vein sampling option is used, the arterial line is substituted by a similar line in a vein on the back of the hand, and the hand is wrapped in a heating blanket that should be set at about 40–45 °C. But again, the two catheters must each be on a different arm.

The two catheters are inserted using sterile technique and, if desired, local anesthetic as well. For the artery, you might use a 20 gauge, and for the vein, an 18 gauge catheter. Both lines need to be kept sterile throughout the study and flushed with saline to prevent clotting. They are both obviously well taped into place and are controlled by stopcocks.

Almost always we will use a forehead sensor-based pulse oximeter to monitor arterial O_2 saturation and heart rate, plus a three lead EKG for rate, rhythm, and pattern through catheter placement and the study as well. The arterial catheter is usually connected to a disposable pressure transducer to monitor/record systemic arterial pressure both for safety and if needed as data.

When ready, the subject dons a noseclip and is connected via a clean mouthpiece to a non-rebreathing valve which supplies inspired gas as desired and which directs all expired gas to the expired gas circuit.

This circuit needs lots of attention, although it is simple in principle. The objective is only to collect mixed expired gas. This requires an in-series mixing chamber to mix alveolar with deadspace gas every breath, so gas is continually mixed and eluted into the room, avoiding cumbersome and problematic bag collections. *It also requires heating to avoid condensation of exhaled water vapor, and this is critical.* The entire expired collection system from valve box to end of the mixing box is heated—to several degrees above body temperature. To do this, electrical heating tapes are wrapped around the tubing, both the external tubing connecting the subject to the mixing box and structures within the mixing box as well. Finally, as little plastic as possible is used in the expired system because the soluble gases may actually dissolve partially in the plastic. We have used a simple 1.5 in. diameter coiled copper pipe of about 27 ft in length, enclosed in a heated metal box, as the mixing chamber. Alternatively, a baffled cube of about 1 cubic ft works well for exercise when ventilation is high. In sum, the system is a flow-through, heated, metal pipe or baffled box connected by tubing to the subject. The whole system is kept above body temperature everywhere from the mouth to the sample collection point, and temperature is monitored continuously to assure it is above body temperature but not so high as to melt any plastic or cause burns. Samples of expired gas are aspirated into glass syringes as desired (see below), and ventilation can be measured by flowmeters of various types hooked in series with the flow-through system. Figure 6.3 shows photos of the expired gas collection system.

Any other hookups for other signals are completed at this time.

6.4 Infusing the Inert Gas Solution Intravenously

Remember that the MIGET is a steady-state technique, and you want to be quite sure that enough time has passed after starting the infusion to assure this.

The halftime of reaching a steady state in a homogeneous lung is given by the term:

$$T½,\ \text{min} = 0.693/k$$

where $k = \left[\dot{V}_A + \lambda \times \dot{Q}\right] / [\text{FRC} + \lambda \times V_{\text{tiss}}]$

Here $\dot{V}_A$ is total alveolar ventilation, λ is blood/gas partition coefficient, $\dot{Q}$ is cardiac output, FRC is functional residual capacity, and V_{tiss} is lung tissue volume (assume about 600 ml for a normal adult human). **At rest**, in min^{-1}:

$$k \sim [5 + 6 \times \lambda] / (4 + 0.6 \times \lambda)$$

so that for a gas like SF_6 where $\lambda \sim 0.005$, $k \sim 1.25\ \text{min}^{-1}$, and $T½$ is less than a minute. In a normal lung, 99% equilibration would take about 8 min. For a gas like acetone where $\lambda \sim 300$, k approximates $10\ \text{min}^{-1}$, and $T½$ is much less than a minute in a normal lung.

However, in reality, we like to allow 20–30 min so that the external connections (valve box, expired system) also have plenty of time to stabilize.

In patients with lung disease, these rapid equilibration predictions are unreliable. To the extent that there are regions of low ventilation or blood flow in relation to gas volume, or lots of alveolar edema, equilibration in those regions may be much slower.

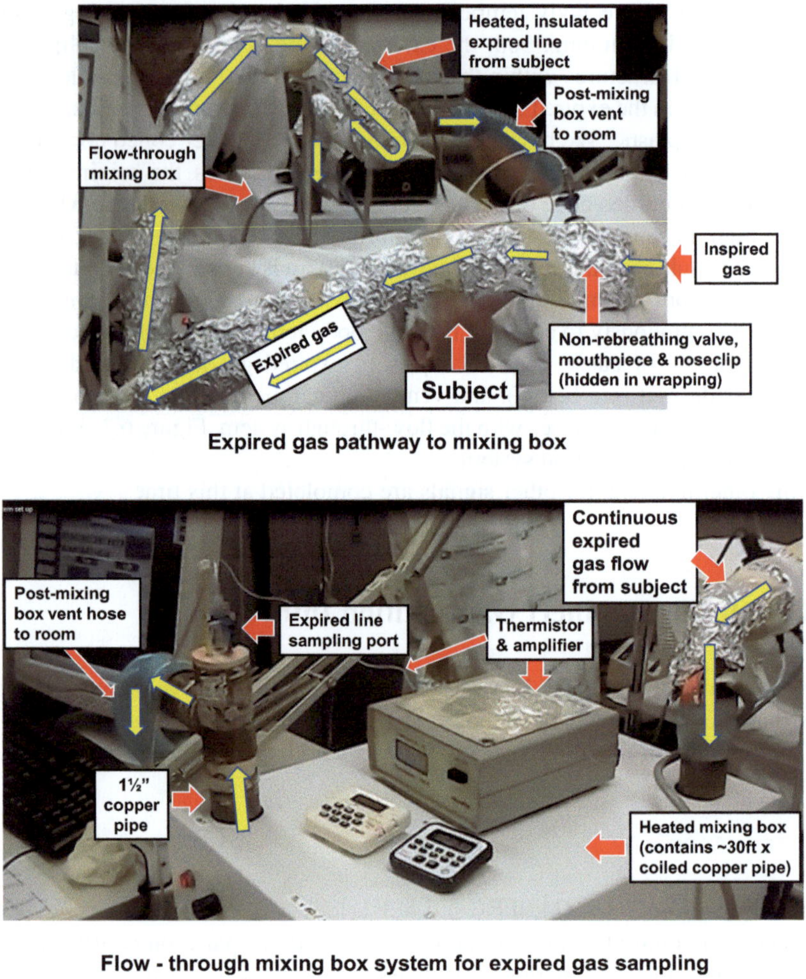

Fig. 6.3 Mixing box system for expired gas sampling. The *right panel* shows a supine subject (lower right) breathing via a mouthpiece through a non-rebreathing valve. Expired gas passes through a wide bore flexible plastic hose (i.e., heated above body temperature and wrapped in aluminum foil) and then into the heated mixing box, better seen in the *left panel*. This box contains a ~30 ft long, spirally coiled 1.5″ diameter copper pipe that leaves the mixing box at the opposite corner, where the now-mixed expired gas can be sampled. *Purple arrows* indicate the expired gas flow path; also seen on top of the mixing box are two timers and a temperature sensor

During exercise, the numerator of the above equation will increase dramatically (both $\dot{V}_A$ and $\dot{Q}$ increase severalfold, while the denominator does not really change). Thus, equilibration is much faster during exercise, and 5 min wait time is sufficient, or even 1–2 min at heavy exercise.

Remember also that the infusion rate is set to numerically equal about 1/3000 of the minute ventilation (i.e., infusion rate, ml/min = 1/3 ventilation, l/min). This has been found to provide sufficient signal-to-noise ratio in MIGET without over-perfusing the subject with saline and works from rest to maximal exercise. The subject will complain of an ether mouth taste, which will stay with him/her for an hour or two after the study ends.

In sum, the infusion is run for 20–30 min at rest, 2–5 min during exercise (usual method with an arterial line) or 90 min (peripheral vein sampling option), and the study is ready to go. This amount of time gets the pulmonary (20 min) and tissue exchange (90 min) of the six gases into a steady state. What this means is that the retentions and excretions will have become constant over time even if the absolute concentrations are slowly rising.

6.5 Collecting Blood and Expired Gas Samples

When the time comes to collecting blood and expired gas samples, the process is simple but there are a few important points. The concept is to collect the blood and gas samples physiologically simultaneously, as follows: The process should be controlled by the person collecting the blood sample. He/she starts collecting when ready (into a sterile glass 30 ml ungreased but heparinized syringe), counting out the aspirated blood volume ml by ml, making sure that the sample is taken steadily over at least 2–3 breaths. See Fig. 6.4a, b. Arterial and pulmonary arterial sampling

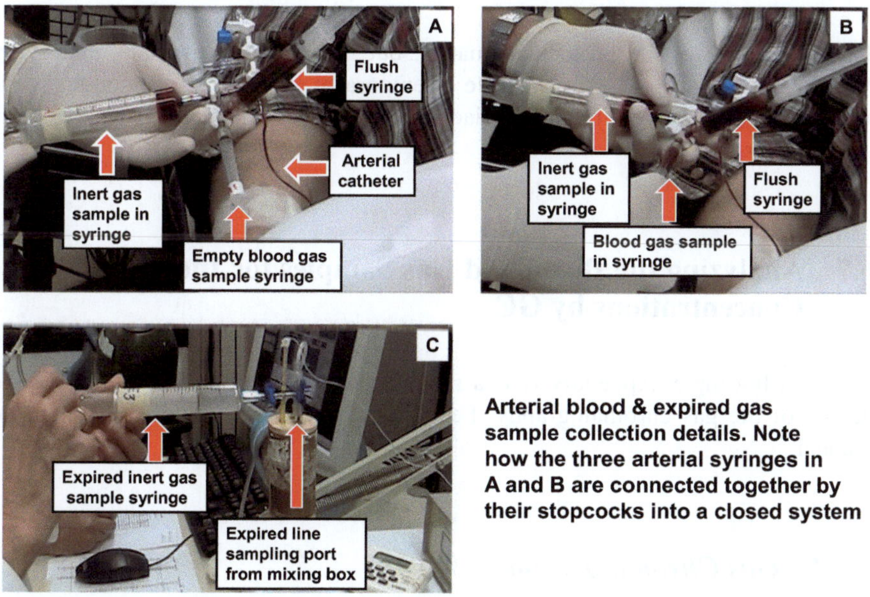

Fig. 6.4 Sample collection details. Panel (**a**) shows the radial arterial catheter as blood is being aspirated into the inert gas sample syringe (top left). Note the manifold consisting of three syringes connected as a closed system to the arterial line: inert gas sample syringe, blood gas sample syringe, and flush syringe. This arrangement avoids contact of blood with air and reduces risk of infection. Panel (**b**) shows the subsequent filling of the blood gas syringe; Panel (**c**) shows how mixed expired gas is sampled at the end of the mixing box system

should occur simultaneously, ml by ml. When half of the sample has been collected, the sampler says "halfway," signaling the expired gas sampling person to start a stopwatch. When the time on the stopwatch (min) equals the volume of the expired tubing plus collecting box (l) divided by the current minute ventilation (l/min), the expired gas is taken into ungreased glass syringes—flushing back and forth between the mixing box and the syringe two to three times and then slowly, over two to three breaths, drawing about 20 ml of expired gas into the syringe (Fig. 6.4c). This delay in collecting expired gas accounts for the transit time from mouth and thus aligns the sampling physiologically in time. It is important to collect the expired gas samples over several breaths as there is the possibility of variation of inert gas concentrations over the course of tidal breaths [39]. Expired line gas temperature should be written down when each sample is collected. In general, 5–8 ml of blood is sufficient per sample. *Any air bubbles in the syringes MUST be allowed to remain in the syringe and NOT be cleared out (opposite instruction to routine collection of blood samples for blood gas measurements, where small bubbles must be rapidly disgorged).*

6.6 Measuring Ancillary Variables Such as Ventilation, Cardiac Output, Temperature, Blood Gases, [Hb], Acid/Base Status, etc.

These are done as the protocol demands, but the important point is that every MIGET blood and gas set must have a temporally aligned set of ancillary variables—blood gases, ventilation, cardiac output, Hb, F_{IO_2}, and acid/base parameters in particular.

6.7 Analyzing the Blood and Gas Samples for Inert Gas Concentrations by GC

This is a big topic. Appendix A is a review [34] written many years ago on the details of the gas chromatography. Feel free to use it as a guide and reference manual.

6.7.1 Gas Chromatograph

A gas chromatograph (GC) is a simple device. It consists of a hollow tube (the column) that is packed with small granules (that enable the temporal separation of the gases in the sample) of any one of many types and connected at both ends to key devices.

At the upstream end, the key device is an inlet system for introducing a gas sample into the column. At the downstream end, the key device is a detector, which measures the gas concentration. There are many variations in the inlet, column, and detector that are chosen by the user to optimize their application. A carrier gas constantly flows through the column to conduct any injected gas sample from the inlet through the column to the detector.

6.7.2 *Inlet System*

The inlet system is a very clever constant volume sample loop valve system: see the attached review in Appendix A for details. You simply connect a syringe of the sample to be analyzed to the inlet port; fill the constant volume loop (any excess sample simply passing through the loop into the room air), and by rotating a valve, sweep just that portion of the sample contained within the constant volume loop into the column by the carrier flow. With a 10–20 ml gas sample in a syringe, and a constant volume sample loop of 0.5–1.0 ml, there is no shortage of sample. The purpose of a constant volume inlet is just that to ensure a precisely constant sample volume to improve reproducibility.

6.7.3 *Column*

Your first question is why is there a (packed) column in the first place? Why not just an inlet (obviously necessary) and a detector (obviously necessary)? The answer is that the column is essential to perform one major task: by choice of packing a granule material and density and size, by choice of column length and diameter, and by choice of carrier gas flow and column temperature (the column is always kept inside an oven), each different gas present in a single sample is transmitted down the column (by the carrier gas flow) at a unique velocity. The column material then separates the gases that began in a single mixed sample—they elute into the detector in temporal sequence. Different velocities basically result from (reversible) bonding between each gas and the packing material, bonding that differs between gases. The other variables above are chosen to empirically achieve the optimal outcome: complete separation of each gas in the mixture in the minimal total analysis time. It is a multivariate optimization game, and there are no shortcuts. You just inject samples and see what you get as you vary all the independent variables mentioned above, one at a time. Be aware that the column is there to not only separate the gases of interest from each other but also to separate each of them from other gases normally present—O_2, CO_2, N_2, and sometimes water vapor. This is critical because these latter gases are present in huge concentrations compared to those used in MIGET which are in the ppm or even ppb range.

6.7.4 Detectors

MIGET uses six gases: SF_6, ethane, cyclopropane, isoflurane (or sometimes enflurane), diethyl ether, and acetone. All except SF_6 contain hydrocarbons and will burn when passed through an already lit flame. This makes these five gases amenable to detection by a flame ionization detector (FID). SF_6 being halogenated is well suited to detection by an electron capture detector (ECD). Note that isoflurane and enflurane are also halogenated and will be detected by ECD as well.

We use two parallel inlet/column/detector paths inside a single GC—one for the five gases other than SF_6 feeding the FID and one for SF_6 feeding the ECD.

The FID is basically a burning candle. The fuel is not wax but a precision-controlled air/hydrogen mixture that is lit and burns continuously. It basically records the current change when each gas within a sample passes, in turn, from the column into the detector and burns, providing a signal that is very sensitive and proportional to gas concentration over a very wide range of several orders of magnitude.

The ECD works by donating electrons normally passing between an anode and cathode to any electron-hungry gas passing through. Halogenated gases are great electron acceptors. This action therefore *reduces* the current between the anode and cathode compared to baseline. This current reduction is simply flipped to give a positive signal, but the implication is clear—above a particular sample concentration, the current falls to *zero* and cannot fall more as gas concentration rises. Thus, the ECD saturates at "high" concentrations and when saturated, obviously cannot be used to measure the concentration. How high? Parts per *million* of SF_6 will saturate it! This is why we usually must dilute expired gas and venous blood samples before introducing them into the GC. Dilution by four- to eightfold is commonly needed and of course must be noted so it can be factored into the final concentration used in the MIGET software.

Note that arterial samples (from lungs without increased SF_6 retention—which is due virtually only to shunt) do *not* require dilution; indeed, it is a challenge to have enough signal for decent measurement. This is because the normal lungs eliminate 99% or more of the SF_6 (remember, $R = \lambda/\left(\lambda + \dot{V}_A/\dot{Q}\right)$ and λ for SF_6 is very, very low at about 0.005).

6.7.5 Running the GC and Putting Samples Through It

Over many years, we have found the following works best for us:

- Hewlett Packard 5890A gas chromatograph.
- Constant volume inlets with 0.5–1.0 ml sample volume.
- Stainless steel columns 1/8 of an inch in diameter packed with Poropak-T granules of 80/100 mesh size.

- 3–6 f. column length for the flame ionization detector (FID).
- 12 f. column length for the electron capture detector (ECD).
- Helium carrier gas for the FID.
- Ultrahigh purity N_2 carrier gas for the ECD.
- Carrier gas flow rates in the range of 20–30 ml/min for each.
- Column temperatures in the range of 130–150 °C.
- Completely separate inlets, columns, and detectors for the FID and the ECD but both of these parallel, independent systems are contained within a single GC, with the columns in a single oven.

What works for you may be different. *It does not matter as long as it works.*

With these parameters, we are able to put each sample through in about 4–5 min. We have a printer/digitizer (called an integrator) attached to each detector (one for the FID and one for the ECD). These do not feed directly into any computer, but they do automatically measure the baseline and peak height for any gas in the sample and print out on the paper both the chromatogram and the table of peak heights and peak areas, so you do not have to do that boring work yourself. The gases elute in the following order on the FID: O_2/N_2, ethane, cyclopropane, ether, acetone, and isoflurane/enflurane. There are a couple of minor blips that one often sees that may reflect CO_2 or water vapor. They never interfere.

Samples obtained from subjects include both arterial (and sometimes venous) blood and expired gas. The chromatograph is a *gas* not liquid system, so we *cannot* put blood into the inlet. It would destroy the inlet/column/detector immediately. We can and do put the expired gas samples directly into the GC.

6.7.6 Expired Gas Sample Processing

Expired gas samples are collected from the expired gas line downstream of the (heated metal) mixing box into clean, ungreased, 20 ml matched plunger/barrel glass syringes closed by a stopcock and kept until measurement. They should be measured as soon as possible, within 30 min ideally. They can be kept at room temperature. The inlet system we have is a Luer system that accepts the male end of any standard stopcock, so sample entry is really simple. The constant volume inlet loop of 0.5–1 ml means introducing about 3–5 ml from the sample syringe (the excess just passing through the loop tubing and back into the room), and then quickly rotating/activating the inlet valve to sweep that sample from the constant volume loop down the column in the carrier gas, and simultaneously pushing the start button on the integrator. What remains in the expired gas syringe should be kept in case a repeat run is desired (stopcock closed very quickly after the sample is injected into the GC). The expired sample can be run through the FID and ECD at the same time, but as they are two independent but parallel systems in one oven, two injections of sample are required, one into each system. The integrators are also separate, so each must be "started" as the corresponding sample is injected. As a practical note, we *always* inject the FID sample first and the ECD sample second—because expired (and mixed venous, if collected) ECD samples contain SF_6 at levels that must be pre-diluted so as not to saturate the ECD detector. We do *not* want to pre-dilute before injecting into the FID—that would be silly.

6.7.7 Blood Sample Processing

Blood samples are a whole other matter and take time, effort, and expertise. In principle, one has to somehow extract the gases out of the blood sample and into a gas phase so it can be injected into the GC just as an expired sample is treated. Here is how we do it:

The blood sample was collected as described above and is inside a closed (by stopcock) glass syringe. It will have a few small bubbles inside, which is fine. They contain some of the inert gas. The stopcock will likely have drops of residual blood/saline in its open arms.

The sample syringe will have been previously (i.e., before sample collection) weighed dry (i.e., plunger + barrel + stopcock + label) and then reweighed after heparin has been added to fill the deadspace of the stopcock + syringe. The difference in weights gives the heparin volume, a needed entry datum for every blood sample.

After collection of the blood sample, the first step is to rinse clean (with tap water) and then dry the outside of the syringe and stopcock ports that usually contain a little blood. Wall air is good to flush through the stopcock to dry it after washing (obviously leaving the stopcock closed to keep the sample closed off from air). The syringe is now weighed again, to calculate the mass of blood in the syringe. From the blood density (measured in a simple measuring cylinder from a pool of all individual samples at the end of the day as weight/volume), the sample blood volume is calculated as weight/density. That blood volume is an important entry datum.

Next, the clean and dry syringe containing the blood sample (with maybe a few bubbles) is taken to a tank of N_2 where a large diameter (at least 0.5 in.) short length of ordinary plastic tubing is attached to the regulator, and in one swift, smooth action, about 10 ml of N_2 is taken into the sample syringe. This must be a *fast* action: Put the stopcock male end into the flowing N_2 stream, open the stopcock to the blood, draw in 10 ml N_2 (NO back and forth, just one smooth withdrawal), and close the stopcock. It is *not* important that this be 10.000000 ml. The volume will expand anyway in the water bath (see below) and is measured after coming out of the water bath (see below). What is *critical* is minimal time the stopcock is kept open for this operation. Less than a second. No back and forth to get the volume "correct" at 10 ml.

Now we have a closed syringe with about 5–8 ml blood + heparin and about 10 ml N_2. That syringe is now placed in a shaking water bath ideally maintained at body temperature of the subject. In any event, that temperature must be recorded and should stay constant for all samples from any one study. The water bath shaking is started. This should not be either too slow or too fast. Too slow and there may not be enough mixing between the N_2 and the blood. Too fast and you will end up with froth and a bad sample. About one shake per second is OK. After a few minutes, the plunger of every syringe in the bath should be rotated briefly to ensure it is not stuck to the barrel. After 40 min equilibrating in the bath, the sample is ready for the next step.

Before this next step, "transfer" syringes should have been prepared and labeled, one such for every sample in the water bath. These are clean, dry, 10 ml glass matched barrel/plunger labeled syringes with their own stopcock. They will have been filled with N_2 a couple of minutes earlier.

All samples in the water bath ready to come out first have their total volume (N_2 + blood) read and recorded. To do this, each syringe, one at a time, should be held horizontally under water in the now-quiet water bath. The plunger should be spun a little to again make sure it is not sticking and compressing its contents, and the volume read off the gradations on the barrel. Next, the syringe is taken out, and *very quickly* wall air is flushed through the open ports of the stopcock to dry it out (a couple of seconds). Immediately, the transfer syringe is hooked up to the blood sample syringe, and after the transfer syringe is emptied of its N_2, it is filled with N_2 from the blood sample syringe. This must take just a few seconds, no more. You must *not get any blood/bubbles from the blood sample syringe transferred along with the* N_2. The best way of assuring this is to hold the blood sample syringe VERTICALLY (stopcock uppermost) and the transfer syringe *horizontally. speed is essential and cannot be overemphasized.* The need for speed is to transfer the sample before it cools down. Figure 6.5 provides a series of photos of the process at key stages.

Once the transfer syringe is filled with the sample gas, it can be laid on the bench and time pressure is off. All such samples in a single water bath batch are best processed together (i.e., transferred) in this way at one time, after which they are measured one at a time in the GC in the same manner as for expired samples into the FID and then the ECD.

It is in this series of steps that the greatest errors can arise, as might be expected. Good quality data demand expertise in this gas extraction process. This can only come with practice.

To summarize to this point, all operations performed so far have gotten you all the expired gas and blood samples analyzed for all six inert gas concentrations. However, you will note there has been no calibration procedure for either the FID or ECD detectors. Why not? Because this is not necessary. The MIGET software only needs *ratios* of gas levels (e.g., expired gas to arterial blood) and thus, as long as the detectors are linear through the origin (which they are in the range in which we use them), no absolute calibration is required, which is good news.

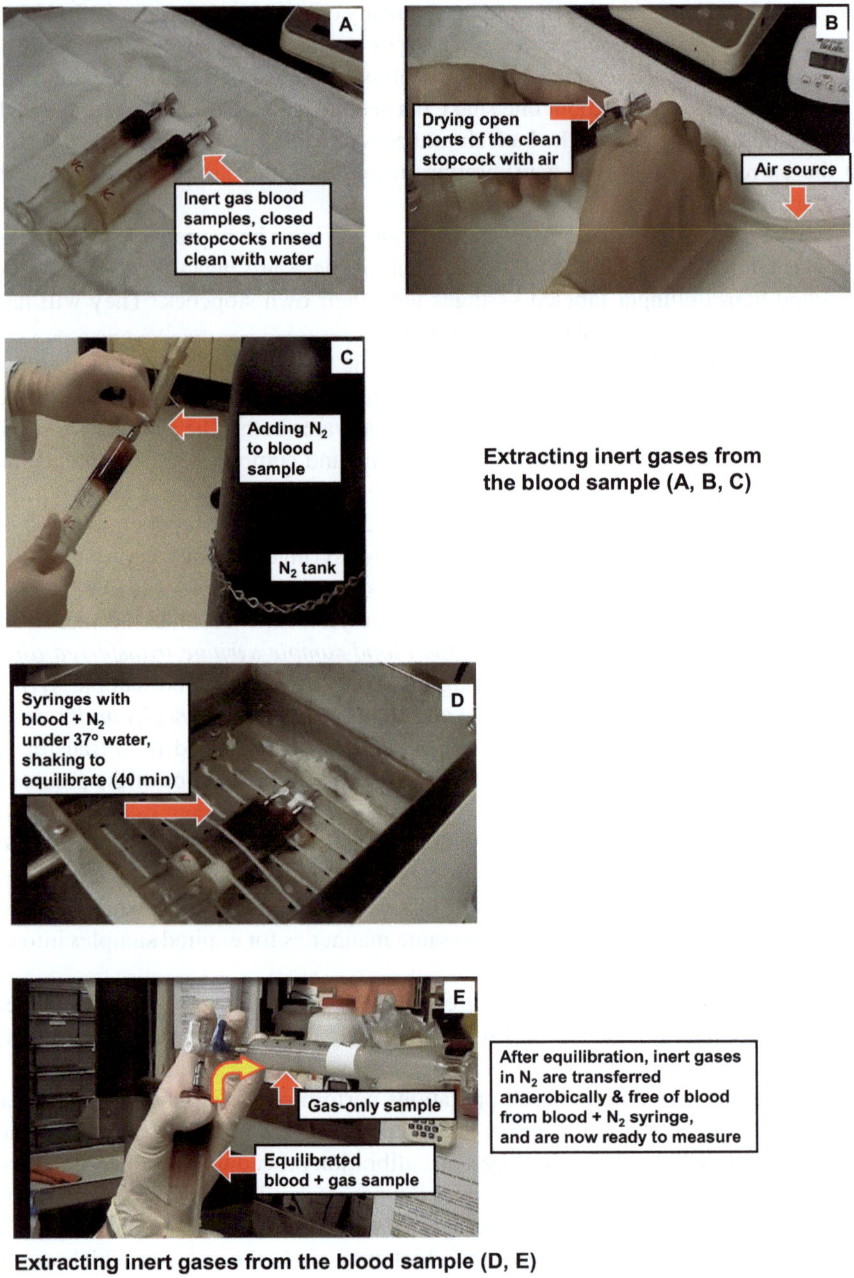

Fig. 6.5 Extracting inert gas from blood sample. Panel (**a**) shows two arterial samples just after collection. Washing the stopcocks with water requires drying with a blast of air (Panel **b**). Then, N_2 is aspirated from a N_2 tank into the sample syringe (Panel **c**) with the stopcock open as short a time as possible. The syringe is then placed fully under water in the water bath, shaken for 40 min (Panel **d**), and finally removed from the water bath. The stopcock is again dried by an air blast, the syringe is then held vertically (Panel **e**), connected to a dry syringe held horizontally, and the gas above the blood transferred into the dry syringe (*purple arrow*). Care is taken to avoid any blood at all reaching the dry syringe

6.8 Measuring Blood/Gas Partition Coefficients of the Gases by GC

The MIGET theory requires the blood/gas partition coefficients to be known for all six gases. If they did not vary across subjects, a standard set of partition coefficients could be selected from the literature and used. Sadly, this is not the case, and so partition coefficients must be measured for each gas in each subject. We measure them (in duplicate as a quality control tactic). This is at once easy and complicated. The concept is easy—conservation of mass as you will see—but the steps are intricate, yet important to perform very well. Here is how this is done.

We always start a MIGET study with control blood samples taken before the inert gas solution is infused—two samples of about 8 ml each. These are indeed used as true controls, as we process them as if they were a blood sample (see above), expecting no inert gas peaks to appear. If peaks do appear, we have to stop and figure why and what to do.

The blood used for these two control samples is now reused to measure the partition coefficients:

1. First, expel any bubbles from the control blood-containing syringes.
2. Now add a small amount of each of the six MIGET gases to a 50 ml dry, glass syringe to provide a single mixture of all six. This is done as follows:

   Gas from the three-gas tank containing SF_6/ethane/cyclopropane is diluted in a dry 50 ml glass syringe to 1 in 10,000. Sounds silly, but otherwise the concentrations are way too high. If you dilute 5 ml gas up to 50 ml using N_2 four times, that will achieve the 1:10,000 target dilution.

   For the three other gases, which exist and are stored in liquid form in individual bottles, we draw the headspace gas above the liquid in the bottle into a 50 ml syringe, using separate syringes for each gas. Then, for isoflurane we dilute 1:20 with N_2. For ether we dilute 1:10, but for acetone, we do not dilute at all. We then transfer 15 ml of the 1:10,000 three-gas mix to a new, empty 50 ml syringe, and add 10 ml of the 1:20 isoflurane gas; 5 ml of 1:10 ether gas; and 10 ml of the acetone gas. This gives a six-gas mix, 40 ml in total, with a reasonable set of concentrations of each.
3. Now add 15 ml of this mixture to each of the two control blood samples in its syringe and equilibrate this blood/gas mix in the shaking water bath for 40 min. After this, move the gas above the blood into a transfer syringe exactly as if it were a regular subject blood sample (see above), and then place the sample in queue to be run through the gas chromatograph to measure the peak heights of the six gases. These two duplicate samples are labeled as $\mathbf{S_1}$ and $\mathbf{S_2}$.
4. *Immediately* after the gas has been transferred (per par immediately above)—before any cooling down of the blood in the syringe—expel any remaining bubbles in the blood-filled syringe through the stopcock into the room. There are always remaining bubbles because if one were to expel them into the transfer syringe, there would be blood in that syringe, which would mess up the GC as mentioned earlier. This gas-bubble-free blood-containing syringe is now cleaned

and dried and weighed. It will have lost a very small amount of blood in the bubble expelling process, but that is not a problem. Note: it had been weighed dry as for all sample syringes earlier in the day. By difference we have the weight of blood in the syringes at this stage.

5. Now, about 15–20 ml new N_2 gas is added to these same blood syringes, and they are stuck back into the water bath for another 40 min shake. After 40 min, their total volumes are measured, they are taken out, their gas is transferred into fresh transfer syringes, and these are labeled $\mathbf{S_1R}$ and $\mathbf{S_2R}$. They are placed in queue for GC analysis.
6. *Immediately* after the gas has been transferred (as in par immediately above)—before any cooling down of the blood in the syringe—the remaining bubbles are expelled through the stopcock into the room. As stated above, there are always remaining bubbles because if one were to expel them into the transfer syringe, there would be blood in that syringe, which would mess up the GC as mentioned earlier. This gas-bubble-free blood syringe/stopcock is now cleaned and dried but does *not* need to be weighed. It will have lost a very small amount of blood in the bubble expelling process, but that is not a problem.
7. You now take a clean dry (non-heparinized) 50 ml syringe with stopcock, fill it to about 40 ml with fresh N_2, and transfer exactly 0.25 ml of this blood into it. This is tricky. You use a 1 ml syringe for accuracy, and connect it to an 18 or 20 gauge long (i.e., spinal) needle. When ready to transfer the 0.25 ml blood, you open the stopcock of the 50 ml syringe (containing the N_2) and insert the spinal needle through it into the inside of the 50 ml syringe and squirt the 0.25 ml blood into it, withdraw the needle and close the stopcock. You do this *fast* but yet *gently*. You *must* make sure the 0.25 ml blood drops into the main part of the 50 ml syringe and that in particular *no blood has reached the plunger or it will migrate up the plunger and be lost*. If that happens, stop and start again with a fresh 50 ml syringe. You cannot recover from this; you have to start over.
8. You take that 50 ml syringe, now containing about 40 ml N_2 and 0.25 ml blood, and immerse it in the water bath. This time you have to hold the plunger so the syringe stays at about 45° (usual sample syringes can be laid flat in the water bath). The reason is again to keep the blood away from the plunger. The majority of the 50 ml syringe must be immersed in the water of the bath, but NOT to the point where the plunger and the barrel meet, which should stay above the water line. You simply have to stand at the bath and hold the syringe in this way—but only for 5 min. After this, read the total volume in the 50 ml syringe and *immediately* transfer some of the gas in that 50 ml syringe into a fresh transfer syringe (after rapidly drying the stopcock of the 50 ml syringe with wall air). Label this transfer syringe as $\mathbf{S_1R^*}$ and the duplicate as $\mathbf{S_2R^*}$.
9. Note that this has been a process requiring THREE sequential water bath shaking periods. The first was to get the six gases INTO the blood and in equilibrium with inert gas in the N_2 gas in the syringe ($\mathbf{S_1}$, $\mathbf{S_2}$ samples). No gas or blood volume numbers needed. The second was to equilibrate blood containing known levels of inert gases (known from $\mathbf{S_1/S_2}$ samples) with new N_2 to get $\mathbf{S_1R}$ and $\mathbf{S_2R}$ samples. Here you *do* need gas and blood volumes as indicated above; let's call them **Vg** and

Vb, respectively. The third water bath shake was with the same blood as indicated by the $\mathbf{S_1R}$ and $\mathbf{S_2R}$ levels, but in a 40 ml N_2 to 0.25 ml blood ratio to end up with $\mathbf{S_1R^*}$ and $\mathbf{S_2R^*}$ samples, again requiring the blood and gas volumes (0.25 ml blood, about 40 ml gas as measured; let's call them **Vg*** and **Vb***, respectively).

One then uses $\mathbf{S_1}$, $\mathbf{S_2}$ and $\mathbf{S_1R}$, $\mathbf{S_2R}$ peaks and **Vg** and **Vb** to compute the partition coefficients of SF_6, ethane, cyclopropane, and isoflurane. One next uses $\mathbf{S_1R}$, $\mathbf{S_2R}$ and $\mathbf{S_1R^*}$, $\mathbf{S_2R^*}$ **peaks,** and **Vg*** and **Vb*** to compute the partition coefficients of ether and acetone:

$$\lambda_1 = \mathrm{Vg}_1/\mathrm{Vb}_1/(\mathrm{S}_1/\mathrm{S}_1\mathrm{R} - 1) \quad \text{for each gas except ether and acetone}$$
$$\lambda_2 = \mathrm{Vg}_2/\mathrm{Vb}_2/(\mathrm{S}_2/\mathrm{S}_2\mathrm{R} - 1) \quad \text{for each gas except ether and acetone}$$

and

$$\lambda_1 = \mathrm{Vg}_1^*/\mathrm{Vb}_1^*/(\mathrm{S}_1\mathrm{R}/\mathrm{S}_1\mathrm{R}^* - 1) \quad \text{for ether and acetone}$$
$$\lambda_2 = \mathrm{Vg}_2^*/\mathrm{Vb}_2^*/(\mathrm{S}_2\mathrm{R}/\mathrm{S}_2\mathrm{R}^* - 1) \quad \text{for ether and acetone}$$

Note the formulas are identical in constitution for all gases. The reason for splitting ether and acetone away from the other four is simply their solubility differences and optimizing the measurement based on the resulting expected differences between $\mathbf{S_1/S_1R}$ and $\mathbf{S_1R/S_1R^*}$.

Where do these equations come from? Mass conservation:

$$\mathrm{Vb} \times \lambda \times \mathrm{S}_1 = \mathrm{Vb} \times \lambda \times \mathrm{S}_1\mathrm{R} + \mathrm{Vg} \times \mathrm{S}_1\mathrm{R}$$

This equation states that the number of molecules of the inert gas (SF_6, ethane, cyclopropane, and isoflurane, each separately considered) in Vb ml of blood *before* the second water bath shake (left side) equals the number of molecules in that same Vb ml blood plus Vg ml gas in the same closed syringe *after* that water bath shake (right side), and it assumes that there is partial pressure equality between blood and gas after the 40 min water bath shake. This equation is simply rearranged to make λ the subject of the formula.

Exactly the same development holds for ether and acetone as the principle is identical, just the choice of blood and gas volumes differ because of the very different solubilities.

Figure 6.6 gives photos of the key steps in solubility measurement, illustrated for the more difficult case of ether and acetone.

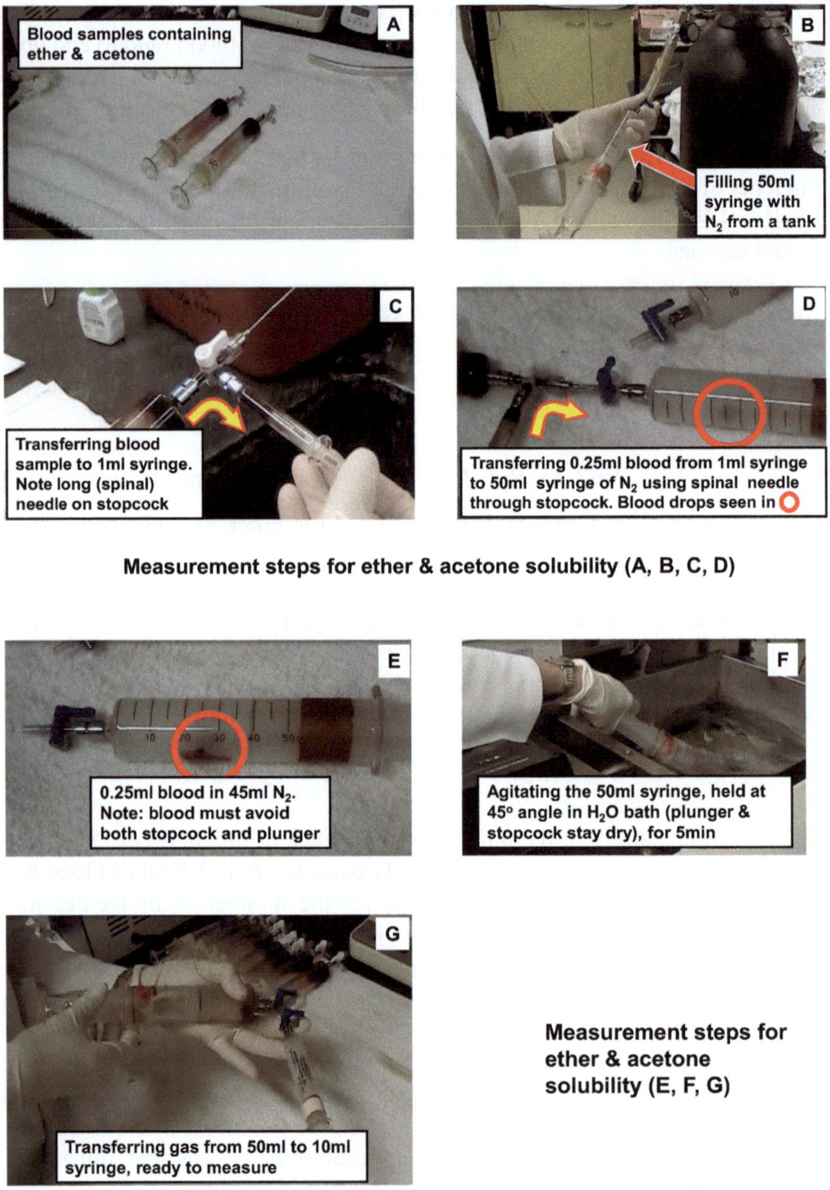

Fig. 6.6 Measurement steps for ether and acetone solubility. Panel (**a**) shows a syringe containing blood already with ether and acetone present. Panel (**b**) shows the filling of a 50 ml glass syringe with N_2 from a tank. Panel (**c**) shows 0.25 ml of the blood from Panel (**a**) being moved into a 1 ml syringe (for accuracy) in a closed system. Panel (**d**) shows that 0.25 ml blood being injected into the 50 ml syringe (i.e., full of N_2), by means of a spinal needle inserted through the stopcock. Panel (**e**) shows the 0.25 ml blood after this transfer lying in the bottom of the horizontal syringe such that no blood has reached the plunger of the syringe, and no blood is in the stopcock. Panel (**f**) shows the operator gently agitating the 50 ml syringe containing the 0.25 ml blood. The syringe is held at about 45° to the vertical to keep the blood from migrating to either the stopcock or plunger. Only 5 min of agitation is needed. Finally, in Panel (**f**) the gas from the 50 ml syringe is transferred into a clean dry syringe for measurement. Both syringes are kept horizontal during the transfer, again to avoid blood reaching the stopcock or plunger

6.9 Measurement of Random Error Levels in a GC System

It is important to occasionally document the (random) experimental error level in your hands, because the entire MIGET depends quite heavily on knowing that value as a coefficient of variation. It is boring but easy to measure. It is important to do this at the concentrations encountered in actual studies, so it is most convenient to do it at the end of an actual study, taking extra blood and gas from the subject after the study is done but *before* the infusion is turned off and *before* catheters are removed. It needs to be done perhaps once a year, after a material change in equipment, or with a new technician.

First, take a single large (50 ml) syringe (matched barrel/plunger) sample of *mixed expired gas* containing the six gases at physiological concentrations and introduce a small amount (5 ml) into the GC and measure the peaks. Repeat nine more times from that one syringe, and for each gas, determine the mean and the SD and divide the SD by the mean. For all FID gases (i.e., except SF_6), this will likely be less than 0.01 and, for SF_6, a bit higher but not much unless there is dilution involved.

Next, do the same thing from a single large pooled *blood* sample containing the six gases at physiological concentrations. Fill a 50 ml heparinized (matched barrel/plunger) syringe with arterial or (mixed) venous blood at the end of the study, and from that fill six heparinized 30 ml regular blood sample syringes with 8 ml each *anaerobically* so you know each syringe has the same starting gas levels. Then process as above for real blood samples, and that will give you six repeats that should be the same. Again, compute the SD/mean from these six samples for each gas. For the five FID gases, you should be able to get 0.03 or so; for SF_6, 0.05 or so.

6.10 Entering the Inert Gas Data and Ancillary Variables onto a Spreadsheet

We use three different spreadsheets.

(A) One is filled out by the GC processing person, and it records the GC-associated variables for each sample—dry syringe weight, heparinized syringe weight, blood-filled syringe weight, total gas volume in the syringe after water bath shaking, and solubility measurement variables of the same type. Attached (Fig. 6.7) is a version of such a sheet to indicate one easy format to use.

(B) A table of solubility peak height numbers for each gas in S_1, S_1R, and S_1R^* and the duplicates (S_2, S_2R, and S_2R^*) and the blood and gas volumes in the S_1R, S_1R^*, and S_2R and S_2R^* syringes. You then calculate the partition coefficients by hand, or by spreadsheet formula, you have set up ahead of time based on the equations given above for the two duplicate sets, and average the two partition coefficients obtained for each gas. They should agree to within about 10% for any one gas—providing a good index of quality control.

Sample, syringe weights/volumes data sheet

Subject id......................

Date:.............PB...... Bath temp.............Density (wt/vol)......................

Sample	A1	A2	A3	A4	A5	A6
Syringe id #						
Dry weight						
+ Heparin						
+ Blood						
Total volume						
Gas volume						
Blood volume						
Heparin volume						

Fig. 6.7 The spreadsheet used to record for each arterial blood sample the syringe weights dry, with heparin, and with blood and the associated volumes of gas, blood, and heparin

(C) The third spreadsheet tabulates all the necessary input data for the MIGET software in the format matching the data entry sequence. Attached (Figs. 6.8 and 6.9) is a version of such a sheet to indicate one easy format to use.

Years of doing this has shown that the effort taken to fill out such forms is worth it. That way, data entry into the computer is much easier as all numbers are together in one place. Even more, errors are easily discerned as there are narrow ranges for most of the numbers from data set to data set that are directly visible as you fill out the form. One number different from the same variable value in another data set is easily seen and should be investigated.

6.11 Entering the Data from the Spreadsheet into the MIGET Software: Either in the Original DOS Format or Now Using a Windows GUI Environment and Running the MIGET Software

Armed with the second spreadsheet (see 6.10 above, Figs. 6.8 and 6.9), you move to the computer where the MIGET software is installed to enter the data and run the program. Entering the data takes perhaps 2 min per data set for an experienced person, and running the program takes a fraction of a second.

Sample data input sheet, first page

DATA SHEET WHEN VENOUS SAMPLES ARE ABSENT

NAME:... DATE:......... PAGE:.......... CV.......
NUMBER OF RUNS:...... PB.......... T,electrodes....... T. water bath.......
NGASES:...... NVAQs:......... Z:....... VQLO:.......... VQHI :..........

SET 1	SF6	ETHANE	CYCLOPROPANE	ISOFLURANE	ETHER	ACETONE
Partition Coefficients						
ARTERIAL Peak height						
EXPIRED Peak height						

VE......... QT......... PB.......... Tb..........Tex..........VGa..........VBa.........VHa.....
VGv...0..... VBv...1.....VHv...0..... Hb....... Hct........ VO2........ VCO2....... TOL 99000
FIO2........ FICO2...0... P50........ PaO2........PaCO2........pHa....... VO2...... VCO2...

SET 2	SF6	ETHANE	CYCLOPROPANE	ISOFLURANE	ETHER	ACETONE
ARTERIAL Peak height						
EXPIRED Peak height						

VE......... QT......... PB.......... Tb..........Tex..........VGa..........VBa.........VHa.....
VGv...0..... VBv...1... VHv...0.... Hb....... Hct........ VO2........ VCO2... TOL 99000
FIO2........ FICO2...0... P50........ PaO2........PaCO2........pHa....... VO2...... VCO2......

Fig. 6.8 Spreadsheet for entering the inert gas data and ancillary variables needed to run the MIGET software. This is the face sheet and includes the first two data sets of a series that can go on to any number

Sample data input sheet, next page

DATA SHEET WHEN VENOUS SAMPLES ARE ABSENT

NAME:.. DATE:......... PAGE:..........
NUMBER OF RUNS:...... PB.......... T,electrodes...... T. water bath.......
NGASES:...... NVAQs:......... Z:....... VQLO:.......... VQHI :..........

SET 3	SF6	ETHANE	CYCLOPROPANE	ISOFLURANE	ETHER	ACETONE
ARTERIAL Peak height						
EXPIRED Peak height						

VE......... QT......... PB.......... Tb..........Tex..........VGa..........VBa.........VHa.....
VGv...0..... VBv...1.....VHv...0..... Hb....... Hct........ VO2........ VCO2....... TOL 99000
FIO2........ FICO2...0... P50........ PaO2........PaCO2........pHa....... VO2...... VCO2...

SET 4	SF6	ETHANE	CYCLOPROPANE	ISOFLURANE	ETHER	ACETONE
ARTERIAL Peak height						
EXPIRED Peak height						

VE......... QT......... PB.......... Tb..........Tex..........VGa..........VBa.........VHa.....
VGv...0..... VBv...1... VHv...0.... Hb....... Hct........ VO2........ VCO2... TOL 99000
FIO2........ FICO2...0... P50........ PaO2........PaCO2........pHa....... VO2...... VCO2......

Fig. 6.9 Continuation of the spreadsheet shown in Fig. 6.7 to the next two data sets

We have *two* programs (both in FORTRAN) that run in sequence—we have for 40 years called them SHORT and VQBOHR, respectively. SHORT is the program you run first—to enter the data from your paper spreadsheet. It calculates the actual retention and excretion values for each data set and allows you to correct any mistakes and then rerun the set. It also creates, automatically, an output file that can be printed with these numbers, and it also creates, automatically, a data file that will become the input data file for the second program, VQBOHR. You just have to enter unique filenames for these files when prompted.

Why separate the programs into two, rather than do it all at once? To give you the chance to correct data entry errors and to be sure that the data are physiologically reasonable (garbage in gives garbage out).

We now have *two* options for entering these data and running these programs:

(A) The first is the original and uses the old DOS system. It is tried and tested, but if you hit the wrong key on the keyboard during data entry, it may crash and you may have to restart data entry from the last successful data set entered. Very frustrating.

(B) The second method uses a GUI (graphical user interface, Fig. 6.10) that a colleague in Barcelona (Isaac Cano, Ph.D.) very kindly constructed. The program package thus contains the GUI—where you enter data in a Windows screen format and thus do not risk crashing from a data entry error—plus the actual MIGET programs. The GUI is linked to the FORTRAN programs so you can directly run them from the GUI screen. There is a detailed section following that describes this GUI approach. The GUI runs like the old DOS-based system—

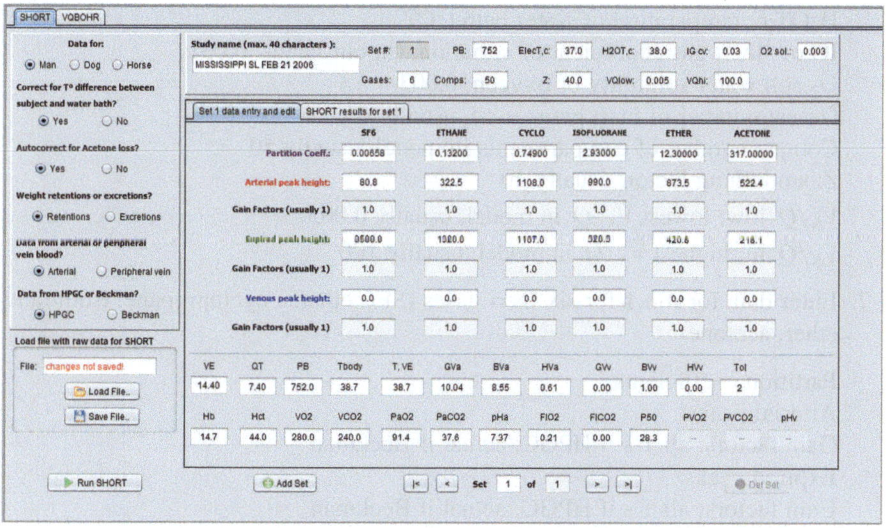

Windows Graphical User Interface for data entry

Fig. 6.10 Photo of the graphic user interface (Windows data entry platform) first SHORT and when you are happy with the data, then VQBOHR. The instructions, kindly provided by A. Cortney Henderson Ph.D., are as follows:

MIGET GUI Program Instructions—A. Cortney Henderson, 12 August 2013

1. Make sure Java is installed on your computer.
2. Double click on MIGET executable jar file using an IBM PC or for Mac users, and run jar file using Parallels.
3. Under SHORT tab, and enter the following for questions in left column:

   Data for: Man (dog, pig...)
   Correct for temperature differences between subject and water bath? Yes/no
   Autocorrect for acetone loss? Yes/no
   Weight retentions or excretions? Retentions/excretions
   Data from arterial or peripheral venous blood? Arterial/peripheral venous
   Data from HPGC or Beckman chromatograph? HPGC/Beckman

4. To load a previously created SHORT input file, under load SHORT parameters file, click "Load from File" button and select file, and then jump to step 13 to run SHORT.
5. Enter study name.
6. Enter variables in top section as follows:

   Number of runs: number of samples to be processed (can be increased by "Add Run" button)
   PB: barometric pressure (mm Hg)
   ElecT,c: temperature of electrodes (°C)
   H_2OT,c: temperature of water bath (°C)
   IG cv: inert gas coefficient of variation, (in our hands, 0.03)
   O_2 sol: water solubility of oxygen, 0.003
   Gases: number of inert gases used, usually 6
   Comps: number of compartments in model, usually 50
   Z: smoothing factor, usually 40
   $\dot{V}_A/\dot{Q}$ low: lowest $\dot{V}_A/\dot{Q}$ in model, usually 0.005
   $\dot{V}_A/\dot{Q}$ hi: highest $\dot{V}_A/\dot{Q}$ in model, usually 100

7. Enter data for run 1 for six inert gases (SF_6, ethane, cyclopropane, isoflurane, ether, acetone):

   Partition coefficients
   Arterial peaks
   Gain factors: all 1's if HPGC; actual if Beckman
   Expired peaks
   Gain factors: all 1's if HPGC; actual if Beckman
   Venous: all 0's unless pulmonary artery samples were taken, then enter peaks
   Gain factors: all 1's if HPGC; actual if Beckman

8. Enter variables in the bottom section as follows:

   $\dot{V}_E$ l/min: expired ventilation corrected to mixing box temperature (ATPS from metabolic cart with inspired temperature set to mixing box temperature)
   $\dot{Q}_T$ l/min: cardiac output, measured for that data set
   PB: barometric pressure (mm Hg)
   T_{body}: body temperature (°C)
   GV_a: volume of gas added to arterial samples
   BV_a: volume of blood in arterial samples
   HV_a: volume of heparin in arterial samples
   GV_v: volume of gas added to mixed venous samples, 0 or actual if samples
   BV_v: volume of blood in mixed venous samples, 1 or actual if samples
   HV_v: volume of heparin in mixed venous samples, 0 or actual if samples
   Tol: tolerance, 99,000
   Hb g/dl: hemoglobin concentration (g/dl)
   Hct %: hematocrit
   $\dot{V}_{O_2}$: oxygen consumed (ml/min)
   $\dot{V}_{CO_2}$: carbon dioxide produced (ml/min)
   P_{aO_2}: arterial oxygen partial pressure (mm Hg)
   P_{aCO_2}: arterial carbon dioxide partial pressure (mm Hg)
   pH_a: pH of arterial blood sample
   F_{IO_2}: fraction of inspired oxygen (air = 0.21)
   F_{ICO_2}: fraction of inspired carbon dioxide, 0
   Standard P_{50}: the P_{O_2} at which hemoglobin is 50% saturated, normal is 26.8

9. Click "Add Run" to add runs and "Del Run" to delete runs.
10. Enter data for other runs.
11. Arrow through runs, and click on tabs to preview each run.
12. At the bottom of the left column under "Load SHORT parameters file," click "Save to File" button and specify filename as shortInput_initialsdate*.txt.
13. Click "Run SHORT" button.
14. In the output window, check box for "Bypass SHORT output to VQBOHR."
15. Click "Save Short text output," and specify filename as shortOutput_initialsdate*.txt.
16. Click "Save Short output for VQBOHR," and specify filename as vqbohrInput_initialsdate*.txt and close window.
17. Click on VQBOHR tab.
18. Enter the following for questions in the left column:

   Check "Bypass SHORT output to VQBOHR."
   Are these sets retention or excretion weighted? Always select retention-weighted.
   Enter set numbers to be skipped: If you want to include all set numbers for analysis, leave this blank.
   Do you want O_2/CO_2 calcs, D/Q infinite? Yes or no.
   Do you want Bohr integration calculations? Yes or No.

19. Click "Run VQBOHR."
20. Arrow through to see results, ventilation-perfusion curves, and retention-excretion curves for each set.
21. Click "Save VQBohr output" and specify filename as vqbohrOutput_initialsdate*.txt.
22. In Excel, click File → Open, select vqbohrOutput_initialsdate*.txt, select Delimited, Next, check box next to "Space," click Next, and click Finish. Now scroll down to copy/paste cells with $\dot{V}_A$, $\dot{Q}$, and $\dot{V}_A/\dot{Q}$ values.

In both cases (A and B), you first need to select among several options for how the programs are to be run. You are prompted in the DOS system; there are check boxes on the GUI for the same options.

Describing the data entry in any more detail here in the text is abstract and a bit futile; it is much better to sit down and do it with the actual programs.

6.12 Looking at and Analyzing the Results: For Data Errors and Physiological Meaning—First, SHORT, Second, VQBOHR

In looking at the output of SHORT for any one data set as you are entering it, you obviously are trying to check for both data errors and unphysiological numbers. Perhaps the best to look at are the retention and excretion numbers for the six gases. They should be monotonically increasing in value from SF_6 up to acetone (left to right across the screen), and excretion numbers must be less than retention numbers for each gas. Duplicate sets should give very similar numbers in each cell.

There are too many permutations of errors to describe them all, but experience will help. Obviously data entry errors need to be corrected and SHORT rerun for that data set.

In looking at the output of VQBOHR, you have by definition passed through the "quality control" examination of the data in SHORT and are more focused on the physiology. The output variable **to** look at first is the RESIDUAL SUM OF SQUARES (RSS). That is usually the final quality control checkpoint.

RSS over many data sets should follow the chi-square distribution for six degrees of freedom and is interpreted probabilistically as for any statistical parameter. RSS should be less than ~24 about 99% of the time, less than ~10 about 90% of the time, and less than ~5.3 about 50% of the time.

This range guides your interpretation of RSS. It is a sensitive indicator of data errors. If RSS is large (>20) for any data set, immediately inspect the individual gas errors in the table above the RSS and see which of the six gases has the largest error. If there is one gas way off, that is an easy pointer to go back to the raw data, and check for errors all along the way for that gas in that sample, starting with the GC output, and then the spreadsheets and SHORT. But note, any one RSS of, say, ten is NOT immediately suspect from the chi-square table, as such a value is expected once in ten sets from normal, acceptable, and random error.

After checking RSS for all data sets and rerunning any sets that had an error causing large RSS, scroll through the parameter tables describing the results. First, the fractional distribution of ventilation and blood flow into decades of $\dot{V}_A/\dot{Q}$ range, and see if they make sense for the type of subjects and conditions studied based on past experience. At the same time, also look at the three moments of both the ventilation and perfusion distributions and make sure they make sense and are relatively similar from duplicate to duplicate.

Next quickly scroll past the retention curve plots and the $\dot{V}_A/\dot{Q}$ distributions themselves. Assuming you have corrected any data errors and the study went well, they should look physiological (monotonic, good visual fits to data) and reproducible from set to set within a given condition.

Finally look at the O_2/CO_2 table and check that the predicted arterial blood gases reasonably agree with the measured values (also printed in the output). As with the RSS, you have to look at this statistically over many data sets to really interpret them—due to random errors in the whole inert gas method and in the blood gases, predicted and measured never agree precisely, but over a sufficient number of data sets, they should agree on average.

The expected degree of agreement between measured and predicted arterial P_{O_2} is very dependent on where on the O_2Hb curve the P_{O_2} lies. On the flat part of the O_2Hb curve, agreement is not often that close—many mm Hg. But on the steep part ($P_{O_2} < 50$–60 mm Hg), agreement is usually within 2–3 mm Hg, which is acceptable.

Deciding when measured P_{O_2} is systematically less than predicted (versus randomly lower from error), indicating diffusion limitation, can really only be made statistically from several data sets unless the differences are very large such that the 95% confidence limits are exceeded for the difference between measured and predicted P_{O_2} values. In essence, if SD for P_{O_2} is about 3 mm Hg such that 2SD is 6 mm Hg, *any one* predicted P_{O_2}—measured P_{O_2} difference needs to be more than 12 mm Hg to have confidence that the difference represents diffusion limitation and not random errors. Much smaller differences are of course detectable from a group of several data sets.

Having inspected all the inert gas and O_2/CO_2 related outputs, you are ready to create your spreadsheet of scientific results and analyze the outcome of your study, as a bona fide MIGET expert.

6.13 Additional Issues Raised by Others, Technical Notes, etc.

For those of you who have managed to read through to this point, it will have become clear that we have opted to tell a story in a logical manner uncontaminated by issues raised by others. In this section, we bring together a host of ideas contributed by several interested colleagues as well as some raised by ourselves.

6.13.1 Theoretical and Mathematical Issues

The basic issue driving this discussion has been to define the information content of MIGET as well as possible: What can it do, what is beyond its capacity? We tackled this by basic theory and by methods of linear and quadratic programming [8, 12–14, 21, 30–32, 38], while others also chipped in: [4, 16, 18, 27, 28, 37]. The simplest high-level conclusion of all this work was that distributions with up to three separate modes were identifiable although the fine detail of such modes was beyond resolution.

In the same domain, Rees et al. sought to simplify the use of MIGET by fitting data to a 5-$\dot{V}_A/\dot{Q}$ compartment model of the lungs [22, 23]. These provide useful information, and it becomes a matter of choice as to the preferred approach [29].

6.13.2 Conducting Airway Influences on MIGET

The consequences of the conducting airways for MIGET have sustained discussion for many years. The traditional view of course is that those airways do not participate in gas exchange, and for the respiratory gases O_2 and CO_2, that remains an acceptable dogma. However, for highly water-soluble gases such as acetone (and especially the even more soluble gas ethanol), there may be some gas exchange across the conducting airway wall. This may be true comparing an inhalation with its subsequent exhalation: acetone reaching the airway wall in bronchial arterial blood may elute into the airway lumen during inspiration (of acetone-free air) and then move back into the airway wall tissue during expiration. However, remember that MIGET is a steady-state technique where over the course of minutes, gas exchange is in a steady state. It does not address breath-by-breath events. MIGET estimates of conducting airway deadspace depend heavily on acetone values in expired gas and have always shown numbers similar to those produced by CO_2, which is about 50 times less soluble than acetone. That said, Mike Hlastala's group feels that significant soluble gas exchange does occur in the conducting airways and provided theoretical calculations in support [2]. If this is to be further explored experimentally, investigators must avoid an unusual but important detail—using heparin (to anticoagulated blood samples) that contains significant acetone concentrations, as we found many years ago [20]. In sum, what soluble gas exchange occurs in the conducting airways seems to have little influence on overall steady-state gas exchange.

6.13.3 Series and Parallel Deadspace: Implications for MIGET

As a branching tree, the conducting airway anatomy imposes both parallel and serial gas flow which results in gas mixing patterns that can have complex effects on gas exchange in general, not just for MIGET.

Depending on many structural and functional properties of the airways, *diffusive* gas mixing between and along airways may not be complete. If so, the high molecular weight gases may exchange less than low molecular weight gases. In MIGET, the molecular weight range is from 30 to about 200, which may translate to an observable effect of diffusive exchange on the retention-solubility curves. It would at the same time affect the least squares solutions at the core of MIGET, such that the residual sum of squares would exceed expectations (because MIGET ignores molecular weight and assumes diffusion equilibration in all aspects of gas exchange), and moreover in a specific manner: High molecular weight gas retentions would be systematically greater than the least squares best fit predicted values, and low molecular weight gas retentions would be less than predicted. We have rarely seen such behavior, although Hlastala predicted [11] the effect would be equivalent to a 5 mm Hg alveolar-arterial difference in P_{O_2}.

Diffusive gas mixing is not the only way the series/parallel airway structure may complicate gas exchange. *Convective* mixing could play a separate role, and this would be blind to molecular weight differences across gases. If an airway divides into two daughter airways, one of which ventilates a region with a high $\dot{V}_A/\dot{Q}$ ratio and the other of which ventilates a region with a low $\dot{V}_A/\dot{Q}$ ratio, the alveolar gas during exhalation from each will mix in the parent airway so that on the ensuing inhalation, each region receives contaminated inspired gas in a manner that MIGET software cannot take into account. This could occur at every generation and branch point and become very complex. The innate nature of such mixing will be to improve overall inert gas elimination and underestimate overall deadspace [19]. Just how much depends on the specifics of the parent-daughter mixing parameters across the entire airway tree, but since MIGET-estimated deadspace and that measured by earlier "conventional" methods with CO_2 are hardly different and are similar to expectations from anatomic measures of conducting airway volume, the issue may not be of great quantitative significance.

We had a concern that in-series $\dot{V}_A/\dot{Q}$ inequality might violate the mathematical foundation of MIGET, which is based only on parallel inequality, and we thus calculated retention and excretion curves in such serial models. We then fed them into the MIGET software and showed that the parallel MIGET analysis worked perfectly even when series inequality was present [9, 33]. This means that for any given in-series pattern of inequality, there exists a mathematically equivalent parallel model that has identical retention and excretion curves across the six-gas spectrum. In other words, the MIGET software will not fail when in-series inequality is present but at the same time cannot separate serial from parallel inequality or say which is present. It will always come up with a parallel $\dot{V}_A/\dot{Q}$ distribution model that will account for the combined effects of serial and parallel inequality.

6.13.4 Mass Spectrometry Versus Gas Chromatography for Inert Gas Measurement

Gas chromatography is slow, especially when blood levels need to be measured using syringe headspace equilibration. It is however inherently linear, very sensitive in the face of low gas concentration, and relatively inexpensive, and the chromatograph itself is mechanically robust and low in maintenance. Mass spectrometry is an attractive alternative and has been proposed. Mastenbrook et al. first tried this in 1982 [17]. So too did Shimadzu et al. [26] after which Baumgardner et al. and others further demonstrated feasibility in 2000 [3], and 2007 [7], followed up by Kretzschmar et al. in 2013 [15]. Measurement is faster; headspace equilibration of samples in blood is not needed, but concentrations in the ppm range pose an ongoing challenge, as does cost. Duenges et al. [6] showed that shunt calculated using this method of analysis was less than that calculated using the Riley technique, and at this stage, it does not yet seem fully validated as a replacement for chromatography.

6.13.5 Inert Gas Solubility in Blood and Mixing of Blood in the Right Ventricle

Over the years, it became apparent that assuming for each MIGET gas an average blood solubility across subjects was dangerous as there is a range of solubility values among subjects for each gas. It also became apparent that common anesthetic gases such as nitrous oxide and halothane could interfere with the MIGET chromatography especially because the anesthetic gases were used in concentrations orders of magnitude higher than those used in MIGET. Dueck et al. figured out protocols to get around this [5]. Yamaguchi wondered if MIGET gases had systematically different solubilities in mixed venous (acidic, desaturated) blood compared to arterial blood, and if so would that interfere with assumption that for any one subject there was one solubility value for each gas? While small systematic differences in solubility were found, the overall conclusion was that they were so small as to have negligible effect on MIGET data analysis [40, 41]. Young and Wagner [42] compared the solubilities of the inert gases in blood to the solubilities in homogenates of canine lung tissue. All were slightly but significantly more soluble in blood than lung tissue with the exception of SF_6.

Guitart et al. [10] used MIGET to show that in rats, inert gases were not fully mixed in the right ventricle compared to more distal sampling locations, highlighting the importance of sampling in the correct vascular location when conducting MIGET studies.

In summary, many specific but minor issues have surfaced in the application and interpretation of MIGET data, but they all have small, second-order, effects on the outcomes that do not materially affect the method or its use.

6.13.6 *Use of MIGET to Validate Other Techniques*

MIGET has been used to validate other techniques. Altemeier et al. [1] used MIGET to validate $\dot{V}_A/\dot{Q}$ measurements derived from microspheres. They did this by predicting gas exchange from their microsphere measurements for oxygen, carbon dioxide, and inert gas. They found excellent correlations ($R = 0.95-0.99$) between their calculated retention and excretions and MIGET. Sato et al. [25] compared shunt measured by MIGET to oxygen-measured shunt and concluded that P_{O_2} electrode hysteresis could account for more of the discrepancy: although the results from the two methods were highly correlated ($r = 0.97$), that slope was significantly less than unity with oxygen shunt overestimating relative to MIGET in the low range of shunt values (<12%) but shunt when shunt was high. Recently Sá et al. [24] compared LogSD, Q and LogSD, V_A measures of heterogeneity those obtained in a single slice of the lung from proton MRI to derived from MIGET. The results were promising with $r = 0.89$ between MRI and MIGET.

References

1. Altemeier WA, Robertson HT, Glenny RW. Pulmonary gas-exchange analysis by using simultaneous deposition of aerosolized and injected microspheres. J Appl Physiol (1985). 1998;85: 2344–51.
2. Anderson JC, Hlastala MP. Impact of airway gas exchange on the multiple inert gas elimination technique: theory. Ann Biomed Eng. 2010;38:1017–30.
3. Baumgardner JE, Choi IC, Vonk-Noordegraaf A, Frasch HF, Neufeld GR, Marshall BE. Sequential V(A)/Q distributions in the normal rabbit by micropore membrane inlet mass spectrometry. J Appl Physiol (1985). 2000;89:1699–708.
4. Clark JS, Lin YJ, Criddle MJ, Cutillo AG, Bigler AH, Farr FL, Renzetti AD Jr. Cardiac output and mixed venous oxygen content measurements by a tracer bolus method: theory. J Appl Physiol (1985). 1997;83:884–96.
5. Dueck R, Rathbun M, Wagner PD. Chromatographic analysis of multiple tracer inert gases in the presence of anesthetic gases. Anesthesiology. 1978;49:31–6.
6. Duenges B, Vogt A, Bodenstein M, Wang H, Bohme S, Rohrig B, Baumgardner JE, Markstaller K. A comparison of micropore membrane inlet mass spectrometry-derived pulmonary shunt measurement with Riley shunt in a porcine model. Anesth Analg. 2009;109: 1831–5.
7. Dunges B, Karmrodt J, Baumgardner JE, Markstaller K. Ventilation-perfusion distributions in the lungs. A novel technique for rapid measurement. Anaesthesist. 2007;56:612–6.
8. Evans JW, Wagner PD. Limits on VA/Q distributions from analysis of experimental inert gas elimination. J Appl Physiol Respir Environ Exerc Physiol. 1977;42:889–98.
9. Fortune JB, Wagner PD. Effects of common dead space on inert gas exchange in mathematical models of the lung. J Appl Physiol Respir Environ Exerc Physiol. 1979;47:896–906.
10. Guitart R, Alfaro V, Roca-Acin J, Palacios L. Blood sampled in rats from right ventricle may not always be truly representative of a mixed venous sample. Lab Anim. 1996;30:273–8.
11. Hlastala MP, Scheid P, Piiper J. Interpretation of inert gas retention and excretion in the presence of stratified inhomogeneity. Respir Physiol. 1981;46:247–59.
12. Kapitan KS, Wagner PD. Information content of multiple inert gas elimination measurements. J Appl Physiol (1985). 1987;63:861–8.

13. Kapitan KS, Wagner PD. Linear programming analysis of VA/Q distributions: average distribution. J Appl Physiol (1985). 1987;62:1356–62.
14. Kapitan KS, Wagner PD. Linear programming analysis of VA/Q distributions: limits on central moments. J Appl Physiol (1985). 1986;60:1772–81.
15. Kretzschmar M, Schilling T, Vogt A, Rothen HU, Borges JB, Hachenberg T, Larsson A, Baumgardner JE, Hedenstierna G. Multiple inert gas elimination technique by micropore membrane inlet mass spectrometry—a comparison with reference gas chromatography. J Appl Physiol (1985). 2013;115:1107–18.
16. Lim LL. A statistical model of the VA/Q distribution. J Appl Physiol (1985). 1990;69:281–92.
17. Mastenbrook SM Jr, Dempsey JA, Massaro TA. Ventilation-perfusion ratio distributions by mass spectrometry with membrane catheters. J Appl Physiol Respir Environ Exerc Physiol. 1982;53:770–8.
18. Munakata M. Measurement of V/Q distribution by inert gas elimination method. Hokkaido Igaku Zasshi. 1982;57:577–88.
19. Petrini MF, Robertson HT, Hlastala MP. Interaction of series and parallel dead space in the lung. Respir Physiol. 1983;54:121–36.
20. Powell FL, Lopez FA, Wagner PD. Effects of acetone in heparin on the multiple inert gas elimination technique. J Appl Physiol (1985). 1985;58:1143–7.
21. Ratner ER, Wagner PD. Resolution of the multiple inert gas method for estimating VA/Q maldistribution. Respir Physiol. 1982;49:293–313.
22. Rees SE, Kjaergaard S, Andreassen S, Hedenstierna G. Reproduction of inert gas and oxygenation data: a comparison of the MIGET and a simple model of pulmonary gas exchange. Intensive Care Med. 2010;36:2117–24.
23. Rees SE, Kjaergaard S, Andreassen S, Hedenstierna G. Reproduction of MIGET retention and excretion data using a simple mathematical model of gas exchange in lung damage caused by oleic acid infusion. J Appl Physiol (1985). 2006;101:826–32.
24. Sa RC, Henderson AC, Simonson TS, Arai TJ, Wagner H, Theilmann RJ, Wagner PD, Prisk GK, Hopkins SR. Measurement of the distribution of ventilation-perfusion ratios in the human lung with proton MRI: comparison with the multiple inert gas elimination technique. J Appl Physiol (1985). 2017;123:136–46.
25. Sato S, Hlastala MP, Hildebrandt J. Intrapulmonary shunt in excised dog lobes: comparison of oxygen and inert gas methods. J Appl Physiol (1985). 1993;74:951–8.
26. Shimazu T, Kishikawa M, Sugimoto T, Yukioka T, Johnson AA Jr, Mason AD Jr, Pruitt BA Jr. Application of a gas chromatography-mass spectrometer (GC-MS) to the multiple inert gas elimination technique: multiple inert gas measurement with a GC-MS at trace level. Kokyu To Junkan. 1989;37:1083–7.
27. Stewart WE, Mastenbrook SM Jr. Graphical analysis of multiple inert gas elimination data. J Appl Physiol Respir Environ Exerc Physiol. 1983;55:32–6.
28. Stewart WE, Mastenbrook SM Jr. Parametric estimation of ventilation-perfusion ratio distributions. J Appl Physiol Respir Environ Exerc Physiol. 1983;55:37–51.
29. Wagner PD. Assessment of gas exchange in lung disease: balancing accuracy against feasibility. Crit Care. 2007;11:182.
30. Wagner PD. Calculating the distribution of ventilation-perfusion ratios from inert gas elimination data. Fed Proc. 1982;41:136–9.
31. Wagner PD. Estimation of distributions of ventilation/perfusion ratios. Ann Biomed Eng. 1981;9:543–56.
32. Wagner PD. Susceptibility of different gases to ventilation-perfusion inequality. J Appl Physiol Respir Environ Exerc Physiol. 1979;46:372–86.
33. Wagner PD, Evans JW. Conditions for equivalence of gas exchange in series and parallel models of the lung. Respir Physiol. 1977;31:117–38.
34. Wagner PD, Lopez FA, Wagner PD, Lopez FA. Gas chromatography techniques in respiratory physiology. In: Otis AB, editor. Techniques in the life sciences. County Clare, Ireland: Elsevier; 1984. p. 403/1–403/24.

35. Wagner PD, Naumann PF, Laravuso RB. Simultaneous measurement of eight foreign gases in blood by gas chromatography. J Appl Physiol. 1974;36:600–5.
36. Wagner PD, Smith CM, Davies NJ, McEvoy RD, Gale GE. Estimation of ventilation-perfusion inequality by inert gas elimination without arterial sampling. J Appl Physiol (1985). 1985;59: 376–83.
37. Wang XH, Poon CS. Spectral vs. compartmental averaging of VA/Q distributions: confidence limits. J Appl Physiol (1985). 1991;70:1290–9.
38. West JB, Wagner PD, Derks CM. Gas exchange in distributions of VA-Q ratios: partial pressure-solubility diagram. J Appl Physiol. 1974;37:533–40.
39. Whiteley JP, Gavaghan DJ, Hahn CE. A tidal breathing model for the multiple inert gas elimination technique. J Appl Physiol (1985). 1999;87:161–9.
40. Yamaguchi K, Mori M, Kawai A, Asano K, Takasugi T, Umeda A, Kawashiro T, Yokoyama T. Effects of pH and SO_2 on solubility coefficients of inert gases in human whole blood. J Appl Physiol (1985). 1993;74:643–9.
41. Yamaguchi K, Mori M, Kawai A, Asano K, Takasugi T, Umeda A, Yokoyama T. Effects of SO_2 and pH on blood-gas partition coefficients of inert gases. Adv Exp Med Biol. 1990;277: 215–24.
42. Young IH, Wagner PD. Solubility of inert gases in homogenates of canine lung tissue. J Appl Physiol Respir Environ Exerc Physiol. 1979;46:1207–10.

Chapter 7
Review of the MIGET Literature

Abstract Since its original reporting in the 1970s, the multiple inert gas elimination technique (MIGET) has been used in hundreds of experiments around the world performed to understand pulmonary gas exchange. These include studies of normal subjects at rest, during exercise, and in different environments, patients with a variety of cardiopulmonary diseases, and patients in the intensive care unit, with many supporting animal disease model studies and measurements in the comparative field, including studies of birds and lizards exercising on treadmills. This chapter briefly summarizes the results of these many studies, grouped into domains of interest. When we started this section of the book, we expected that this discussion would be straightforward. However, we were surprised by the breadth of studies from great many centers. It is inevitable that some work will have been missed, although every attempt has been made to be as inclusive as possible.

7.1 Introduction

MIGET has been used to study pulmonary gas exchange in great many studies and has been the subject of several review articles [1, 130, 132, 151, 217, 230, 300, 308, 309, 384, 385, 387, 388, 391, 392, 394–396] that discuss both technical aspects and also application to study pulmonary gas exchange in health and disease. Data derived from MIGET studies have also formed the basis of modeling studies that inform limitations to oxygen transport [43] and the effects of maldistribution of hematocrit [426] on gas exchange. MIGET has been shown to have a high degree of reliability [404] and has been cross-validated with inert gas washout [275] and microspheres [8, 295]. In the section that follows, we briefly review manuscripts that have used MIGET in the study of pulmonary gas exchange.

S.R. Hopkins, P.D. Wagner, *The Multiple Inert Gas Elimination Technique (MIGET)*, Methods in Physiology, https://doi.org/10.1007/978-1-4939-7441-2_7

7.2 Pulmonary Gas Exchange in Healthy Human Subjects and Animals

7.2.1 *Assessing $\dot{V}_A/\dot{Q}$ Mismatch in Normal Human Subjects and Animals*

Early studies in MIGET were used to characterize the ventilation-perfusion distribution in normal humans and animals. The first of these evaluated 12 normal subjects aged 21–60 years [409]. This work demonstrated that in the youngest subjects (aged 21–24), the recovered distributions were lognormal and that the LogSD of both was less than 0.45 indicating little $\dot{V}_A/\dot{Q}$ mismatch. There was no shunt while breathing air, but these subjects developed shunt with breathing 100% oxygen (more about this below). The older subjects had more $\dot{V}_A/\dot{Q}$ mismatch than the younger subjects, and this is also discussed further below. Very quickly the field branched out to evaluate pulmonary gas exchange in healthy humans under a variety of experimental conditions such as with exercise [101, 115, 365] or at high altitude [116, 401]. Data obtained under baseline conditions, such as preceding exercise or in normobaria, confirmed the original findings of a narrow unimodal $\dot{V}_A/\dot{Q}$ distribution without appreciable $\dot{V}_A/\dot{Q}$ inequality or shunt in healthy subjects.

Work in healthy dogs [408] and pigs [165] showed a similar amount of $\dot{V}_A/\dot{Q}$ mismatch, although in horses some authors reported a greater extent of $\dot{V}_A/\dot{Q}$ mismatch and a small shunt [140], whereas others reported very narrow $\dot{V}_A/\dot{Q}$ distributions and no appreciable shunt [156, 334, 402]. In the sections that follow, the use of MIGET to study normal physiology is highlighted in several areas.

7.2.2 *In Healthy Aging*

It is well known that even in healthy individuals, pulmonary gas exchange efficiency worsens with age. For example, there is a reduction in arterial P_{O_2} [118, 340] and an increase in the alveolar-arterial difference for oxygen (A-aDO_2) [122] with increasing age. MIGET studies have shown that these changes occur as a result of increasing $\dot{V}_A/\dot{Q}$ mismatch [44]. As mentioned previously in Table 2.2, the normal range for LogSD,Q is ~0.30–0.60 and for LogSD,V_A is 0.30–0.65. On average, up until age 50 years, the mean LogSD,Q and LogSD,V_A is <0.45 and does not change with increasing age (Table 7.1). However, above age 50 both LogSD,Q and LogSD,V_A rise and may approach or even exceed the upper limit of normal. For this reason the upper limit (+2SD) at age <30 years for LogSD,Q is ~0.60, whereas at age 60+, +2SD would increase to 0.70. Similar logic applies to LogSD,V_A with an upper limit of 0.65 for ages <30 years and increasing to 0.73 in ages 60+. It is notable that there is no evidence for the development of shunt or diffusion limitation with increasing age in healthy subjects.

Table 7.1 Effect of age on P_{aO_2} and MIGET measures of heterogeneity

Age group (years)	N	Mean age (years)	FVC (% pred)	P_{aO_2}	LogSD,Q	LogSD,V_A
18–29	25	23 ± 3	104 ± 9	101 ± 6	0.38 ± 0.09	0.41 ± 0.12
30–39	14	35 ± 3	107 ± 12	100 ± 8	0.38 ± 0.09	0.45 ± 0.14
40–49	8	45 ± 3	102 ± 14	101 ± 9	0.38 ± 0.08	0.46 ± 0.11
50–59	13	54 ± 3	108 ± 12	103 ± 8	0.42 ± 0.09	0.46 ± 0.19
60+	4	66 ± 5	101 ± 12	85 ± 9	0.55 ± 0.15	0.53 ± 0.14

Data from [44]; FVC = forced vital capacity

7.2.3 *With Exercise at Sea Level*

Some of the earliest work involving MIGET in healthy humans explored the effects of exercise on pulmonary gas exchange. MIGET quantifies the extent of $\dot{V}_A/\dot{Q}$ inequality, intrapulmonary shunt, and deadspace ventilation from the 50 compartment models where a $\dot{V}_A/\dot{Q}$ ratio of less than 0.005 is assigned to the shunt compartment and $\dot{V}_A/\dot{Q}$ ratios above 100 assigned to deadspace. An important advance for using MIGET to the study of pulmonary gas exchange with exercise was the ability to indirectly measure diffusion limitation as the amount of A-aDO_2 not accounted for by $\dot{V}_A/\dot{Q}$ mismatch and shunt. As discussed previously, the recovered $\dot{V}_A/\dot{Q}$ distribution and shunt is used to predict a P_{aO_2}, the P_{aO_2}(p), and an AaDO_2—the AaDO_2(p). The difference between the predicted (p) and observed (o) values is the estimate of diffusion limitation. In addition, where diffusion limitation is present, it is possible to quantify a diffusing capacity for oxygen (D_{LO_2}) as was also outlined previously.

Numerous MIGET studies have shown that $\dot{V}_A/\dot{Q}$ inequality accounts for almost all of the AaDO_2 [44, 101, 116, 162, 291, 365] at rest, aside from a small contribution from venous admixture (i.e., physiological shunting) as a result of the bronchial and Thebesian circulation returning deoxygenated blood to the left ventricle (see discussion below). With increasing exercise intensity, numerous MIGET studies have shown that $\dot{V}_A/\dot{Q}$ mismatch increases as evidenced by an increase in the LogSD,Q and LogSD,V_A, although the extent of $\dot{V}_A/\dot{Q}$ mismatch varies widely between individuals [26, 101, 158, 162]. $\dot{V}_A/\dot{Q}$ inequality continues to account almost exclusively for the AaDO_2 by until $\dot{V}_{O_2}$ is greater than ~3 l/min. Although it is highly variable between individuals, it is after this point that that some individuals show a difference between the measured AaDO_2 and that predicted by MIGET, thus implicating diffusion limitation of oxygen. At $\dot{V}_{O_2max}$, depending on the level of aerobic fitness, $\dot{V}_A/\dot{Q}$ inequality may comprise anywhere from 30% to 100% of the AaDO_2 [162, 291]. It is important to note that none of the gas exchange studies using MIGET have documented any significant intrapulmonary shunt during exercise in healthy humans [116, 158, 162, 183, 258, 265, 401] reviewed in [163, 164].

MIGET shows that when increased $\dot{V}_A/\dot{Q}$ mismatch is observed during exercise, there are alterations in $\dot{V}_A/\dot{Q}$ mismatch in recovery which persists beyond the point

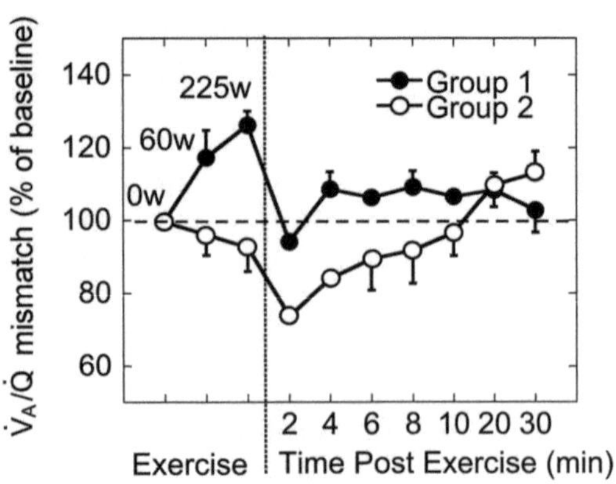

Fig. 7.1 Ventilation-perfusion mismatch during recovery from heavy exercise. Subjects who experience an increase in $\dot{V}_A/\dot{Q}$ mismatch with progressive exercise (Group 1) have a different pattern of gas exchange with increased $\dot{V}_A/\dot{Q}$ mismatch during recovery than those who do not increase $\dot{V}_A/\dot{Q}$ mismatch with exercise (Group 2) (w = watts). Redrawn from [321]

at which ventilation and cardiac output normalize to pre-exercise levels [321] as seen in Fig. 7.1. Exercise in hypoxia (discussed further below) increases the extent of the $\dot{V}_A/\dot{Q}$ mismatch, which can then be alleviated by breathing 100% oxygen [116]. Although the extent of $\dot{V}_A/\dot{Q}$ mismatch increases with increased exercise intensity on incremental exercise protocols, it can be increased by increasing exercise duration even at relatively low exercise intensities [158]. Together these findings have led to the suggestion that the mechanism of the increase in $\dot{V}_A/\dot{Q}$ mismatch is interstitial pulmonary edema acting to compress small airways and blood vessels [101]. This was further evaluated with MR imaging studies showing that the extent of the increase in $\dot{V}_A/\dot{Q}$ mismatch with exercise was associated with the increase in spatial heterogeneity of pulmonary blood flow following prolonged heavy exercise [41]. Notably, however, similar changes in the spatial distribution of ventilation were not seen with a similar exercise protocol [355], suggesting that these changes are more likely on the vascular side rather than the airways.

7.2.4 Sex-Based Differences in Pulmonary Gas Exchange During Exercise

Even when differences in body size are taken into account, females have smaller lung volumes [47, 332, 359, 418] and airway diameters [222, 228] and lower expiratory flow rate [332, 418] and diffusing capacity [48], leading to considerable interest in any sex-based differences in gas exchange (see [121, 159, 335] for review). The early studies of pulmonary gas exchange during exercise using MIGET were conducted largely using male subjects, and recent studies exploring sex-based differences have largely relied on measuring arterial blood gases or arterial saturation using pulse oximetry, and only one has incorporated MIGET. This study [258] matched age, lung size, and mass-specific $\dot{V}_{O_2max}$ in men and women. P_{aO_2}, SaO_2 or A-aDO_2, and AaDO_2(o-p) did not differ between sexes during normoxic and hypoxic ($F_{IO_2} = 0.125$) exercise, but the LogSD,Q was *lower* in the women than the men indicating less exercising $\dot{V}_A/\dot{Q}$ mismatch. Although the calculated diffusing capacity for oxygen D_{LO_2} in hypoxia was less in women, it is thought that it is the ratio of diffusional to perfusional conductances (i.e., the ratio of diffusing capacity (D_L) and pulmonary capillary blood flow ($\dot{Q}$) • the slope of the oxygen hemoglobin dissociation curve (β)) that determines the completeness of diffusion equilibration [322]. Since the $D_L/\beta \cdot \dot{Q}$ was not different between men and women, this potentially explains the lack of differences in $AaDO_2$(o-p). These data suggest that in females there may be factors that are beneficial for gas exchange that may compensate for lung size. However, the understanding of sex-based differences (if any) in pulmonary gas exchange during exercise remains limited.

7.2.5 *Gas Exchange During Exercise in Elite Athletes*

For many years researchers have been fascinated by the apparent paradox that some highly aerobically trained humans and some animals such as horses and racing greyhound dogs experience a reduction in P_{aO_2} and arterial oxygen saturation [67, 120, 291, 407] during exercise. This condition, termed exercise-induced arterial hypoxemia (EIAH), is reviewed in [68]. In this context, an $AaDO_2$ greater than 25 Torr during exercise is considered to be a mild gas exchange impairment, whereas $AaDO_2$ greater than 35–40 Torr is consistent with a severe gas exchange impairment [68]. An important consequence of EIAH for athletic performance is that even small amounts of EIAH have a significant detrimental effect on limiting O_2 transport and utilization during maximal exercise [119, 272]. The cause of EIAH may be multifactorial, including relative hypoventilation, and in the horse this is a major contributor to EIAH [86, 402]. However, in humans P_{aO_2} is only loosely associated with alveolar ventilation as inferred from the P_{aCO_2} [154] and explains ~20% of the variation in P_{aO_2} between subjects, leaving ~80% of the variability in P_{aO_2} attributable to differences in gas exchange efficiency and the $AaDO_2$. MIGET studies have allowed further insights into EIAH, by evaluating the contributors to the large $AaDO_2$ and the significant impairment in gas exchange efficiency seen in EIAH.

Although $\dot{V}_A/\dot{Q}$ inequality increases within an individual with increasing exercise intensity [101, 115, 162], aggregated data from multiple MIGET studies [101, 158, 258, 291] show that there is no significant difference in the extent of $\dot{V}_A/\dot{Q}$ inequality between highly aerobic subjects and those of low to average aerobic ability as measured by the LogSD$\dot{Q}$ [154, 159] (Fig. 7.2). In fact, some elite athletes may have very little increase in $\dot{V}_A/\dot{Q}$ mismatch if any, with exercise as documented by Rice et al., [291]. Thus, it is not surprising that during maximal exercise, the contribution to $\dot{V}_A/\dot{Q}$ mismatch to the $AaDO_2$ is similar between highly fit and less fit individuals. In subjects with minimal gas exchange impairment, irrespective of fitness ($AaDO_2 < 25$ Torr), $\dot{V}_A/\dot{Q}$ inequality and very small

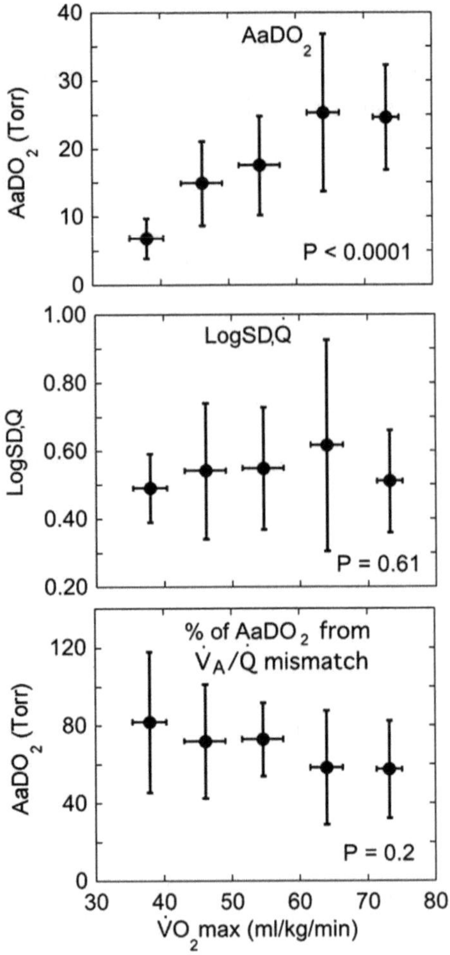

Fig. 7.2 Ventilation-perfusion mismatch during exercise in subjects grouped by $\dot{V}_{O_2max}$. Although the $AaDO_2$ increases with increasing $\dot{V}_{O_2max}$, the extent of $\dot{V}_A/\dot{Q}$ mismatch and the contribution of $\dot{V}_A/\dot{Q}$ mismatch to the $AaDO_2$ are similar between the groups of subjects (Data from [26, 101, 157, 158, 162, 265, 291, 412])

contributions of shunt (discussed below) explain virtually all of the $AaDO_2$. However, in subjects with severe gas exchange impairment ($AaDO_2 > 35$ Torr), $\dot{V}_A/\dot{Q}$ mismatch accounts for less than 40% of the $AaDO_2$ with the remainder being attributed to diffusion limitation of oxygen transport. This is shown in Fig. 7.3. Thus MIGET studies have shown that diffusion limitation is the primary gas exchange defect that distinguishes those who develop EIAH from those who do not. It should be noted that as mentioned previously in the section on exercise and gas exchange, a consistent finding is that intrapulmonary shunt makes a minimal contribution to the $AaDO_2$, and this finding extends to individuals with EIAH [158, 162, 291].

MIGET evidence for diffusion limitation [significant increase in $AaDO_2$(o-p)] of pulmonary oxygen transport is rare in individuals exercising at <3 l/min $\dot{V}_{O_2}$ and is consistent with the finding that EIAH is unusual in subjects who do not exercise at

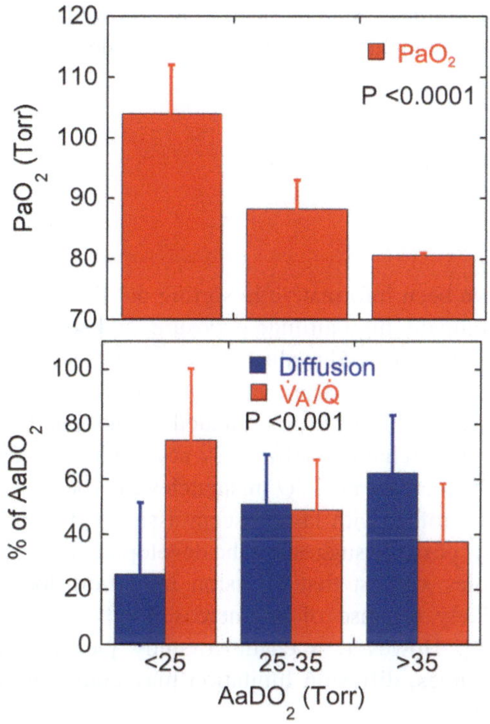

Fig. 7.3 P_{aO_2} and the contribution of $\dot{V}_A/\dot{Q}$ mismatch and diffusion limitation to the $AaDO_2$ during heavy to maximal exercise in subjects with different degrees of gas exchange impairment. In subjects with little gas exchange impairment ($AaDO_2 < 25$ Torr), P_{aO_2} is maintained during maximal exercise, and the $AaDO_2$ is largely due to $\dot{V}_A/\dot{Q}$ mismatch. This trend is reversed in subjects with severe gas exchange impairment; subjects with moderate gas exchange impairment have equal contributions from both $\dot{V}_A/\dot{Q}$ mismatch and diffusion limitation. The differences between subject groups are statistically significant P_{aO_2}, $p < 0.0001$, and $\dot{V}_A/\dot{Q}$ mismatch and diffusion limitation both $p < 0.001$ [26, 101, 157, 158, 162, 265, 291, 412]

this intensity. In subjects capable of greater aerobic work, EIAH become more common. The mechanism of the diffusion impairment documented by MIGET is thought to be rapid capillary transit times, and in highly trained athletes, the $AaDO_2$(o-p) is correlated with whole-lung red blood cell transit times measured using ^{99m}technecium-labeled red blood cells used as a marker for capillary transit times [157]. Subjects who develop diffusion impairment often do so at submaximal levels of exercise [291], but these are almost always highly fit individuals where an oxygen consumption of 3 l/min is as little as 50% of $\dot{V}_{O_2max}$. This might argue against rapid red cell transit as the mechanism of diffusion limitation because capillary recruitment is presumably not maximal and transit time is not minimal but if diffusional, and perfusional conductances (i.e., D_L and $\beta \bullet \dot{Q}$) are recruited with different time constants, and this may lead to transient diffusion limitation during submaximal exercise. Notably, Rice et al., showed that subjects with EIAH have a lower $D_L/\beta \cdot \dot{Q}$ compared to those who do not develop EIAH [291].

7.2.6 High Altitude/in Hypoxia

7.2.6.1 Rest

MIGET studies have been informative in sorting out the changes in gas exchange efficiency that accompany high-altitude exposure, and some of the first studies in humans using MIGET [101, 365] evaluated the effect of simulated high altitude on ventilation-perfusion matching. The overall extent of resting $\dot{V}_A/\dot{Q}$ mismatch is increased with increasing altitude (or simulated altitude) and is the major contributor to the resting gas exchange inefficiency at altitude [26, 265, 401] (Fig. 7.4). The magnitude of the increase in $\dot{V}_A/\dot{Q}$ mismatch varies between subjects, elevation reached, and ascent profile, with faster ascent associated with greater ventilation-perfusion inequality, possibly suggesting the development of subclinical pulmonary edema. These studies suggest that diffusion limitation does not contribute the resting $AaDO_2$, likely because of an increased diffusing capacity in hypoxia [26, 115–117, 412]. However, a chamber study (see below) showed that at extremely high altitudes, diffusion limitation may contribute a small amount in some individuals [412].

7.2.6.2 Exercise in Hypoxia/High Altitude

As previously mentioned, MIGET studies have shown that $\dot{V}_A/\dot{Q}$ mismatch increases with increasing exercise intensity in normoxia [162] and is also further increased by exercise in hypoxia [101, 116, 401]. Thus, ventilation-perfusion inequality as a contributor to inefficient gas exchange is likely present in most individuals exercising at altitude. However, since the net effect of ventilation-perfusion inequality on the $AaDO_2$ depends on the overall P_{AO_2} of lung units on the oxygen hemoglobin dissociation curve, although the extent of ventilation-perfusion inequality increases markedly, the effect on the $AaDO_2$ decreases to an even greater extent. Thus, MIGET has shown that, in contrast to the resting data, $\dot{V}_A/\dot{Q}$ mismatch plays a minor role in gas exchange inefficiency during hypoxic exercise. Conversely, as seen in Fig. 7.5, MIGET studies suggest that diffusion limitation becomes an increasingly larger contributor to the $AaDO_2$ during exercise at altitude, despite an overall increase in D_{LO_2} [116, 183, 265]. The exception to this is at extreme simulated altitude [412], where Wagner and coworkers showed larger amount of $\dot{V}_A/\dot{Q}$ inequality at rest and during exercise consistent with the development of pulmonary edema as discussed below.

Naeije et al., [242] studied the effect of chemoreceptor stimulation on gas exchange in acute hypoxia using almitrine, a peripheral chemoreceptor agonist.

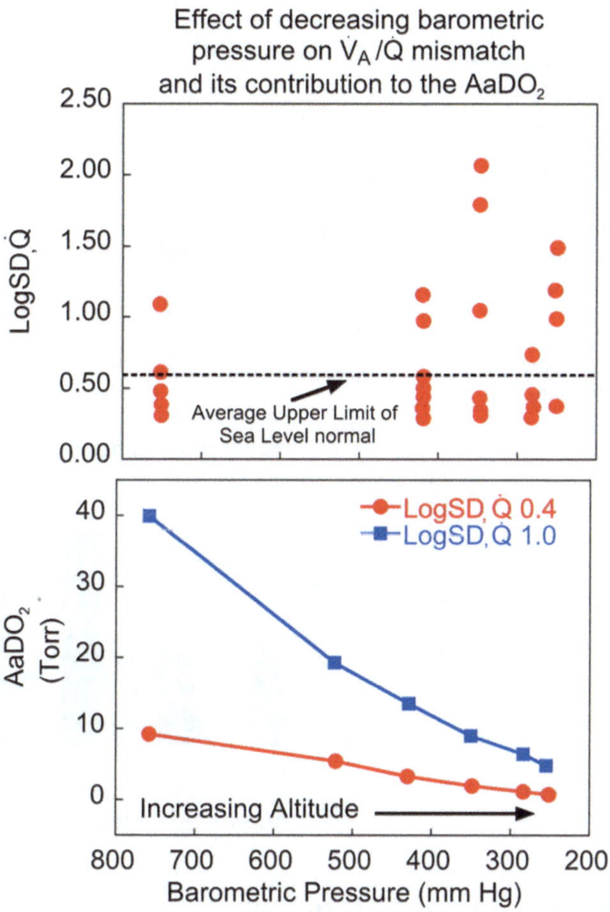

Fig. 7.4 The effect simulated altitude on ventilation-perfusion mismatch and its contribution to the $AaDO_2$. The change in LogSD,*Q* with decreasing barometric pressure is variable between individuals (*top*) with the *dotted line* indicating the upper bounds of normal sea-level values. For a fixed amount of $\dot{V}_A/\dot{Q}$ mismatch, the $AaDO_2$ decreases with decreasing barometric pressure (*bottom*). The net result is a reduction in the $AaDO_2$ attributable to ventilation-perfusion inequality. Figures modified from [386]

Almitrine was suggested to increase arterial O_2 saturation in the absence of changing minute ventilation during hypoxic exercise; thus, gas exchange efficiency was evaluated. In normoxic exercise, almitrine increased P_{aO_2} and decreased P_{aCO_2} consistent with hyperventilation, but $\dot{V}_A/\dot{Q}$ mismatch was unchanged. Similar findings were observed during hypoxic exercise and there was no evidence for altered lung diffusion. It was concluded that the beneficial effects of this drug on gas exchange were via increased ventilation and a left shift of the oxyhemoglobin dissociation curve, without any changes in pulmonary gas exchange per se.

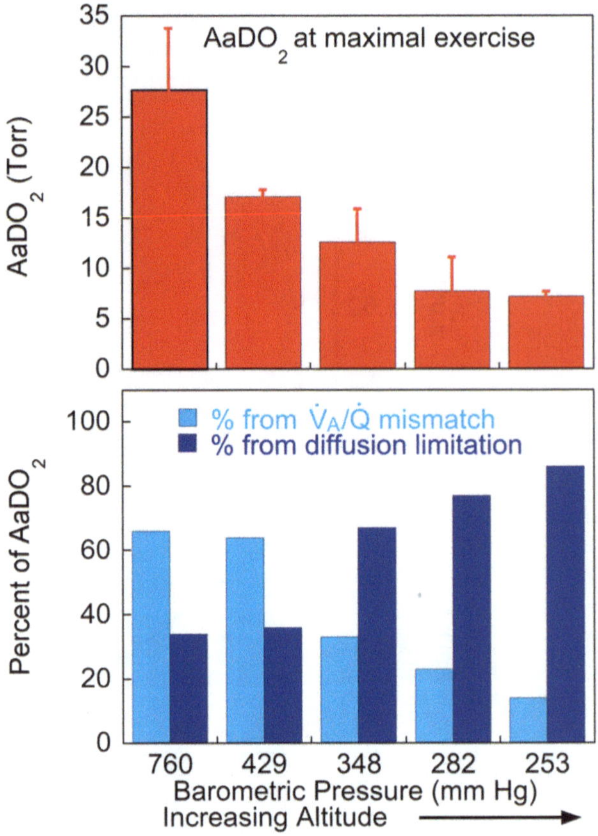

Fig. 7.5 The effect of maximal exercise at simulated altitude on the $AaDO_2$ and the relative contributions to the $AaDO_2$ from diffusion limitation and ventilation-perfusion mismatch. With increasing altitude the contribution of $\dot{V}_A/\dot{Q}$ mismatch to the $AaDO_2$ decreases and the $AaDO_2$ decreases. However, diffusion limitation becomes more pronounced. Data from [412]

7.2.6.3 With Chronic Exposure to Hypoxia

Few MIGET studies have evaluated the effects of chronic exposure to high altitude/hypoxia on pulmonary gas exchange. In neonatal piglets, Gibson et al., [103] showed that although 2 weeks of exposure to normobaric hypoxia ($F_{IO_2} = 0.10$) caused an increase in pulmonary vascular resistance compared to control animals, there were no differences in pulmonary gas exchange as assessed by the LogSD,Q and shunt.

In humans, there are data at extreme "altitude" from one study conducted as part of a series of perhaps the most ambitious physiology studies in human ever conducted, Operation Everest II [412]. In this study, subjects were taken over 40 days through a simulated ascent of Mount Everest inside a hypobaric chamber, by mimicking the barometric pressure of a typical ascent profile before ending at the "summit." After a rapid transition from a higher barometric pressure (lower "altitude") to a Pbar of 340 mm Hg, resting $\dot{V}_A/\dot{Q}$ inequality as measured by the LogSD,Q was increased suggesting the development of pulmonary edema. This elevated resting $\dot{V}_A/\dot{Q}$ inequality was also seen at the "summit" (Pbar = 240 mm Hg).

The slope of the relationship between LogSD,Q and $\dot{V}O_2$ during exercise was increased with increasing altitude. Diffusion limitation became progressively more important with decreasing barometric pressure and was present even during relatively low oxygen consumptions. At the extreme conditions of the "summit," diffusion limitation was apparent in several of the subjects at rest.

Studies performed during acclimatization at a constant altitude show that the contribution of diffusion limitation to the $AaDO_2$ during exercise decreases with acclimatization [26] compared to acute hypoxia, presumably on the basis of reduced $\dot{Q}$ and a consequent increase in $D_L/\beta{\bullet}\dot{Q}$. However, the extent of ventilation-perfusion inequality and shunt (which was minimally present if at all at sea level) was unchanged.

7.2.7 High-Altitude Pulmonary Edema

High-altitude pulmonary edema (HAPE) is a potentially fatal, rare disease, characterized by a non-cardiogenic high-permeability edema, that develops in otherwise healthy individuals following 24–72-h exposure to altitudes above 2400 m (~8000 ft). A hallmark of HAPE is that after treatment with supplemental oxygen, descent from altitude and pulmonary vasodilators, recovery is extremely rapid, and the previously affected individual quickly returns to normal pulmonary function. It should be noted that the single biggest predictor of developing HAPE is a previous history of HAPE [178]. HAPE-susceptible subjects have greater pulmonary arterial pressures in hypoxia compared to HAPE-resistant subjects, greater resting pulmonary vascular resistance, and higher pulmonary arterial and capillary wedge pressures during exercise. Since interstitial edema is hypothesized to cause the increase in $\dot{V}_A/\dot{Q}$ inequality with exercise in normal subjects, MIGET has been used to evaluate whether HAPE-susceptible individuals (i.e., those with previous history of HAPE) also experienced increased $\dot{V}_A/\dot{Q}$ inequality, to a greater extent than control subject particularly during hypoxic exercise. Podolsky et al., [265] studied seven subjects with a prior history of HAPE with MIGET, at rest; 35%, 65%, and 85% of $\dot{V}_{O_2max}$; at sea level, and at 3810 m, in normoxia ($P_{IO_2} = 148$ Torr), and hypoxia ($P_{IO_2} = 91$ Torr), at both locations. They were compared to nine control subjects who had an extensive history of altitude travel without developing altitude-related illness. Both groups of subjects developed increased $\dot{V}_A/\dot{Q}$ mismatch during normoxic and hypoxic exercise at sea level as measured by the LogSD,Q. When lung size was considered as a covariate, the increase in LogSD,Q (and other MIGET indices of dispersion) was significantly greater in the HAPE-susceptible subjects than the control subjects. After 2 days acclimatization at altitude although $\dot{V}_A/\dot{Q}$ mismatch increased with exercise both in normoxia and hypoxia, there was no longer a difference between subject groups. However, there was a close relationship between pulmonary vascular pressures and the extent of $\dot{V}_A/\dot{Q}$ mismatch as measured by the LogSD,Q [265] (Fig. 7.6). The authors suggested that HAPE

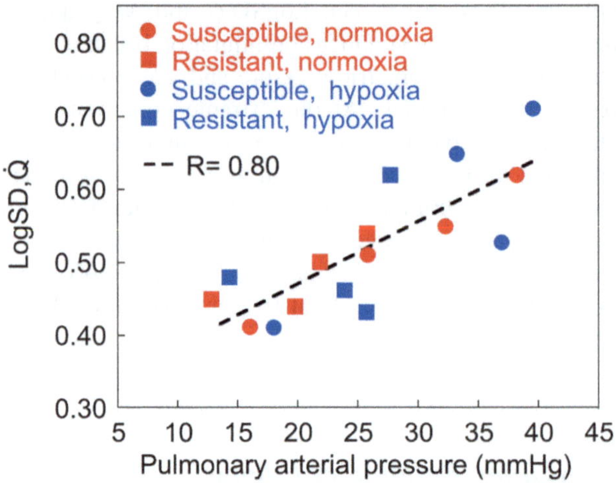

Fig. 7.6 The relationship between pulmonary arterial pressure and LogSD,Q at sea level in normoxia and hypoxia in patients who have previously suffered from high-altitude pulmonary edema (susceptible) and high-altitude control subjects who have never been ill with HAPE (resistant). LogSD,Q is correlated with pulmonary arterial pressure. Data from [265]

may represent the extreme endpoint, exaggerated by hypoxia, of the same process that is responsible for increased $\dot{V}_A/\dot{Q}$ inequality during sea-level exercise—an increase in fluid efflux and the development of interstitial edema.

7.2.8 With Hypoxic Pulmonary Vasoconstriction

Several studies both in humans and in animals have evaluated the effects of hypoxic pulmonary vasoconstriction on pulmonary gas exchange. In humans Melot et al., [233] measured the effects of acute hypoxia ($F_{IO_2} = 0.125$) on pulmonary arterial pressures and gas exchange in healthy subjects. Hypoxia increased pulmonary arterial pressure from 13 mm Hg to 19 mm Hg, and this response was partially blocked by nifedipine (16 mm Hg). $\dot{V}_A/\dot{Q}$ mismatch as measured by the LogSD,V_A increased from 0.34 ± 0.02 at baseline to 0.45 ± 0.05 and further increased to 0.51 ± 0.03 with nifedipine. There were no significant changes for LogSD,Q. However, when blood flow and ventilation were constrained to the baseline normoxic values in the distributions recovered during hypoxia ("normalization procedure"), a slight improvement in $\dot{V}_A/\dot{Q}$ matching was apparent, which was blunted during hypoxia after nifedipine. It should be noted, however, that most studies using a similar hypoxic stimulus in normal subjects either show no change in $\dot{V}_A/\dot{Q}$ matching [265, 291] or a slight improvement [26], although this was not tested statistically in this last study. Other work under anesthesia in humans with healthy lungs [114] showed that when one lung was hypoxic, administration of inhaled nitric oxide to the opposite lung enhanced the shift of blood flow away from the hypoxic lung. Similarly, in dogs, Hopkins et al., [161] showed that the effect of whole-lung inhaled nitric oxide on gas exchange depended on the underlying gas exchange defect. In the presence of shunt created by an airway obstruction, inhaled nitric oxide improved pulmonary gas exchange, presumably by diverting blood away from the unventilated lung regions. However, in the presence of poorly ventilated lung regions, inhaled nitric oxide worsened gas exchange, likely because it lifted any compensatory hypoxic pulmonary vasoconstriction.

In an interesting animal study, Domino et al., [77] ventilated the right lung of dogs with 100% oxygen, while the left lower lobe of the lung was ventilated with either 100% oxygen or 5% oxygen. By selective cannulations, MIGET was performed on the whole lung, the right lung, and the left lower lobe. Considering the entire lung, left lower lobe hypoxia had no effect on $\dot{V}_A/\dot{Q}$ heterogeneity although the mean $\dot{V}_A/\dot{Q}$ ratio was increased. However left lower lobe hypoxia increased LogSD,Q of the left lower lobe, while it decreased that of the right lung, suggesting that the regional effects of hypoxic pulmonary vasoconstriction were compensated for by improved $\dot{V}_A/\dot{Q}$ matching elsewhere in the lung. This work was later extended [72] to evaluate the effect of oleic acid-induced pulmonary edema on the ability of hypoxic pulmonary vasoconstriction to alter blood flow: the extent of $\dot{V}_A/\dot{Q}$ mismatch in the left lower lobe was increased more by hypoxia than in normal lungs, and the percentage of blood flow to shunt and low $\dot{V}_A/\dot{Q}$ regions was increased as a percentage of cardiac output but not in absolute terms because left lower lobe blood flow was reduced. However, the hypoxic pulmonary vasoconstrictive response was similar between 1 h and 24 h after lung injury indicating no further deterioration in the response.

Walther et al., [417] used a similar dog model to evaluate the interaction between posture (prone or supine) and the response to regional alveolar hypoxia. Local hypoxia increased blood flow to regions of high $\dot{V}_A/\dot{Q}$ ratio (10–100), and this effect was larger in supine posture. These differences are possibly explained by the larger cardiac output in prone posture or the interaction of vascular asymmetry between right and left and posture.

Escourrou et al., [89] evaluated the effect of a calcium channel blocker nitrendipine on pulmonary gas exchange during hypoxia in piglets. Pulmonary vascular resistance increased in hypoxia but this response was eliminated by nitrendipine. Nitrendipine improved P_{aO_2} and reduced P_{aCO_2} in hypoxia. There was no shunt in any condition, but perfusion to low $\dot{V}_A/\dot{Q}$ regions increased in hypoxia and was unchanged by nitrendipine for reasons that are unclear. Conversely, Hirano et al., [150] found that the administration of verapamil, another calcium channel blocker, worsened pulmonary $\dot{V}_A/\dot{Q}$ relationships by increasing regions of low $\dot{V}_A/\dot{Q}$ ratio, but in animals that increased cardiac index with the drug, P_{aO_2} was not decreased.

Finally Romaldi et al., evaluated the effects of almitrine, a drug thought to enhance hypoxic pulmonary vasoconstriction, on pulmonary gas exchange in dogs. Almitrine increased pulmonary artery pressure and pulmonary vascular resistance when it was administered in hypoxic (12% O_2) but not in hyperoxic

(100% O_2) conditions. However there was no effect on pulmonary gas exchange as measured by MIGET. Further studies using almitrine [65, 66, 234, 242, 290] are discussed in the sections that follow.

7.2.9 With Hypocapnia and Hyperventilation

As discussed above, the pulmonary circulation responds to alveolar hypoxia by vasoconstriction in a manner that is consistent with optimizing pulmonary gas exchange and $\dot{V}_A/\dot{Q}$ relationships. However, it is much less well appreciated that CO_2 may also affect $\dot{V}_A/\dot{Q}$ relationships either by acting on the pulmonary circulation to vasoconstrict or vasodilate vessels in regions of high or low local CO_2, respectively, or on the airways to bronchodilate or bronchoconstrict in regions of hypercapnia and/or hypocapnia, respectively. There have been several MIGET studies that have evaluated the effects of hyperventilation and the addition of inspired CO_2 producing hypercapnia on $\dot{V}_A/\dot{Q}$ matching.

In contrast to the fairly clear-cut effects of increased inspired CO_2 acting to improve $\dot{V}_A/\dot{Q}$ matching, hypocapnia induced by hyperventilation affects $\dot{V}_A/\dot{Q}$ matching in a manner that is affected by tidal volume, acid/base status, duration of hyperventilation, and the presence or absence of PEEP. Both pigs [75] and dogs [80] demonstrate a worsening of $\dot{V}_A/\dot{Q}$ relationships with hypocapnia under some situations. For example, in dogs some studies have shown hypocapnia results in decreased $\dot{V}_A/\dot{Q}$ matching, as measured by the LogSD,V_A and dispersion indices, when it is induced by increased respiratory rate but not when induced by increased tidal volume [80]. The effect of hypocapnia is thought to be mediated by changes in pH rather than effect of local CO_2 per se [79]. Hyperventilation-induced hypocapnia also results in a deterioration in $\dot{V}_A/\dot{Q}$ relationships and an increase in the LogSD,Q in dogs with oleic acid-induced lung injury [78]. However in the presence of metabolic acidosis, this deterioration is not seen [76]. Another work has reported that in dogs a high $\dot{V}_A/\dot{Q}$ mode is seen in animals in the presence of high tidal volume-induced hypercapnia, when PEEP is also applied [374, 375], but that this resolves over time and $\dot{V}_A/\dot{Q}$ matching is improved [375].

7.2.10 With Hypercapnia and Increased F_{ICO_2}

Studies using HCl infusion to induce metabolic acidosis in dogs [95] found no effect of acidosis on $\dot{V}_A/\dot{Q}$ mismatch but an improvement in the $AaDO_2$, thought to occur because of the Bohr effect. However, studies using normally ventilated anesthetized dogs have demonstrated that breathing small concentrations of inspired CO_2 (2–4%) improves pulmonary gas exchange by a reduction in $\dot{V}_A/\dot{Q}$ mismatch as measured by a decrease in the LogSD,V_A without a significant change in the LogSD,Q [350]. The effect of CO_2 on $\dot{V}_A/\dot{Q}$ matching was similar in hyperoxia ($F_{IO_2} = 0.5$) with only LogSD,V_A showing a significant decrease, but in hypoxia ($F_{IO_2} = 0.15$) both LogSD,Q and LogSD,V_A declined. The addition of inspired CO_2 also reduced deadspace measured by MIGET in normoxia and hyperoxia, but this effect was not seen in hypoxia [350].

The effect of CO_2 on $\dot{V}_A/\dot{Q}$ matching does not appear to be related to nitric oxide since it was not affected by inhibition of nitric oxide synthase [39]. Also, MIGET studies, where CO_2 was added to the inspired gas late in the respiratory cycle and changes in blood pH were prevented, suggest that the effect of increased CO_2 on $\dot{V}_A/\dot{Q}$ mismatch is not a primary result of an induced systemic acidosis [40], but rather largely due to local effects on the lung with acidosis playing a secondary role. This is presumed to be mainly via effects on the airways; however, CO_2 also has effects on the distribution of pulmonary blood flow [350]. Studies that used carbonic anhydrase inhibition with acetazolamide (Diamox®) suggest that the effects of CO_2 are mediated by carbonic anhydrase [349], which augments the rate of any compensatory ventilation as well as perfusion responses to induced or spontaneous changes in local blood flow and/or ventilation affecting $\dot{V}_A/\dot{Q}$ matching [348].

However, all of the MIGET data on the effects of CO_2 on $\dot{V}_A/\dot{Q}$ matching are not concordant. For example, in dogs who are ventilated with high tidal volumes and PEEP, a reported improvement in $\dot{V}_A/\dot{Q}$ matching with the addition of CO_2 to inspired gas [374] was, in a subsequent study, attributed to a time-dependent change [375] leading to the loss of a high $\dot{V}_A/\dot{Q}$ mode and improved $\dot{V}_A/\dot{Q}$ matching independent of CO_2. In addition, in animals with oleic acid-induced lung injury, the addition of inspired CO_2 worsened $\dot{V}_A/\dot{Q}$ relationships [78], although the relationship to normal physiology in this context is unclear, as intense inflammation may impair the normal response. The use of MIGET to understand pulmonary gas exchange in the presence of lung injury is discussed further below.

7.2.11 With Diamox

Acetazolamide (Diamox®) is a carbonic anhydrase inhibitor, commonly taken to prevent and relieve the symptoms of acute mountain sickness. Ingestion of acetazolamide increases ventilation and P_{aO_2} and improves arterial oxygen saturation. In dogs, administration of acetazolamide increased the response time for changes in regional ventilation when rapid lobar changes in regional perfusion were induced by vessel occlusion. This led to a transient increase in $\dot{V}_A/\dot{Q}$ mismatch measured with MIGET at 2 min, but not 10 min after the changes in perfusion [348]. MIGET in humans studies have shown that acetazolamide decreases the overall extent of $\dot{V}_A/\dot{Q}$ mismatch both at rest and during exercise in normoxia and hypoxia compared to placebo [183]. Acetazolamide has also been shown to reduce diffusion limitation assessed by MIGET during acute hypoxic exercise [183]. The mechanism is suggested to occur through a reduction in β and improving $D/\beta\dot{Q}$ although it is unknown if the effect is sustained with chronic ingestion, such as would occur with travelers to high altitude. Thus, although acetazolamide may affect regional ventilatory response time kinetics, MIGET studies suggest that the overall effect on gas exchange is beneficial.

7.2.12 Altered Pulmonary Blood Flow, Blood Volume, or Hematocrit

The influence of altered pulmonary blood flow on gas exchange has been studied with MIGET. Domino et al., [74] used the split lung model described earlier to evaluate the effects in increasing local blood flow on $\dot{V}_A/\dot{Q}$ mismatch. Blood flow to the left lower lobe was altered by means of an arteriovenous fistula and partially diverting blood from the right pulmonary artery, while minute ventilation was kept constant. $\dot{V}_A/\dot{Q}$ mismatch as measured by the LogSD,V_A was markedly increased in the left lower lobe from 1.30 ± 0.20 to 1.76 ± 0.14 with a threefold increase in local blood flow, and the extent of the response was highly correlated with the extent of the blood flow increase. Another study [256] from the Seattle group using this model showed that when blood flow was reduced to the lower left lobe, $\dot{V}_A/\dot{Q}$ mismatch in this region was also increased as reflected in an increase in the LogSD,Q.

Lynch et al., [213] used MIGET to study the effect of changes in cardiac output on intrapulmonary shunt in dogs. Shunt was induced by creating an oleic acid, pulmonary edema, which resulted in a bimodal $\dot{V}_A/\dot{Q}$ distribution with one compartment of shunt and the other being normal. Perhaps unsurprisingly, the extent of the shunt increased with increasing blood flow although the general shape of the $\dot{V}_A/\dot{Q}$ distribution was unchanged, leading the authors to suggest interpretation of changes in shunt must be interpreted considering any overall changes in cardiac output.

Fortune et al., [94] created hemorrhage in dogs and studied pulmonary gas exchange at baseline and then after the animals were bled and then the heparinized blood reinfused. Hypotension created regions of high $\dot{V}_A/\dot{Q}$ ratio, but these were not alleviated by reinfusion, possibly because of leukocyte margination and aggregation. Balgos et al., [15] transfused packed red blood cells to increase hematocrit from 43% to 57% and 68% and then later performed plasma exchange transfusion to restore hematocrit back to 43% to evaluate the effect of hematocrit on $\dot{V}_A/\dot{Q}$ mismatch. MIGET showed that increasing hematocrit did not increase in $\dot{V}_A/\dot{Q}$ mismatch and LogSD,Q and LogSD,V_A were unchanged. The mean of the distributions was shifted to a higher $\dot{V}_A/\dot{Q}$ ratio because of an overall decrease in cardiac output, which was suggested to be a result of increased blood viscosity from polycythemia. Since the overall decrease in cardiac output with the highest level of polycythemia was rather large, and comparable to changes evaluating the effects of blood flow alone discussed above, it may have been that polycythemia defended $\dot{V}_A/\dot{Q}$ matching or alternately that the regional changes discussed above are rather small and possibly undetectable when the whole lung is considered.

Deem et al., [57] studied the effects of serial hemodilution in Flemish Giant rabbits. Isovolemic anemia was associated with significantly increased P_{aO_2} and reduced $\dot{V}_A/\dot{Q}$ mismatch as measured by the dispersion index DispR-E; although the LogSD,Q and LogSD,V_A showed similar directional changes, they were not statistically significant. Expired nitric oxide was increased by anemia, and it was suggested that this might be the basis of improved $\dot{V}_A/\dot{Q}$ mismatch with anemia. Interestingly this finding was also observed in animals with acute lung injury induced by venous gas embolization [58].

7.2.13 *Splenectomy in Animals*

To determine the importance of the spleen in O_2 transport during exercise in the dog, which rapidly autotransfuses itself with stored red cells via splenic contraction, Hsia et al., [170] used MIGET before and after recovery from splenectomy. Splenic contraction increases both hematocrit and circulating blood volume. The latter raises cardiac filling pressures and augments cardiac output. Postsplenectomy, exercise capacity was impaired, cardiac output was lower, $\dot{V}_A/\dot{Q}$ relationships were unaffected, but diffusing capacity was lower. This is consistent with what is seen in the thoroughbred racehorse, which also relies on its spleen for rapid autotransfusion during exercise. In the thoroughbred racehorse, the spleen may be even more important for exercise capacity; splenectomy greatly impaired treadmill running speed, and this was mediated by reduced cardiac output from lower circulating blood volume [383] and not from failure to increase [Hb].

7.2.14 *With Blood Substitutes*

A single study has evaluated the effect of perfluorocarbon emulsions used as blood substitutes on pulmonary gas exchange and oxygen transport. Johnson et al., [181] evaluated different doses of perfluorooctylbromide emulsion in normovolemic, anemic dogs (hematocrit ~ 25%). Administration of perfluorooctylbromide at 12 ml/kg increased oxygen solubility by a factor of 2 but had no effect on ventilation-perfusion relationships.

7.2.15 *With Ultrasound Contrast Agents*

Kirkton et al., [189] evaluated the effect of Imagent®, a contrast agent used in cardiac ultrasound imaging, on hemodynamics and pulmonary gas exchange in dogs with pulmonary hypertension. Doses up to 16 mg/kg had no effects on any measured variable including pulmonary gas exchange although the 16 mg/kg dose transiently increased pulmonary arterial pressure and pulmonary vascular resistance.

7.2.16 With Head-Out Water Immersion

A single study evaluated the effects of head-out water immersion on ventilation-perfusion relationships and pulmonary gas exchange [69]. Two groups of subjects were studied: younger subject without evidence of airway closure at their expiratory reserved volume on land, and older subjects (average age 45) with evidence for airway closure. Younger subjects did not show development of gas exchange abnormalities with immersion, but older subjects developed small amounts of shunt consistent with airway closure during tidal breathing. However, other gas exchange abnormalities such as development of low $\dot{V}_A/\dot{Q}$ regions were not seen.

7.2.17 With High-Frequency Voluntary Ventilation

Another small study also in the category of exploratory science was the study of high-frequency ventilation in trained yogis [98]. These unique subjects were able to sustain a respiratory rate of over 225 breath/min for a duration of 30–60 min allowing MIGET to be performed in the usual manner. Mean ventilation was over 90 l/min, but tidal volume was reduced compared to normal breathing such that P_{aCO_2} and pH were not markedly altered. All three of the subjects had normal $\dot{V}_A/\dot{Q}$ distributions at baseline but developed a region of high $\dot{V}_A/\dot{Q}$ ratio with high-frequency breathing for reasons that are unclear.

7.3 In Lung Disease

7.3.1 COPD

MIGET was developed in the 1970s, before CT methods to quantify emphysema were available and when COPD was commonly described as showing two main phenotypes—the pink puffer and the blue bloater. The former was thought to be associated mostly with emphysema, while the latter was considered related to chronic bronchitis, with most patients exhibiting a combination of both pathologies and some mix of phenotypic appearance. The first MIGET study to explore this was in 1977 [397]. The findings were striking, with three distinct patterns of $\dot{V}_A/\dot{Q}$ distribution discovered, shown in Fig. 7.7. Thus, COPD was not a disease with a

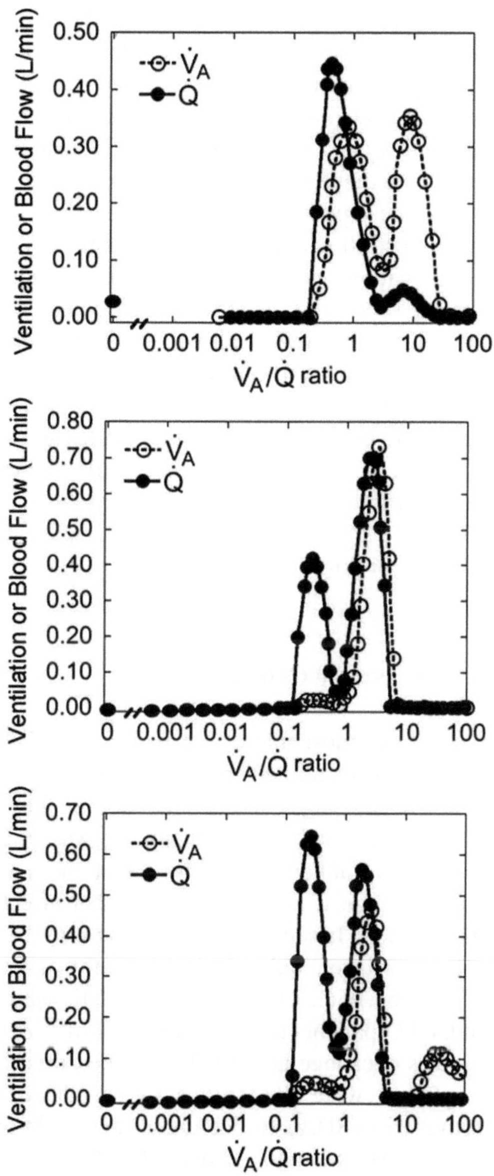

Fig. 7.7 The patterns of gas exchange in patients with COPD. Patients with relatively normal P_{aO_2} (pink puffers) had regions of high $\dot{V}_A/\dot{Q}$ ratio without marked regions of low $\dot{V}_A/\dot{Q}$ ratio (*top*). In some patients with the "blue bloater" phenotype, there are regions of low $\dot{V}_A/\dot{Q}$ ratio (*middle*). Some patients had a combination of regions of both high and low $\dot{V}_A/\dot{Q}$ ratio. In all cases shunt was minimal. Redrawn from [397]

single, typical $\dot{V}_A/\dot{Q}$ pattern. Each pattern contained considerable $\dot{V}_A/\dot{Q}$ inequality—the findings were not subtle. Essentially all pink puffers showed a pattern with a distinct population of high $\dot{V}_A/\dot{Q}$ areas (in addition to a population of normal $\dot{V}_A/\dot{Q}$ areas). However, there were no areas of low $\dot{V}_A/\dot{Q}$ or shunt. Physiologically, this would fit with the regional alveolar wall destruction known to occur in emphysema, because this would reduce capillary number and thus blood flow while alveolar spaces were still ventilated, resulting in a higher than normal $\dot{V}_A/\dot{Q}$ ratio. It was predicted that blue bloaters with their cough, sputum production, and underlying chronic bronchitis would show the mirror image pattern—a population of low $\dot{V}_A/\dot{Q}$ (and a normal $\dot{V}_A/\dot{Q}$ group as well), explained by mucus retention in small airways. However, only 4 of 12 such patients in fact showed this pattern, and many had the high $\dot{V}_A/\dot{Q}$ distribution pattern or a combination of both patterns. Of course these patients likely had some degree of emphysema as well. Only 1 of the more than 20 patients had any shunting of blood through unventilated pathways, and no cause was established for this individual. Atelectasis and consolidation were absent, so maybe an atrial septal defect was to blame. Importantly, the $\dot{V}_A/\dot{Q}$ inequality observed in all patients explained the amount of hypoxemia measured, no matter what distribution pattern was seen. This confirmed that $\dot{V}_A/\dot{Q}$ inequality was the major cause of gas exchange abnormality in COPD and that other causes such as hypoventilation, shunt, or diffusion limitation were not as important.

Since 1977 at least 15 further studies using MIGET in COPD were carried out in many different research centers [2, 17–20, 43, 70, 109, 219, 234, 237, 249, 263, 274, 290, 298, 302, 305, 320, 366, 394, 397] investigating various aspects of pathophysiology.

7.3.1.1 Stable COPD

One of the most interesting papers on COPD is that of Rodriguez-Roisin et al., [305], who examined the relationship between $\dot{V}_A/\dot{Q}$ inequality and spirometry (i.e., GOLD stage) across 150 patients they had accumulated in MIGET studies. They found that $\dot{V}_A/\dot{Q}$ inequality was greater earlier in the disease than were spirometric abnormalities (i.e., that there was obvious $\dot{V}_A/\dot{Q}$ heterogeneity in GOLD Stage I) and in a manner that affected blood flow more than ventilation distribution. It suggests that COPD may be manifest first by distal changes that affect perfusion and not by airway obstruction.

Barbera et al., [17] probed the relationship between pathology and gas exchange in COPD patients by evaluating patients undergoing lung resection. The extent of the gas exchange abnormality as measured by the $AaDO_2$, the LogSD,Q, and the LogSD,V_A were all significantly correlated with the lung morphology, and interestingly LogSD,V_A (reflecting primarily abnormalities in the perfusion distribution) was also significantly correlated with the extent of inflammatory infiltrates of small airways. They later evaluated the relationship between pulmonary vascular structure and $\dot{V}_A/\dot{Q}$ mismatch in patients with mild COPD undergoing lung resection surgery [19]. Patients were grouped according to spirometry and the change in LogSD,Q with oxygen indicating hypoxic pulmonary vasoconstriction and pulmonary vascular responsiveness. The two groups of patients with either airflow obstruction and a high response to oxygen and/or patients with airflow obstruction and a low response to oxygen had narrower lumens and thicker walls of small pulmonary arteries than did patients with normal spirometry. These findings were most marked in the patients with the low response to oxygen. The extent of intimal narrowing was negatively related to P_{aO_2} and positively related to the extent of $\dot{V}_A/\dot{Q}$ mismatch suggesting that the changes in the pulmonary circulation might underlie some of the gas exchange abnormalities in COPD. The pulmonary circulation was further evaluated by Castaing et al., [45] who showed mild worsening in $\dot{V}_A/\dot{Q}$ distributions with hyperoxia suggesting at least some extent of hypoxic pulmonary vasoconstriction.

Since 1977 there have been several studies, mostly directed at effects of therapeutic interventions. Prediletto et al., [274] confirmed the major findings of the 1977 study by Wagner et al., and in addition showed modest improvement with long-term clinical care. Melot et al., [234] found $\dot{V}_A/\dot{Q}$ inequality in COPD improved by the drug almitrine due to enhanced vasoconstriction, which redirected blood flow from low $\dot{V}_A/\dot{Q}$ regions to better ventilated areas. Bratel et al., [35] went the other way, finding that a calcium channel antagonist (felodipine), which lowered pulmonary vascular resistance, worsened $\dot{V}_A/\dot{Q}$ inequality. These two studies demonstrate a significant role for hypoxic pulmonary vasoconstriction in modulating blood flow distribution and hence gas exchange in COPD. However, Moinard et al., [237] failed to see any change in inequality after inhaled NO in COPD patients despite a lowering of pulmonary vascular pressures. Roca et al., [298] failed to find any changes in pulmonary gas exchange in patients with stable COPD given naloxone, which was suggested to lift hypoxic pulmonary vasoconstriction. However pulmonary vascular pressures were also unchanged. Sandek et al., [320] found that 2 months of inhaled steroids failed to improve $\dot{V}_A/\dot{Q}$ inequality. Pillet et al., [263] compared β2 agonist drugs to anticholinergics in a short-term study and found essentially no immediate gas exchange consequences. Sandek et al., [319] studied COPD patients with sleep-disordered breathing and found greater nocturnal sleep disturbances occurring in those with more daytime $\dot{V}_A/\dot{Q}$ inequality. Manier et al., [219] looked at the separate influences of extrapulmonary factors (cardiac output and mixed venous P_{O_2}) on arterial P_{O_2} in COPD patients. They found arterial P_{O_2} higher when mixed venous P_{O_2} was higher—this would be expected from classical predictions of how mixed venous P_{O_2} affects arterial P_{O_2}—but interestingly, when cardiac output was higher, there was greater $\dot{V}_A/\dot{Q}$ inequality, such that arterial P_{O_2} was actually lower than with lower cardiac output.

7.3.1.2 Acute Exacerbations

Yamasawa et al., [423] found, as one would expect, that patients with COPD suffering an acute exacerbation displayed worsened $\dot{V}_A/\dot{Q}$ relationships. Diaz et al., [70] examined the effects of noninvasive ventilation on gas exchange in patients in the hospital for an acute exacerbation and found that despite some improvement in arterial P_{O_2}, there was no change in $\dot{V}_A/\dot{Q}$ relationships. Rather, higher alveolar ventilation explained the gas exchange improvement. Torres et al., [366] used MIGET to study COPD patients recovering from an exacerbation severe enough to require several days of mechanical ventilation. Their focus was on the consequences of weaning, and they found substantial worsening of $\dot{V}_A/\dot{Q}$ relationships when mechanical ventilation was discontinued and patients breathed spontaneously and in spite of maintained minute ventilation and an increase in cardiac output. Barbera et al., [18] studied COPD patients given aminophylline (a bronchodilator) after an exacerbation and found that while the drug improved spirometry as expected, $\dot{V}_A/\dot{Q}$ inequality either failed to improve or in some actually worsened. Robinson et al., [297] sought to explain the cause(s) of CO_2 retention associated with breathing 100% O_2 in COPD patients recovering from an acute exacerbation. The principal cause was simply reduced ventilation, even though $\dot{V}_A/\dot{Q}$ inequality worsened with hyperoxia due to loss of hypoxic pulmonary vasoconstriction.

7.3.1.3 COPD and Lung Volume Reduction Surgery

In 2011, Cremona et al., [50] studied a small number of patients by MIGET before and after lung volume reduction surgery. Surgery resulted in only a small gas exchange improvement (higher arterial P_{O_2} due to less $\dot{V}_A/\dot{Q}$ inequality) despite substantial improvement in lung mechanics. Changes in arterial P_{CO_2} were closely related to mitigation of dynamic hyperinflation and not to changes in $\dot{V}_A/\dot{Q}$ matching. The effects were however quite variable across subjects.

7.3.2 Asthma

Asthma is a disease characterized by clinical variability within subjects over time, by large between-subject variability in natural progression and therapeutic response, and by major advances in drugs used to treat the disease. The use of MIGET to study asthma was reviewed by Rodriguez-Roisin [301] in 1990 and Rodriguez-Roisin and Roca in 1994 [308]. In 1978, Wagner et al., used MIGET in a small group of stable, asymptomatic asthmatics with mild airway obstruction [399] and were surprised to find extensive $\dot{V}_A/\dot{Q}$ inequality and a repeatable $\dot{V}_A/\dot{Q}$ distribution pattern characterized by a distinct mode of low $\dot{V}_A/\dot{Q}$ units ($\dot{V}_A/\dot{Q}$), quite separate in $\dot{V}_A/\dot{Q}$ ratio from units in the normal $\dot{V}_A/\dot{Q}$ range (Fig. 7.8). There was however little or no shunting, and arterial P_{O_2} was surprisingly well preserved (generally >80 mm Hg) in spite of the $\dot{V}_A/\dot{Q}$ inequality. This raised many still unanswered questions:

(A) Why a bimodal distribution rather than one with a single mode that is broad? It suggests that airways are either affected substantially or they are not affected, at least in terms of airflow, rather than reflecting a continuous spectrum of obstruction. Of note, imaging studies [229] have confirmed the bimodal nature of $\dot{V}_A/\dot{Q}$ inequality in human asthma by an obviously independent method, and this is consistent with a pattern of bistable airways with alveoli rapidly

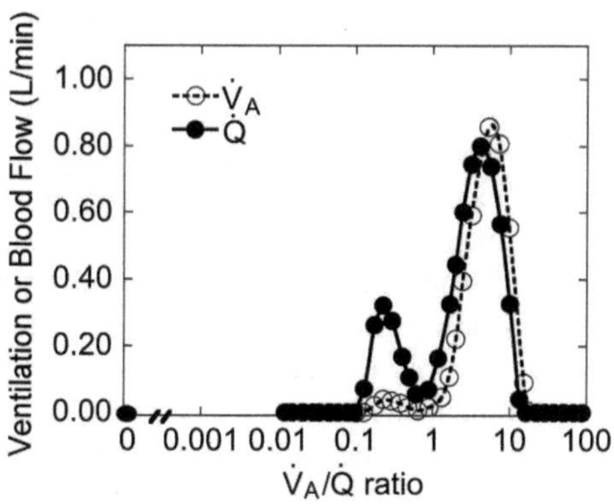

Fig. 7.8 The distribution of $\dot{V}_A/\dot{Q}$ ratios in a patient with asthma. There is a bimodal distribution with regions of low $\dot{V}_A/\dot{Q}$ ratio and regions of normal $\dot{V}_A/\dot{Q}$ ratio. Redrawn from [399]

transiting from open to nearly closed. This is also suggested by MIGET studies in bronchoconstricted pigs that showed that the extent of $\dot{V}_A/\dot{Q}$ mismatch was reduced by higher tidal volumes [191].

(B) How can significant numbers of very low $\dot{V}_A/\dot{Q}$ units persist without becoming totally obstructed, causing atelectasis and shunt? This suggests that collateral ventilation between gas exchange units (acini) may be an important phenomenon in human small airways (the same may be said of COPD when low $\dot{V}_A/\dot{Q}$ areas are found). This was supported by MIGET studies in dogs (known to have substantial collateral ventilation) that had small airways occluded by beads of various sizes [200]. As the size of the bead increased, presumably occluding more proximal airways, where collateral ventilation was less efficient, a bimodal $\dot{V}_A/\dot{Q}$ distribution developed, whereas it was broadened at the smallest bead size. In another dog study [317], broad unimodal distributions were found only in mild bronchoconstriction, and more severe bronchoconstriction resulted in the typical bimodal distribution, emphasizing the importance of collateral ventilation in this species.

(C) Why is arterial P_{O_2} well preserved in the face of substantial $\dot{V}_A/\dot{Q}$ inequality? This was explained by a relatively high cardiac output that maintained a high mixed venous P_{O_2}. Within a study, conducted in dogs [304], the acute effects of inhaled isoproterenol (beta 1 and 2 sympathetic agonist) were studied, and 5 min after inhalation, $\dot{V}_A/\dot{Q}$ inequality was substantially worse despite normalization of spirometry. This immediately suggested, first, that retained secretions in small airways were important effectors of $\dot{V}_A/\dot{Q}$ inequality (rather than just bronchoconstriction) and, second, that the bronchodilator also reversed pulmonary hypoxic vasoconstriction causing vasodilatation and thus increasing perfusion of low $\dot{V}_A/\dot{Q}$ units. Some 10 min later, the $\dot{V}_A/\dot{Q}$ pattern had returned to its baseline appearance with low $\dot{V}_A/\dot{Q}$ areas. Clearly, bronchodilators can normalize airflow but not necessarily correct the gas exchange defects.

Stimulated by these findings, Bylin et al., [42] and Wagner et al., [403] studied asthmatic subjects weekly for 9 weeks testing the effects of the mucolytic/antioxidant N-acetylcysteine (NAC) in a double-blind crossover study (3 weeks run-in, 3 weeks NAC, 3 weeks placebo). The outcome was negative—no effects of NAC on gas exchange, spirometry, or symptoms—but the existence of often substantial $\dot{V}_A/\dot{Q}$ inequality was confirmed in every subject at some time points, with clear week-to-week change within subjects unexplained by external events. Roca et al., [299] studied asthmatics admitted with an acute attack and found that $\dot{V}_A/\dot{Q}$ inequality was unrelated to spirometric abnormality. Furthermore, improvement in spirometry with treatment was much quicker than reduction in $\dot{V}_A/\dot{Q}$ inequality, consistent with the above idea that bronchoconstriction per se is not a major cause of $\dot{V}_A/\dot{Q}$ inequality and that other factors such as small airway mucous plugging may be important. The poor relationship between spirometry and gas exchange was later confirmed by Lagerstrand et al., [193], and Rodriguez-Roisin et al., [303] also studied patients with acute severe asthma in hospital and confirmed the unique bimodal nature of $\dot{V}_A/\dot{Q}$ inequality and virtual absence of shunt. As expected, the degree of inequality was much more severe than observed a decade earlier in asymptomatic asthmatics. Giving 100% O_2 worsened the $\dot{V}_A/\dot{Q}$ distribution and led to small shunts appearing, both the consequence of relaxation of hypoxic vasoconstriction in low $\dot{V}_A/\dot{Q}$ ratio units.

7.3.2.1 Role of Inflammatory Mediators

Rodriguez-Roisin et al., [306] studied the effect of platelet-activating factor, a potent inflammatory mediator, which is known to cause bronchoconstriction and potentiate the effects of methacholine on gas exchange in normal humans. Platelet-activating factor administered by nebulizer caused a fall in P_{aO_2} and an increase in the $AaDO_2$ because of increased $\dot{V}_A/\dot{Q}$ mismatch. They suggested that platelet-activating factor may play a major role as a mediator of inflammation in the lung that might be important for the development of ARDS or asthma.

7.3.2.2 Bronchial Challenge and Gas Exchange

An obvious question was whether bronchial challenge substantially reducing FEV_1 led to acute development of $\dot{V}_A/\dot{Q}$ inequality. Lagerstrand et al., [197] used allergen challenge and found moderate $\dot{V}_A/\dot{Q}$ inequality with no clear relationship to spirometry, increasing the evidence suggesting that gas exchange and airflow impairments in asthma are the result of different pathophysiological changes. Lagerstrand et al., [198] in 1995 found in mild, well-controlled asthmatics with normal spirometry, that modest $\dot{V}_A/\dot{Q}$ inequality was present chronically. Similar outcomes were reported by Munoz et al., [241], both after exercise-induced asthma and after inhaled mannitol. Rodriguez-Roisin et al., [307] also showed worsening gas exchange with methacholine-induced bronchoconstriction in patients with mild stable asthma. Using a combination of imaging and MIGET, Schmekel et al., [327] showed that the particle size of the methacholine aerosol affected the changes in airway conductance, but the gas exchange patterns were similar and lasted longer than the changes in conductance.

7.3.2.3 Gas Exchange Responses to Bronchodilators

Ballester et al., [16] studied inpatients who were admitted to the hospital with an acute severe exacerbation of asthma before and during beta-agonist administration (either intravenous salbutamol or inhaled salbutamol) or 100% oxygen. All patients had marked $\dot{V}_A/\dot{Q}$ inequality that correlated poorly with spirometry. After intravenous salbutamol perfusion to low $\dot{V}_A/\dot{Q}$ areas and LogSD,Q increased significantly, but these changes were much less than with inhaled salbutamol. Breathing 100% oxygen also resulted in a large increase in the LogSD,Q from 1.11 to 1.44. Montserrat et al., [240] studied the effects of (intravenous) aminophylline given to hospitalized patient with acute severe asthma. They found improved spirometry, ventilation, and thus a drop in arterial P_{CO_2} but no improvement in arterial P_{O_2} or $\dot{V}_A/\dot{Q}$ distribution. The dissociation between spirometry and gas exchange in asthma appears to be a strong and consistent observation made by several laboratories and over a range of asthma severity from mild to severe [405].

7.3.2.4 Asthma Animal Model Studies

Watremez et al., [420] used methacholine to create a pig model of asthma and recapitulated the $\dot{V}_A/\dot{Q}$ inequality seen in human asthma. Thus, this model is available for preclinical studies. They used this animal model [419] to show that in severe bronchospasm, at a level in humans which would require intubation and mechanical ventilation, the administration of helium-oxygen mixtures reduced work of breathing and lung resistance and elastance but increased the LogSD,Q and also worsened the metabolic acidosis and hypercapnia for reasons that are unclear. In dogs Ueno et al., [377] evaluated the effect of the antioxidant N-acetylcysteine on gas exchange in dogs that were bronchoconstricted with methacholine. N-Acetylcysteine improved gas exchange and resulted in less $\dot{V}_A/\dot{Q}$ mismatch and a higher P_{O_2}.

7.3.3 Cystic Fibrosis

This form of chronic obstructive pulmonary disease has not been extensively studied using MIGET, but in 1982, Dantzker et al., [55] reported on six adults with cystic fibrosis. In contrast to tobacco-induced COPD, where shunting is rare, substantial shunting (mean 13%) was seen in all patients, and additional, modest regions of low $\dot{V}_A/\dot{Q}$ ratio were observed in half of them. As the authors suggested, the shunt was likely the result of retained secretions obstructing airways. There was no clear evidence of alveolar-capillary O_2 diffusion limitation, but their data suggest that the two most hypoxemic of their patients (arterial $P_{O_2} < 60$ mm Hg) may have shown a small (<5 mm Hg) contribution to hypoxemia from this mechanism. Two subjects also showed high $\dot{V}_A/\dot{Q}$ regions, commonly seen in emphysema. Exercise improved $\dot{V}_A/\dot{Q}$ relationships, and this allowed arterial P_{O_2} to remain at resting values. Lagerstrand et al., [196] found that in well-treated patients, $\dot{V}_A/\dot{Q}$ relationships, while abnormal, were not severely deranged. Two weeks of antibiotic therapy improved $\dot{V}_A/\dot{Q}$ relationships, as did 2 weeks of treatment with inhaled amiloride. Additionally, as what appears to be the rule for most lung diseases, gas exchange abnormalities were not predictable from other measures such as spirometry or CT.

7.3.4 Pulmonary Fibrosis

Surprisingly little research has been done using MIGET in this class of lung diseases. A review of MIGET contributions in this area has been reviewed by Agusti and Barbera [1]. The work performed by Wagner et al., [398] and Agusti et al., [4] showed moderate $\dot{V}_A/\dot{Q}$ mismatch with very low $\dot{V}_A/\dot{Q}$ ratio areas and/or shunt present, together averaging about 15% of the cardiac output. All of the hypoxemia at rest was explained by these changes, and diffusion limitation of O_2 uptake was not present. On exercise, $\dot{V}_A/\dot{Q}$ relationships did not change, yet arterial hypoxemia worsened substantially. The explanation was a combination of the appearance of diffusion limitation and a fall in mixed venous P_{O_2}, which aggravated the hypoxemia already there from $\dot{V}_A/\dot{Q}$ inequality, shunt, and diffusion limitation. This work was confirmed by later studies from the Hedenstierna group [180].

In another study of patients with more severe IPF (D_{LCO} 52% predicted), Agusti et al., [4] found that about 20% of the resting $AaDO_2$ was attributable to diffusion limitation of O_2 uptake—the first pulmonary disease to show this phenomenon after COPD, asthma, pneumonia, and embolism all failed to display any diffusion limitation. Just as with the earlier study, exercise worsened the hypoxemia, and this was due to worsening diffusion limitation combined with a large fall in mixed venous P_{O_2}, which aggravated the hypoxemia. $\dot{V}_A/\dot{Q}$ relationships were not affected by exercise. These three studies thus showed essentially identical gas exchange disturbances, the only difference being resting diffusion limitation in the Agusti study.

In a case report of a patient with severe hypoxemia, Prediletto et al., [273] reported severe diffusion limitation for oxygen and $\dot{V}_A/\dot{Q}$ mismatch. CT scanning confirmed fibrosing alveolitis. In another case report, a patient with pulmonary alveolar microlithiasis [96], a condition characterized by intra-alveolar deposits with a fibrotic response, was studied. MIGET showed elevated $\dot{V}_A/\dot{Q}$ mismatch (LogSD,$Q = 0.94$, LogSD,$V_A = 0.99$) and 16% shunt, but diffusion limitation was not evaluated. Shunt but not $\dot{V}_A/\dot{Q}$ mismatch was improved by CPAP, suggesting that in this unusual case of pulmonary fibrosis, atelectasis and shunt played a role in hypoxemia.

The conclusions we draw from these studies are as follows: (1) $\dot{V}_A/\dot{Q}$ inequality is clearly present, manifest by a combination of very low $\dot{V}_A/\dot{Q}$ regions plus shunt. Presumably, low $\dot{V}_A/\dot{Q}$ areas reflect fibrotic alveoli with low compliance which are therefore poorly ventilated. Shunt may be the manifestation of complete diffusion limitation with capillaries buried so deeply in collagen that no gas exchange takes place. It is also possible that due to loss of alveolar volume in fibrotic alveoli, there may have been atelectasis contributing to the shunt. (2) Diffusion limitation appears at rest when D_{LCO} is about 50% of predicted or lower and during exercise in all patients studied. (3) Exercise does not change $\dot{V}_A/\dot{Q}$ relationships, and the almost universal large fall in arterial P_{O_2} during exercise is explained by diffusion limitation and a low mixed venous P_{O_2}.

Why diffusion limitation does not occur in COPD, even during exercise, when it is present in fibrosis is an interesting question. In both diseases, D_{LCO} is often greatly reduced. It is suggested that the reason is that in COPD, D_{LCO} is low due to destruction of alveoli and thus loss of capillaries, while walls of the remaining alveoli are essentially normal. However in fibrosis, while there may be similar capillary destruction in fibrotic alveolar walls, those walls are greatly thickened by the deposited collagen, adding to the difficulty in achieving diffusion equilibration for O_2.

Ogasawara [252] evaluated the pattern of gas exchange in an animal model of interstitial lung disease using dogs that had been exposed to paraquat and developed pneumonitis. The histology was not well matched with the patterns of gas exchange, which consisted of variable shunt (0–30%) and moderate $\dot{V}_A/\dot{Q}$ mismatch with low $\dot{V}_A/\dot{Q}$ regions.

7.3.5 During Exercise in Patients with Lung and Chronic Disease

Several studies have explored $\dot{V}_A/\dot{Q}$ relationships during exercise in patients with lung disease or other chronic disease that results in pulmonary gas exchange abnormalities such as liver cirrhosis. An overview of some of these studies is given in reviews by Wagner in 1977 [393] and 1992 [394].

In a detailed evaluation of a small group of patients with stable COPD, Agusti and colleagues [2] showed that, somewhat unexpectedly, $\dot{V}_A/\dot{Q}$ heterogeneity improved with exercise and previously bimodal distributions became unimodal.

Although there was no diffusion limitation at rest, there was evidence for diffusion limitation during exercise accounting for approximately 20% of the $AaDO_2$. The net result was that the $AaDO_2$ was unchanged in these patients between rest and exercise. Interestingly, adding nifedipine to release hypoxic pulmonary vasoconstriction worsened $\dot{V}_A/\dot{Q}$ relationships, and the LogSD,Q was increased compared to exercise without the drug, suggesting that hypoxic pulmonary vasoconstriction was having some effect to optimize pulmonary gas exchange. Barbera et al., [20] evaluated the relationship between pathologic features in the lungs of patients with COPD undergoing lung resection at rest and during submaximal exercise. Also in this subject population, P_{aO_2} increased during exercise and $\dot{V}_A/\dot{Q}$ mismatch lessened. While at rest, the emphysema score and extent of bronchiolar abnormalities correlated significantly with P_{aO_2} and $\dot{V}_A/\dot{Q}$ mismatch, this relationship was not seen during exercise. However this is not a universal finding as Dantzker and D'Alonzo [54] reported that in their study of patients with severe COPD, $\dot{V}_A/\dot{Q}$ heterogeneity was unchanged with exercise, and P_{O_2} fell most likely because of an inadequate ventilatory response that led to increased CO_2.

In addition to the above study, the group in Barcelona also studied patients with pulmonary fibrosis [4]. Patients with pulmonary fibrosis had evidence for diffusion limitation at rest comprising ~20% of the $AaDO_2$ and moderate $\dot{V}_A/\dot{Q}$ mismatch with ~3% of cardiac output perfusing low $\dot{V}_A/\dot{Q}$ units. In a similar manner to the COPD patients, $\dot{V}_A/\dot{Q}$ mismatch was unchanged by exercise; however, diffusion limitation increased and contributed ~40% to the $AaDO_2$, which was markedly elevated at 49 Torr. There were modest amounts of shunt 2–3%, which was not significantly changed with exercise.

This group also evaluated pulmonary gas exchange during exercise in patients with liver cirrhosis [5]. In this group of patients, there was moderate $\dot{V}_A/\dot{Q}$ mismatch at rest (LogSD,Q = 0.79), which was not increased with exercise. Shunt was low at 1% and not significantly changed with exercise, and there was no evidence of diffusion impairment. The pattern of gas exchange in these may well be different in patients with more severe disease and larger intrapulmonary shunts and/or diffusion limitation at rest [49, 141].

Finally, McEvoy and colleagues in 1986 [227] investigated pulmonary gas exchange at rest and during exercise in patients with mitral stenosis, where it might be expected that the pulmonary venous congestion would result in alveolar flooding and thus increased regions of low $\dot{V}_A/\dot{Q}$ ratio and shunt. Unexpectedly, while there was an increase in $\dot{V}_A/\dot{Q}$ inequality, this was due to the development of regions of high (not low) $\dot{V}_A/\dot{Q}$ ratio, and there was no increase in shunt. This led the authors to postulate that interstitial edema might cause redistribution of pulmonary blood flow, although this possibility seems unlikely given more recent data in normal subjects that suggests that interstitial edema has the potential to cause either regions of low $\dot{V}_A/\dot{Q}$ ratio [276] or a generalized broadening of the $\dot{V}_A/\dot{Q}$ distribution [158].

7.3.6 Pulmonary Embolism and Pulmonary Hypertension

One might think that in pulmonary thromboembolic disease with otherwise normal lungs, ventilation-perfusion relationships would just reflect areas of reduced blood flow in embolized regions, giving rise to high $\dot{V}_A/\dot{Q}$ areas, while the rest of the lung displayed normal $\dot{V}_A/\dot{Q}$ ratios. In pulmonary arterial hypertension (PAH), as long as there was no concomitant airway or alveolar disease, there would be little $\dot{V}_A/\dot{Q}$ inequality from pulmonary arterial wall thickening unless there was also reduction in vessel lumen and thus blood flow. Even then, lumen diameters would need to be distributed in a nonuniform manner. As for pulmonary embolism, reduced lumen size would reduce blood flow and create a $\dot{V}_A/\dot{Q}$ distribution with high $\dot{V}_A/\dot{Q}$ areas in the affected regions and normal $\dot{V}_A/\dot{Q}$ ratios elsewhere. However, several additional factors could be involved.

First, the non-affected regions, because of their lower vascular resistance, would receive more than their usual amount of blood flow (diverted from affected, high-resistance vessels), which would result in a modest reduction in their $\dot{V}_A/\dot{Q}$ ratios (if cardiac output were maintained). Thus, while there might be discernible $\dot{V}_A/\dot{Q}$ inequality and hypoxemia from these lower than normal $\dot{V}_A/\dot{Q}$ ratios in unaffected lung, the hypoxemia could be easily prevented by modest increases in total alveolar ventilation to return $\dot{V}_A/\dot{Q}$ ratios to normal in the unaffected regions.

Additionally, with pulmonary hypertension of any cause, right atrial pressures may be elevated, and a patent foramen ovale might result in a right-to-left intracardiac shunt. Furthermore, pulmonary embolism is known to interfere with surfactant metabolism, and it is possible that some (micro)atelectasis may develop as a result. Also, the reduced airway P_{CO_2} in high $\dot{V}_A/\dot{Q}$ ratio regions may cause bronchoconstriction in those areas, reducing their ventilation and redistributing it to non-embolized regions, raising their $\dot{V}_A/\dot{Q}$ ratio back toward normal. Then, much as is believed to be the case with high-altitude pulmonary edema, pulmonary hypertension might cause local edema, especially in dependent regions, impairing gas exchange by creating areas of low $\dot{V}_A/\dot{Q}$. Finally, cardiac output is frequently reduced from right ventricular stress or failure, and this will cause mixed venous $P_{\bar{v}O_2}$ to fall, which will aggravate any hypoxemia caused by the $\dot{V}_A/\dot{Q}$ inequality itself. Thus, there may be multiple, interacting factors affecting $\dot{V}_A/\dot{Q}$ distribution and arterial oxygenation in a complex manner. The possibility that vascular obstruction in affected areas raises perfusion of non-affected vessels also means that red cell transit times in high-perfusion areas are shorter and that diffusion limitation of O_2 uptake might occur.

Not much work has been done using MIGET in PAH; more has been done after pulmonary thromboembolism, and in several reports, both kinds of patients have been combined. Some of the MIGET findings in pulmonary thromboembolism are reviewed here [104, 411]. The first MIGET study in human pulmonary hypertension was published in 1979 by Dantzker and Bower [52], with seven patients with chronic disease of both types. Arterial P_{O_2} varied greatly among patients (from 46 to 90 mm Hg), but all showed significant $\dot{V}_A/\dot{Q}$ inequality and a very high alveolar-arterial P_{O_2} difference. Arterial P_{O_2} predicted from the measured $\dot{V}_A/\dot{Q}$ inequality agreed closed with measured arterial P_{O_2}, disproving the hypothesis that diffusion limitation of O_2 exchange was a factor in arterial oxygenation. However, the $\dot{V}_A/\dot{Q}$ patterns were surprising in that no areas of high $\dot{V}_A/\dot{Q}$ were observed in any patient, and deadspace was essentially normal. In contrast, every patient showed small shunts and/or areas of extremely low $\dot{V}_A/\dot{Q}$ (0.01 or lower) (Fig. 7.9a). The authors had no clear explanation for this pattern, but did not raise the possibility of atrial septal defect since prior catheterization failed to show such defects. Rather, they considered that dependent lung edema from high vascular pressure may be responsible, although chest X-ray did not reveal evident edema. Importantly, arterial hypoxemia was found to be aggravated by reduced mixed venous P_{O_2}. The absence of high $\dot{V}_A/\dot{Q}$ areas could have been the result of technical error if significant loss of exhaled soluble gas occurred before sampling—something that will occur if stringent precautions to avoid this are not taken. Then in 1983, the same group studied two patients with acute massive pulmonary embolism [51]. These patients did show high $\dot{V}_A/\dot{Q}$ regions (Fig. 7.9b) and not areas of low $\dot{V}_A/\dot{Q}$—however, they both had substantial shunting (20% and 29% of cardiac output) explaining their hypoxemia, but the basis of the shunts (foramen ovale, edema, atelectasis, or other causes) was not established. In 1983, Melot et al., [236] studied two very hypoxemic patients with severe PAH. In both, the $\dot{V}_A/\dot{Q}$ distribution was right shifted (due to the combination of hyperventilation and low cardiac output). Both also showed the presence of significant regions of reduced $\dot{V}_A/\dot{Q}$ ratio, and one had a large (20%) shunt. In both, diffusion limitation was excluded because arterial P_{O_2} was very closely predicted from the $\dot{V}_A/\dot{Q}$ inequality

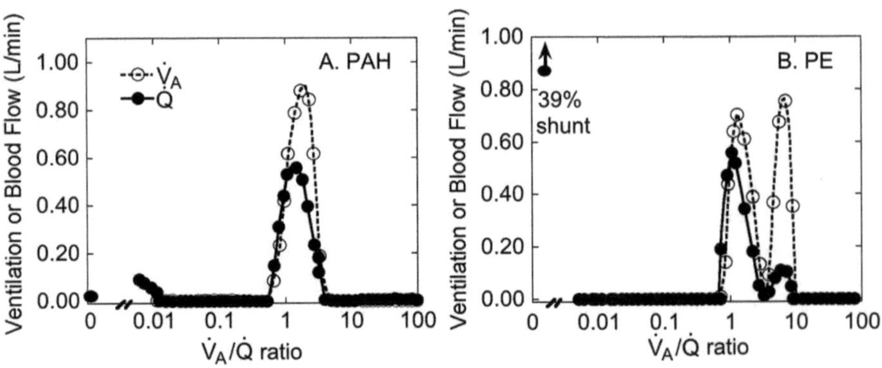

Fig. 7.9 The distribution of $\dot{V}_A/\dot{Q}$ ratios in a patient with pulmonary arterial hypertension (PAH) in (**a**) and with pulmonary embolism (PE) in (**b**). Surprisingly there are no regions of high $\dot{V}_A/\dot{Q}$ ratio, and deadspace was normal in patients with PAH. However, there were regions of low $\dot{V}_A/\dot{Q}$ ratio. These findings were confirmed in subsequent studies, although the pattern is highly variable across studies. Shunt is not uncommon and may be substantial; also, regions of both very low and very high $\dot{V}_A/\dot{Q}$ ratio are not uncommon. In the patient with PE, regions of high $\dot{V}_A/\dot{Q}$ ratio are apparent and there is a large shunt. The reasons for the development of shunt in PE are unclear. Redrawn from [51, 52]

plus shunt. Nifedipine worsened $\dot{V}_A/\dot{Q}$ relationships but increased cardiac output such that arterial P_{O_2} actually improved.

In 2007, Bratel et al., [37] studied a similar cohort, again seven patients. While they too saw small (4%) shunts, there were no low $\dot{V}_A/\dot{Q}$ areas observed. However, all of their patients had extensive regions of high $\dot{V}_A/\dot{Q}$ ratio in line with physiological expectation. The authors were very familiar with the need to keep exhaled gas above body temperature and dry to avoid soluble gas loss and a technical error explanation for these findings is unlikely. IV prostacyclin acutely worsened $\dot{V}_A/\dot{Q}$ relationships and increased shunt and cardiac output, and as a result, arterial oxygenation was unaffected. Two years later, Kapitan et al., [186] reported on 25 patients with chronic large vessel thromboembolism and found hypoxemia, hypocapnia, a large alveolar-arterial P_{O_2} difference, and substantial $\dot{V}_A/\dot{Q}$ inequality. No shunt or low $\dot{V}_A/\dot{Q}$ regions were found, nor distinct areas of high $\dot{V}_A/\dot{Q}$, but the $\dot{V}_A/\dot{Q}$ distribution was much broader than normal. There was increased deadspace (i.e., ventilation of unperfused lung). Predicted and measured arterial oxygenation agreed closely, further suggesting that diffusion limitation was not a factor in the hypoxemia, but as with prior studies, cardiac output and mixed venous P_{O_2} were found to be reduced, aggravating the hypoxemia caused by the $\dot{V}_A/\dot{Q}$ inequality. The same group studied several patients about a year after thromboendarterectomy and found near normalization of oxygenation, cardiac output, and $\dot{V}_A/\dot{Q}$ relationships [187]. In 1992, Manier and Castaing [218] studied 16 patients with acute pulmonary embolism and found that the $\dot{V}_A/\dot{Q}$ distribution pattern varied with cardiac output such that low $\dot{V}_A/\dot{Q}$ areas were often seen when cardiac output was preserved, but high $\dot{V}_A/\dot{Q}$ areas were evident when cardiac output was reduced. But once again, $\dot{V}_A/\dot{Q}$ inequality was universal, as was hypoxemia, and overall, mixed venous P_{O_2} was low, contributing significantly to the arterial hypoxemia. When they administered dobutamine, cardiac output increased, arterial P_{O_2} rose, but $\dot{V}_A/\dot{Q}$ inequality was acutely worsened by increased perfusion of low $\dot{V}_A/\dot{Q}$ areas. Finally, Voswinckel et al., [382] used an inhaled vasodilatory prostaglandin and also inhaled NO in 39 patients with PAH and studied 5 of them by MIGET. They observed shunting (about 6%), presence of low $\dot{V}_A/\dot{Q}$ areas, and presence of high $\dot{V}_A/\dot{Q}$ areas, with little effect of either vasodilator on gas exchange or $\dot{V}_A/\dot{Q}$ inequality.

Bratel et al., [34, 35] evaluated the effects of felodipine, a pulmonary vasodilator, on hemodynamics and pulmonary gas exchange in patients with COPD-associated pulmonary hypertension. In the acute study [34], intravenous felodipine had no effect on pulmonary arterial pressure but reduced both pulmonary and systemic vascular resistance, as cardiac output was increased. The drug caused increased shunt and increased $\dot{V}_A/\dot{Q}$ mismatch because of the development of regions of high $\dot{V}_A/\dot{Q}$ ratio, and arterial oxygenation was lower compared to baseline. Since there were beneficial effects on pulmonary vascular resistance, felodipine was continued orally in these patients, and they were reevaluated after 3–5 months of therapy [35]. There was marked patient heterogeneity in the response to felodipine, and some, but not all, showed a continued decrease in pulmonary vascular resistance. However, when these subjects were exercised, pulmonary vascular resistance was reduced. In this longer-term study, the increase in shunt with the drug was not significant, but the increased $\dot{V}_A/\dot{Q}$ mismatch was significant, and the increase in regions of high $\dot{V}_A/\dot{Q}$ ratio persisted. The outcome was that the authors felt that the primary goal of decreasing pulmonary arterial pressure and vascular resistance was not achieved using this drug, although there were clearly responders and nonresponders to the therapy.

A final study of felodipine evaluated the effects of felodipine on pulmonary gas exchange in patients undergoing long-term oxygen therapy [36]. In a similar manner to the previous studies, an acute infusion of felodipine lowered pulmonary vascular resistance, increased cardiac output, and worsened pulmonary gas exchange by increasing shunt. In contrast to the development of high $\dot{V}_A/\dot{Q}$ regions in the studies without long-term oxygen therapy, regions of low $\dot{V}_A/\dot{Q}$ ratio accounted for the increased $\dot{V}_A/\dot{Q}$ mismatch observed. Also in contrast to the studies without oxygen therapy, after ~14 weeks of oral felodipine, pulmonary gas exchange was restored back to baseline, and the beneficial effects on pulmonary vascular resistance persisted. Based on this, the authors suggested that felodipine be considered for long-term adjunct therapy in combination with oxygen therapy.

In patients with pulmonary fibrosis-associated pulmonary hypertension, Ghofrani and colleagues [102] compared the effect on the pulmonary circulation and gas exchange of sildenafil to inhaled nitric oxide and infused epoprostenol. All three drugs decreased pulmonary vascular resistance. While epoprostenol worsened gas exchange by increasing $\dot{V}_A/\dot{Q}$ mismatch and shunt, both inhaled nitric oxide and sildenafil did not. In addition, sildenafil improved arterial P_{O_2} significantly leading the authors to suggest that it may enhance local vasoregulatory mechanisms in the pulmonary circulation. In a follow-up safety study [110], this group evaluated the effects of another pulmonary vasodilator, bosentan, an endothelin antagonist on pulmonary gas exchange in idiopathic pulmonary fibrosis patients. Only one patient showed any change in gas exchange (increased shunt and low $\dot{V}_A/\dot{Q}$ regions) measured by MIGET following acute administration of bosentan. These findings led to the conclusion that, similar to the results with sildenafil, worsening pulmonary gas exchange abnormalities were unlikely in patients with pulmonary fibrosis with this drug.

Clearly, gas exchange in pulmonary hypertension, whether idiopathic, post-thromboembolic, or secondary to lung disease in origin, is complex, but several studies described above allow some general conclusions:

1. There is usually arterial hypoxemia. This is surprisingly variable in degree from little to severe.
2. There is usually hyperventilation with hypocapnia, and so the alveolar-arterial P_{O_2} difference is substantial.
3. $\dot{V}_A/\dot{Q}$ inequality is universally found, but the pattern is highly variable between patients and across studies. In some cases, there is simply widening of the $\dot{V}_A/\dot{Q}$ distribution without shift; in others, when cardiac output is low, the whole $\dot{V}_A/\dot{Q}$ distribution is shifted toward higher $\dot{V}_A/\dot{Q}$ ratios; shunt is not uncommon and

may be substantial. Whether this is due to a patent foramen ovale, to scattered atelectasis, or to local alveolar edema has not been determined. Distinct regions of both very low and very high $\dot{V}_A/\dot{Q}$ ratio are not uncommon; the former may be due to local edema; the latter may be due to vascular obstruction.
4. Diffusion limitation of O_2 uptake has never been observed in pulmonary hypertension per se but, of course, is present when pulmonary hypertension is secondary to pulmonary fibrosis.
5. Reduction in cardiac output, causing a fall in mixed venous P_{O_2}, considerably worsens the hypoxemia initially caused by the $\dot{V}_A/\dot{Q}$ inequality.
6. The use of pulmonary vasodilators has a variable effect on gas exchange depending on selectivity. Nonselective pulmonary vasodilators tend to worsen gas exchange, whereas more modern drugs such as bosentan and sildenafil either improve gas exchange or have no effect.

The above human research has been complemented by several animal studies of acute autologous clot or glass bead venous embolism [56, 59–62] that in essence recapitulated the basic observations described above. In general they found that the size of the embolus mattered, with larger-sized emboli associated with high $\dot{V}_A/\dot{Q}$ areas and smaller emboli more with shunt and low $\dot{V}_A/\dot{Q}$ areas, much as Young et al., found in 1980 [425]. These authors also found that vasodilators had little effect on $\dot{V}_A/\dot{Q}$ relationships despite reducing vascular resistance, while vasoconstrictors worsened $\dot{V}_A/\dot{Q}$ relationships. Strong hypoxic pulmonary vasoconstriction also had deleterious effects [63]. Kawai et al., [188] used glass bead embolism in dogs and found both low and high $\dot{V}_A/\dot{Q}$ regions developed. They also found that a serotonin antagonist mitigated the low $\dot{V}_A/\dot{Q}$ regions, but had no effect on the development of high $\dot{V}_A/\dot{Q}$ regions. This is consistent with the presumed mechanical basis for high $\dot{V}_A/\dot{Q}$ area creation (i.e., vascular obstruction) but an inflammatory and/or pressure-mediated basis for the generation of low $\dot{V}_A/\dot{Q}$ areas. Truog et al., [369] gave TNF-alpha to neonatal piglets and saw increased pulmonary vascular resistance but little gas exchange effect using MIGET. In contrast, Walmrath et al., [415] studied blood-free perfused rabbit lungs, using a thromboxane analog to raise vascular resistance, and found substantial $\dot{V}_A/\dot{Q}$ inequality and shunt, mitigated to some extent by inhaled vasodilatory prostaglandins. This result may in part be due to the fragility of isolated rabbit lungs.

7.3.7 Consolidation and Atelectasis

Alveolar consolidation (alveolar gas space filling with cellular debris in pneumonia) and collapse (atelectasis) would both be expected to result in unventilated but perfused lung regions and thus shunting. In canine right lower lobar pneumonia induced by direct bronchial installation of pneumococcal bacteria, that was exactly what was found at 48 h after giving the bacteria [408]. Twenty-four hours later, after antibiotic treatment, shunt had fallen considerably and given way to regions of low $\dot{V}_A/\dot{Q}$ ratio. This was interpreted as partial alveolar clearing and corresponding partial restoration of ventilation. In rats, breathing 100% oxygen has been shown to produce shunt, consistent with the development of atelectasis, in animals that previously had regions of low $\dot{V}_A/\dot{Q}$ ratio [371].

Redding et al., [288] induced left lower lobar collapse in piglets and found shunting occurred as expected. Of more interest, left lower lobe blood flow, which was 30% of the cardiac output before collapse, fell by 70% to just 9% after collapse, indicating strong hypoxic pulmonary vasoconstriction materially mitigating the size of the shunt and the ensuing gas exchange defect. When the piglets were given 12% O_2 to breathe, hypoxic vasoconstriction in the rest of the lung caused fractional blood flow in the collapsed left lower lobe perfusion to more than double.

Mols et al., [238] compared strategies to keep atelectatic alveoli open either by sustained inflation (with 20 cm H_2O pressure) maneuvers and PEEP alone or in combination in isolated rabbit lungs. The extent of shunt and low $\dot{V}_A/\dot{Q}$ regions significantly differed between the sustained inflation alone and sustained inflation with PEEP, suggesting that alveolar recruitment by sustained inflation maneuvers requires that alveoli are stabilized by sufficient PEEP and that repeated sustained inflations on PEEP may not offer any advantages to appropriate PEEP alone. They also found [239] that the lowest level of PEEP to keep the lung open could be evaluated by compliance measures within tidal volume breaths without the need for total lung capacity maneuvers.

7.3.8 Following Pneumonectomy

Pneumonectomy not only removes functioning gas exchange tissue but may also result in hyperinflation and distortion of remaining airways, blood vessels, and lung tissue. These structural effects have the potential to affect gas exchange. Of course, the amount of lung resected and which lobes of the lung are removed from the asymmetrical chest wall cavity will be factors involved. Accordingly, Connie Hsia and Bob Johnson carried out a series of treadmill exercise studies over several years in different groups of pneumonectomized foxhounds using MIGET. Studies could only be done after surgery, but sham surgery done in control dogs at rest and during exercise shows minimal $\dot{V}_A/\dot{Q}$ inequality, no hypoxemia, and no diffusion limitation.

After left lung removal (42% of the lung mass, [167]), exercise led to substantial $\dot{V}_A/\dot{Q}$ inequality plus significant diffusion limitation, the latter dominating during hypoxic exercise as expected. In another group, right lung removal (58% of lung mass, [169]) was studied at 2 and 12 months after surgery. At 2 months, exercise limitation, high pulmonary vascular resistance, $\dot{V}_A/\dot{Q}$ inequality, and diffusion limitation were seen during exercise, but by 12 months, all had significantly improved but were still abnormal. There was also evidence of gas-phase diffusion limitation, as retention values of high molecular weight gases were systematically higher than those of low molecular weight gases, correcting for differences in solubility. This effect, while likely insignificant for a low molecular weight gas like O_2, probably reflects alveolar hyperinflation. In another study, after right lung removal, an inflatable balloon was inserted into the right chest cavity and compared after long-term inflation or deflation to examine how mediastinal shift and lung strain of the left lung would be affected [172]. Interestingly, allowing strain (balloon deflated) promoted diffusive gas exchange; $\dot{V}_A/\dot{Q}$ relationships were unaffected by balloon inflation state. They interpreted these findings to suggest that pneumonectomy causes regeneration of alveolar surface but not of conducting airways. Hypoxia during lung growth leads to larger lungs, which may improve exercise capacity. They then removed 70% of the lungs in another group of foxhounds [168]. This was done by two different resection plans designed to prevent or allow distortion of the remaining lung through mediastinal shift. After recovery from surgery, there were substantial pulmonary hypertension and hypoxemia during exercise, worse when mediastinal shift occurred. The major effect was of worsening $\dot{V}_A/\dot{Q}$ relationships, especially with distortion. While O_2 diffusion limitation was seen, it played a minor role in the hypoxemia. This contrasts with less lung removal studies above where diffusion limitation appeared to be as, or maybe more, important than $\dot{V}_A/\dot{Q}$ inequality. It suggests that alveolar tissue regeneration was more complete than changes in conducting vessels. The rich data sets from these studies led [173] to combine them and assess the relationship between O_2 diffusing capacity determined by MIGET and that determined from the conventional D_{LCO} converted to D_{LO_2}. There was certainly noise in the comparison, but on average, the value derived from MIGET differed from that derived from D_{LCO} by only 2%.

Johnson et al., [182] compared gas exchange after pneumonectomy in humans and foxhounds. MIGET data were from the above studies in the dogs; there were no MIGET studies in humans, but there was no exercise-induced hypoxemia in the humans to explain. The principal finding was that the $\dot{V}_A/\dot{Q}$ inequality and diffusion limitation were more extensive in dogs than could be imagined in normoxemic humans, even when large volumes of lung had been removed in humans. In contrast, in humans, maximal cardiac output was reduced, and the suggestion was made that the pericardial space/cardiac fossa anatomy may have been compromised by surgery in humans and may be the more important factor.

In 2007, Hsia et al., [171] compared $\dot{V}_A/\dot{Q}$ relationships and diffusion limitation during hypoxic exercise in foxhounds raised at sea level compared to littermates raised at 3800 m. The studies were done 2.5 years after the altitude-raised dogs were returned to sea level. While during exercise $\dot{V}_A/\dot{Q}$ relationships were unaffected by altitude, diffusing capacity was improved and the alveolar-arterial P_{O_2} difference was lower.

This animal model was also used to study the effects of crocetin, a carotenoid compound that was suggested to be beneficial in enhancing oxygen diffusion during exercise [406]. However although there were small improvements in $\dot{V}_A/\dot{Q}$ mismatch of questionably physiological significance (LogSD,Q fell from 0.48 to 0.43), there were no effects on oxygen diffusion.

7.4 In Other Systemic Diseases

7.4.1 Liver Disease

Gas exchange impairments in liver cirrhosis have been reviewed by Wagner [390]. Hypoxemia, sometimes severe, is not uncommon in chronic liver disease even when spirometry suggests normal airway structure and function. Moreover, arterial P_{O_2} frequently improves from upright to supine position. Several studies have used MIGET to understand why hypoxemia occurs without spirometric abnormality (Fig. 7.10). Edell et al., [85] found both low $\dot{V}_A/\dot{Q}$ areas and shunts to be present, sometimes taking almost 30% of the cardiac output. These two abnormalities accounted for most, but not all, of the observed arterial hypoxemia. Thus, a significant contribution from diffusion limitation was inferred. Most authors attribute the gas exchange defects to large arteriovenous communications within the lung that do not allow complete (or in some cases, any) gas exchange to occur. The postural changes suggest that those communications exist in the dependent lung regions. These authors also noted an elevated cardiac output (at rest) which served to buffer the arterial P_{O_2} from further fall by helping elevate mixed venous P_{O_2}. However, Agusti et al., [3] who used propranolol to reduce cardiac output showed no change in arterial P_{O_2} after administration of the drug. This was because shunt was reduced and the $\dot{V}_A/\dot{Q}$ distribution was shifted to the right likely because of the lower cardiac output, although changes in vascular tone could not be ruled out. Melot et al., [232] compared cirrhotic patients with hypoxemia to those with normal arterial P_{O_2}. Hypoxemic patients showed mostly low $\dot{V}_A/\dot{Q}$ areas (with only small shunts), and together these accounted for most of the hypoxemia. There were additional small contributions to arterial P_{O_2} from anemia (reducing P_{O_2}) and an elevated Hb P_{50} and hyperventilation (both raising P_{O_2}), but there was no O_2 diffusion limitation observed. They specifically looked for impaired hypoxic pulmonary vasoconstriction as a factor in the blood flow through hypoxic regions, but did not find evidence. Eriksson et al., [88] also found a combination of shunting and

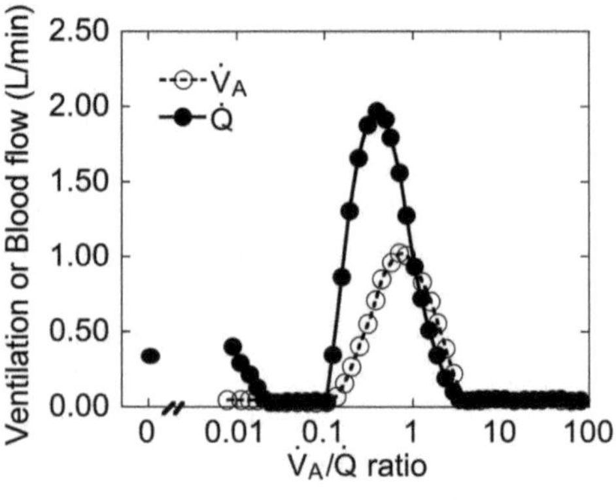

Fig. 7.10 Ventilation-perfusion relationships in a patient with hypoxemic liver disease. LogSD,Q is moderately elevated (~0.63), and there are regions of low $\dot{V}_A/\dot{Q}$ ratio as well as significant shunt (~4%). Redrawn from [3]

low $\dot{V}_A/\dot{Q}$ regions in such patients and no evidence of O_2 diffusion limitation. Interestingly, 2–6 months after liver transplantation, $\dot{V}_A/\dot{Q}$ relationships normalized and shunt disappeared. In 1993, Van Obbergh et al., [380] confirmed the presence of shunts (with small regions of low $\dot{V}_A/\dot{Q}$ as well) and again disappearance after liver transplantation. In 1991, Hedenstierna et al., [141] confirmed Eriksson's main findings—a combination of shunt and low $\dot{V}_A/\dot{Q}$ regions in cirrhotic patients—but in this study, there was evidence for O_2 diffusion limitation as an additional contributing factor to the hypoxemia. They suggested that the diffusion problem might arise from inequality in distribution of diffusing capacity versus blood flow. This could fit with arteriovenous communications as the basis for hypoxemia, as such channels would have relatively high blood flow and being larger than capillaries, with thicker walls, a low diffusing capacity.

Crawford et al., [49] published a case study in 1995, applying both MIGET and lung scintigraphy using radiolabeled macroaggregates, at rest in both normoxia and while breathing 100% O_2. Perfusion of low $\dot{V}_A/\dot{Q}$ regions was observed but was small (2% on air, 1% during hyperoxia). In normoxia, shunt (by MIGET) was about 19%, while that by macroaggregate transmission was 41%. During hyperoxia, shunt by the venous admixture equation for O_2 was 18%, that by MIGET increased to 25%, and that from macroaggregates was unchanged, at 40%. In normoxia, measured arterial P_{O_2} was significantly lower than predicted by MIGET, supporting a role for diffusion limitation as a factor contributing to hypoxemia. However in hyperoxia, MIGET predicted a P_{O_2} less than measured, a unique observation. The authors wrote a long explanation implying diffusion limitation for SF_6, the least soluble inert gas used in MIGET. However, the necessary aberration in SF_6 retention would have to be relatively enormous and would cause the MIGET software to indicate a poor fit to the data, which was not the case. That the shunt by aggregate transmission was much larger than those by MIGET or O_2 equations was explained on the basis of a range of particle sizes so that the 40% value includes both smaller and larger channels. Only the larger channels are hypothesized to impair gas exchange.

In 1995, Mayoral and Sabate [225] analyzed MIGET data from three studies of patients with cirrhosis, focusing on the combined data of those with hypoxemia. They found substantial $\dot{V}_A/\dot{Q}$ inequality (16% blood flow in low $\dot{V}_A/\dot{Q}$ regions), substantial shunt (10%), and strong evidence for O_2 diffusion limitation. Reviews by Lotterer in 1994 and 1997 [208, 209] and by Robert et al., in 1999 [294] came to the same general conclusions.

In summary, liver disease appears to produce gas exchange abnormalities in the absence of demonstrable lung disease. The common features are a combination of low $\dot{V}_A/\dot{Q}$ ratio regions and unventilated (shunt) regions, plus a contribution from incomplete O_2 diffusion equilibration. These three factors are rather variable within and between studies. The structural basis for all three appears to be dilated arteriovenous vascular channels that due to high blood flow, thick walls, and low or zero ventilation causing impaired diffusion, partial gas exchange, or even zero gas exchange. The commonly found increase in arterial P_{O_2} moving from upright to supine suggests a concentration of these channels in the lung bases. Most impressively, liver transplantation has consistently been found to normalize gas exchange (in patients without lung disease from other causes).

7.4.2 Heart Failure

There appears to be just one study of patients with chronic heart failure [30] and that was limited to just five patients. These were all patients with reduced ejection fraction, with heart failure based on ischemic heart disease, and were clinically stable without evidence of (acute) pulmonary edema. Arterial P_{O_2} varied between 59 and 99 mm Hg across patients, and a generally small degree of $\dot{V}_A/\dot{Q}$ inequality was seen in each, despite no clinical evidence of pulmonary edema or of lung disease. With cardiac output averaging only 3.1 l/min, mixed venous P_{O_2} was only 29 mm Hg (normal 40 mm Hg). This aggravated the arterial hypoxemia and explained why patients with small amounts of $\dot{V}_A/\dot{Q}$ mismatch could still have significant arterial hypoxemia. Presumably, $\dot{V}_A/\dot{Q}$ inequality reflected small amounts of alveolar and/or interstitial edema. Following intravenous nitroprusside infusion, $\dot{V}_A/\dot{Q}$ relationships worsened with perfusion of poorly ventilated or completely unventilated regions increasing from 1.5% to 9.4% of the cardiac output. This was probably due to abolition of hypoxic vasoconstriction by nitroprusside in those areas. Despite a large relative increase in cardiac output, from 3.1 to 4.8 l/min, mixed venous P_{O_2} increased by only 4 mm Hg after nitroprusside, an amount insufficient to prevent hypoxemia from worsening as a result of increased blood flow in low $\dot{V}_A/\dot{Q}$ areas.

7.4.3 Obesity

We are taught that obesity pushes the diaphragm up and, when reclining, compresses the chest wall. As a result, FRC is reduced, and there may be airway closure during the breathing cycle to impair gas exchange and cause hypoxemia. However, that is not what is always found, even in morbid obesity. Some morbidly obese subjects maintain normal gas exchange, while others develop small shunts and some $\dot{V}_A/\dot{Q}$ inequality [293], but overall, the effects do not seem as large as one might have expected. Bariatric surgery was found to mitigate the gas exchange defects [293]. The same authors [292] compared gas exchange in morbidly obese subjects in the upright and supine positions and surprisingly found no deterioration in the supine position.

7.4.4 *Hemodialysis in Chronic Renal Failure*

Hemodialysis has been shown to cause acute hypoxemia in patients with chronic renal failure. Romaldini et al., [310] used MIGET in such patients to better understand the reason. The complex findings were interesting: $\dot{V}_A/\dot{Q}$ relationships actually improved. The reason for hypoxemia was mostly a fall in alveolar ventilation caused by significant removal of CO_2 by the dialysis system. However, because there was no significant addition of O_2 from the dialysis system, the fall in alveolar ventilation caused a fall in alveolar and thus arterial P_{O_2}. Cardiac output was slightly lower, which contributed to the drop in arterial P_{O_2} (via reduction in mixed venous P_{O_2}), while the improvement in $\dot{V}_A/\dot{Q}$ relationships worked to buffer that drop. There was no evidence for O_2 diffusion limitation. Precisely the same conclusions were reached by Ralph et al., [285] in a canine model of hemodialysis.

7.5 Children

7.5.1 *Children with Normal Lungs*

Application of the MIGET technique in children was reviewed by Hedenstierna and Freyschuss [135]. In the one reported study [136], nine children 7–15 years old who were undergoing cardiac catheterization for clinical indications, but who had no evidence of pulmonary disease, were studied using MIGET. The clinical diagnoses were of heart valve lesions or coarctation of the aorta. No patients had pulmonary hypertension or intracardiac shunting. In the older children, results were as previously found for healthy adults—a single narrow $\dot{V}_A/\dot{Q}$ distribution centered on a $\dot{V}_A/\dot{Q}$ ratio of about 1.0. However in three of the four youngest children, a separate mode of high $\dot{V}_A/\dot{Q}$ areas was observed. This is a very difficult population to study, and the findings were not known until after the catheter work had been completed, and so the physiological basis for such high $\dot{V}_A/\dot{Q}$ regions remains unknown. Other research has narrowed the causes of high $\dot{V}_A/\dot{Q}$ areas to (a) COPD, (b) pulmonary embolism, (c) high levels of PEEP, and (d) other scenarios where Zone 1 conditions exist (e.g., severe acute blood loss). The only one of these that could be invoked in this pediatric population is pulmonary embolism, but this was not mentioned or evaluated in the paper.

7.5.2 *Children with Lung and Liver Disease*

While the abovementioned study appears to be the only one in healthy children, investigators have also used MIGET to evaluate children with lung and liver disease. In one such study, Freyschuss et al., [97] evaluated $\dot{V}_A/\dot{Q}$ relationships before and after an exercise test. All 11 children had smooth unimodal $\dot{V}_A/\dot{Q}$ distributions before exercise with a LogSD,Q in the high normal range (0.49–0.58) and an elevated LogSD,V_A (0.64–0.70). In approximately half of the subjects, exercise did not provoke symptoms and spirometry was unchanged. However, although the $\dot{V}_A/\dot{Q}$ distribution remained unimodal, there was an increase in LogSD,Q postexercise (to 0.70), but this was not statistically significant. The other half of the subjects developed evidence of exercise-induced bronchospasm, and these subjects developed a bimodal $\dot{V}_A/\dot{Q}$ distribution following exercise with a significant increase in both LogSD,Q and LogSD,V_A. Shunt was not a feature of either group either before or after exercise. The authors hypothesized that the high $\dot{V}_A/\dot{Q}$ mode was caused by increased intrathoracic pressure from local hyperinflation impeding regional blood flow.

In another study, this same group of investigators induced bronchospasm by histamine challenge in five children with a history of asthma and evaluated gas exchange with MIGET [144]. Before provocation, the children had largely normal $\dot{V}_A/\dot{Q}$ distributions with unimodal distributions, although LogSD,Q and LogSD,V_A were greater than the normal children reported above. There was also small but detectable shunt (~2%). After provocation with histamine, the LogSD,Q and the LogSD,V_A were significantly increased, and in some subject shunt was increased to ~4%. The increase in $\dot{V}_A/\dot{Q}$ heterogeneity was caused by the appearance of bimodal $\dot{V}_A/\dot{Q}$ distributions which were driven the development of a high mode, but there were no significant regions of low $\dot{V}_A/\dot{Q}$ ratio. The authors suggested that histamine provocation caused local hyperinflation leading to compromised $\dot{V}_A/\dot{Q}$ relationships.

In the final study reported in children, Van Obbergh et al., [380] evaluated three children before and after liver transplantation for severe liver disease. Like their adult counterparts, these children demonstrated severe hypoxemia that was almost entirely the result of a large shunt (27–56%) with minimal regions of low $\dot{V}_A/\dot{Q}$ ratio. Postoperatively, there was a progressive decrease in shunt and at the end of the follow-up period shunt ranged from 0% to 8% of cardiac output.

7.5.3 During Growth and Development

Escourrou et al., [90] evaluated the effects of normal lung growth on pulmonary gas exchange using piglets ranging in age from 12 to 65 days. P_{aO_2} was lower in young animals, and $AaDO_2$ was higher. Shunt and deadspace were not correlated with age, but the LogSD,Q decreased in older animals. The mean modes of the perfusion and ventilation distributions did not differ significantly between age groups. Unexpectedly, there was a suggestion that in younger animals there was diffusion limitation but the reasons for this are unclear.

7.6 Anesthesia and Surgery

7.6.1 Anesthesia in Humans

General anesthesia, even in patients with normal lungs, has long been known (e.g., see [130]) to cause sometimes severe, and rapidly developing, gas exchange disturbances, as was pointed out in 1984 [400]. As with acute lung injury, this topic has been a fertile ground for studies of gas exchange by MIGET. There are many risks that come with general anesthesia, risks that may affect gas exchange. These include alveolar collapse or overdistension, under- or overventilation, retention of airway secretions, change in vascular and airway tone, changes in F_{IO_2}, changes in cardiac output, and the so-called second gas effects of large fluxes of soluble anesthetics such as nitrous oxide. These all provide potential challenges to gas exchange and its evaluation that together become very complex. In particular, high F_{IO_2} and the second gas effect can substantially mask $\dot{V}_A/\dot{Q}$ inequality by shoring up arterial P_{O_2}. The reader is directed to several reviews of gas exchange during anesthesia [128–134].

To use MIGET in anesthetized patients is itself a technical challenge because the six gases used are present by design in just parts per million, while gaseous anesthetics are present in orders of magnitude higher concentrations. This makes for analytical difficulty especially in chromatography, the backbone method of MIGET. To this end we worked out chromatography conditions to get around this problem [82].

One of the first studies in 1983 involved epidural anesthesia [212] and found no significant changes in $\dot{V}_A/\dot{Q}$ relationships, a finding which they later confirmed [111]. The same year, Hedenstierna and colleagues [138, 139, 210] studied older patients undergoing leg vascular surgery, hip arthroplasty, or abdominal surgery and found both shunt and considerable $\dot{V}_A/\dot{Q}$ heterogeneity in each type of surgery. The same group evaluated patients undergoing halothane anesthesia and found modest shunts (~10%) that increased with increasing F_{IO_2} and fell again with reduced F_{IO_2}. There was no mention of alveolar collapse, rather that there may be "shunt vessels" subject to reversible hypoxic vasoconstriction [211]. They later extended this work [33] to show that supine awake older subjects developed bimodal distributions with a low $\dot{V}_A/\dot{Q}$ mode that converted to shunt with induction of anesthesia and mechanical ventilation and were reduced with the addition of PEEP. With PEEP there was development of a high $\dot{V}_A/\dot{Q}$ model that they attributed to the development of Zone 1 conditions. Animal (dog) studies [143] showed that development of high $\dot{V}_A/\dot{Q}$ regions was related to open corner vessels in the presence of alveolar vessel collapse.

Then in 1986, Hedenstierna et al., [142] published a landmark paper using MIGET and CT simultaneously in patients with normal lungs undergoing elective surgery under halothane anesthesia. The findings were not subtle. Remarkably, shunts as high as 17% appeared within 15 min of induction (Fig. 7.11), and CT imaging identified atelectasis in dependent lung regions as their likely cause. The

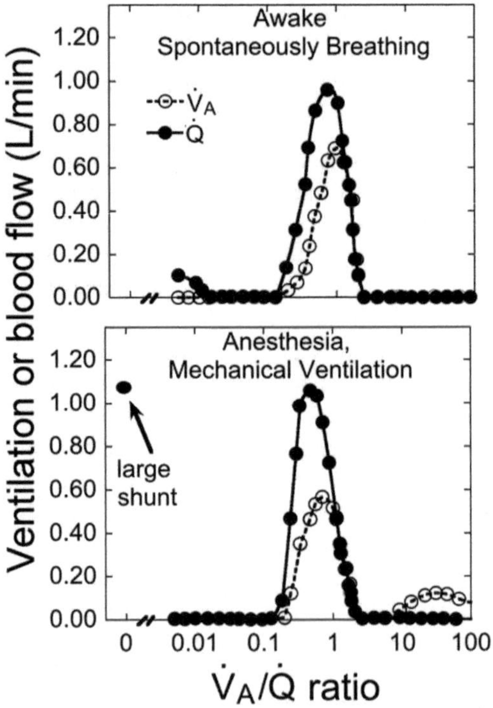

Fig. 7.11 The effects of anesthesia and mechanical ventilation on the distribution of $\dot{V}_A/\dot{Q}$ ratios in a supine subject. After induction of anesthesia and mechanical ventilation, a large shunt of ~17% of cardiac output is present. CT scanning obtained shortly thereafter showed regions of atelectasis. Redrawn from [142]

extent of atelectasis by CT showed a close relationship with MIGET shunt ($R = 0.93$). Seven of their eight patients also showed high $\dot{V}_A/\dot{Q}$ areas, which may well have reflected overexpanded alveoli from the mechanical ventilation, likely from high ventilator pressures that still did not prevent rapid onset of alveolar collapse. Additional work by these authors [361, 362] in both spontaneously breathing and mechanically ventilated patients (halothane anesthesia again) confirmed these findings (rapid atelectasis and shunt onset) adding observations of sometimes considerable low $\dot{V}_A/\dot{Q}$ region development. In addition, muscle paralysis worsened atelectasis by CT and shunt by MIGET, while PEEP reduced atelectasis but not always shunt.

The same group next studied these phenomena during ketamine anesthesia [363] and found the same outcome, albeit less severe than with halothane. However, muscle paralysis again worsened atelectasis and gas exchange in a correlated manner. The authors suggested altered chest wall mechanics were the key pathway to alveolar collapse and thus shunt and hypoxemia. Klingstedt et al., [192] also showed that the extent of $\dot{V}_A/\dot{Q}$ mismatch was worsened by the lateral position, compared to supine. In 1989, Gunnarsson et al., [107] compared nitrous oxide anesthesia to that by enflurane and found the two gases produced similar degrees of atelectasis, shunt, and $\dot{V}_A/\dot{Q}$ inequality, with evidence for absorption atelectasis when nitrous oxide was used. In 2002, Loeckinger et al., [203] compared sevoflurane to isoflurane in patients undergoing coronary artery surgery and found that while both agents caused $\dot{V}_A/\dot{Q}$ heterogeneity during anesthesia, only sevoflurane also depressed arterial P_{O_2}. However, in normal supine awake subjects, thoracoabdominal restriction using a corset [360] reduced lung volumes but did not cause atelectasis or affect gas exchange as measured by MIGET.

In 1988, Dueck et al., [81] related shunt development during anesthesia to lung volume and found that the lower the lung volume, the higher the shunt. Again, the changes from awake were substantial, and the data supported the general thought that loss of lung volume from chest wall mechanical changes was to blame. Next, Gunnarsson et al., [108] studied the interaction between age and gas exchange impairment under anesthesia with halothane or enflurane. Age worsened the gas exchange defect caused by anesthesia because of greater $\dot{V}_A/\dot{Q}$ heterogeneity and not because of greater shunt, which was nonetheless elevated in all age groups. They confirmed the close correlation between extent of atelectasis by CT and shunt by MIGET. Studies then turned to strategies to prevent atelectasis under anesthesia. Rothen et al., [314–316] found that if F_{IO_2} was held to 0.4, hyperinflation maneuvers eliminated atelectasis for the 40 min studied [313], but that if F_{IO_2} was 1.0, atelectasis recurred within 5 min after a hyperinflation maneuver. The deleterious effect of high F_{IO_2} on gas exchange was confirmed in a pig model by Loeckinger et al., [205]. Rothen et al., [312] went on to show that both atelectasis and airway closure contributed to gas exchange abnormalities under anesthesia.

In 1999, Tenling et al., [357] found that thoracic epidural anesthesia (TEA) given to patients as an adjunct to general anesthesia for cardiac surgery to control postoperative pain did not worsen gas exchange associated with general anesthesia. Additionally, TEA allowed earlier extubation and improved $\dot{V}_A/\dot{Q}$ relationships. Reber et al., [287] also showed that lumber epidural anesthesia did not worsen with gas exchange or lung aeration (measured by CT) nor did lithotomy position affect gas exchange. Hachenberg et al., [112, 113] confirmed rapid development of shunt in patients undergoing cardiac surgery for mitral valve disease or coronary artery disease. They found the magnitude of disturbance similar in the two groups. Further work by Tenling et al., [356] showed persistence of atelectasis and shunt into the day after surgery. Interestingly, in contrast to the findings in normal subjects and other patient groups, Gunnarsson et al., [109] found that although $\dot{V}_A/\dot{Q}$ mismatch worsened from the preanesthetic baseline with anesthesia in COPD patients, this was not associated with significant increases in shunt or with large increases in the amount of atelectasis detecting on CT scanning. Dueck et al., [83] showed that arterial blood gases grossly underestimated the severity of the underlying $\dot{V}_A/\dot{Q}$ disturbance induced in patients who were smokers with COPD of varying severity undergoing halothane anesthesia, but that the development of CO_2 retention reliably identified these individuals.

7.6.2 Anesthesia in Nonhuman Species

Note that all of the above work (both human and animal) on the effects of anesthesia came from the Hedenstierna laboratory in Sweden (except for that done by Dueck in San Diego and Loeckinger in Austria), truly a significant contribution to the literature. In horses, the Hedenstierna group [248] found very large dependent region shunts in anesthetized horses. Peep applied at the trachea did not reinflate these regions, but if PEEP was applied just to the airways of the collapsed regions, reinflation was possible and shunt fell substantially. Marntell et al., [220] also found that $\dot{V}_A/\dot{Q}$ inequality and shunt (atelectasis) developed in horses as in man, but that acepromazine added to the drug regime improved oxygenation. The same authors [221] confirmed that during anesthesia in horse, 100% inspired O_2 increased shunt compared to room air breathing and that this shunt persisted after return to room air, implicating absorption atelectasis in its genesis. Dodam et al., [71] showed that in horses undergoing anesthesia, clenbuterol, a beta-2 agonist, decreased pulmonary vascular resistance but worsened hypoxemia, because of increased shunt. Magnusson et al., [214] tried to prevent atelectasis and thus shunt by continuous positive airway pressure ventilation in a pig model of open-heart surgery with cardiopulmonary bypass. This strategy unfortunately failed.

A recent case study from the Hedenstierna laboratory [251] showed that unlike the work in horses, the giraffe showed little development of shunt under anesthesia despite much larger lungs. The reason for this is unclear but may include the short duration of the anesthesia; however, as mentioned above, humans typically develop shunt within 15 min of induction of anesthesia [142], so this is less likely. Other possible explanations include that the animal was in the semi-lateral recumbent posture, was spontaneously ventilating, and was not breathing hyperoxic gas, or this may be a feature of the giraffe lung.

7.6.3 Before and After AV Malformation Surgery

Diseases causing pulmonary arteriovenous malformations such as hereditary hemorrhagic telangiectasia (HHT) result in arterial hypoxemia from intrapulmonary vascular shunts. Andrivet et al., [10] used MIGET to study a patient with HHT before and 4 months after therapeutic embolization of radiologically visible malformations. Both percent shunt and cardiac output at rest fell after surgery: shunt from 31% to 19% of cardiac output and cardiac output itself from a very high value of 12.4 to 7.4 l/min. Thus absolute shunt flow fell by 63% from 3.8 to 1.4 l/min. However, arterial P_{O_2} increased by only 15 mm Hg, from 64 to 79, which, considered alone, would not have indicated a strong therapeutic response. These data are explained by the beneficial effect of the high cardiac output in the presurgical state on arterial oxygenation via preservation of mixed venous P_{O_2}.

7.6.4 *Orthopedic Surgery*

MIGET has been used to evaluate aspects of gas exchange during orthopedic surgery. Hip arthroplasty is known to result in hypoxemia, which can either be a transient phenomenon or last for a prolonged period of time. Hedenstierna et al., [139] used MIGET to evaluate gas exchange in eight patients undergoing hip arthroplasty. Anesthesia alone induced minor changes in gas exchange with the development of small amount of shunt (~5%). In addition 4/8 patients developed a small low $\dot{V}_A/\dot{Q}$ mode. After the insertion of the acetabular prosthesis, there was a small increase in shunt (~1%), and the extent of shunt was also increased after the insertion of the femoral prosthesis in 4/8 subjects. There were no discernable trends in deadspace during the study, but there was the development of regions of high $\dot{V}_A/\dot{Q}$ ratio (i.e., $\dot{V}_A/\dot{Q}$ 10–100). The authors suggested that the changes resulted from gravitationally based redistributions in blood flow away from nondependent lung.

7.6.5 *Coronary Bypass and Other Cardiac and Vascular Surgery*

Oku et al., [257] wondered if transection of the vagus nerve and sympathetic nerve chain to the lung, such as often happens during thoracic surgery, had an effect on pulmonary gas exchange. Their work in dogs showed that blood flow in the denervated lung was reduced and this had the effect of creating regions of high $\dot{V}_A/\dot{Q}$ ratio, but the disturbance to pulmonary gas exchange was relatively minor.

Hacheberg et al., [112] evaluated MIGET in patients undergoing mitral valve surgery; they hypothesized that increased $\dot{V}_A/\dot{Q}$ mismatch might be present because of their greater risk for postoperative respiratory issues. However, they found that the major gas exchange abnormality was the development of shunt and this was true with the induction of anesthesia and after the patients were removed from extracorporeal circulation.

Anjou-Linskog and colleagues published several papers, which used MIGET to evaluate pulmonary gas exchange in patients in the early postoperative period after cardiac surgery. The first [13] evaluated the effects of nitroglycerin infusion on gas exchange after bypass grafting. At baseline postsurgery and before nitroglycerine, $\dot{V}_A/\dot{Q}$ mismatch was increased in this group of patients, and shunt averaged 6% of cardiac output. Nitroglycerin decreased cardiac output, mean arterial and pulmonary arterial pressure, consistent with the known vasodilator effects of this drug, and increased shunt, without an overall change in the extent of $\dot{V}_A/\dot{Q}$ mismatch. This change with nitroglycerin persisted even when patients [11] were pre-ventilated with 100% oxygen to lift any hypoxic pulmonary vasoconstriction. The higher levels of oxygen induced a small decrease in pulmonary arterial pressure and an increase in cardiac output, and regions of either low $\dot{V}_A/\dot{Q}$ ratio or shunt were increased consistent with the development of atelectasis. Later this group evaluated the effects of ventilation with 100% oxygen compared to 30% oxygen in a population of patients post bypass surgery [12]. Again higher levels of oxygen-induced regions of either low $\dot{V}_A/\dot{Q}$ ratio or shunt were increased consistent with the development of atelectasis. That development of atelectasis was responsible for the shunt was confirmed in an animal model [216] using MIGET combined with CT scanning. Magnusson et al., [214] showed in an animal model that the shunt that occurred with bypass could be improved with vital capacity maneuvers. They also showed that CPAP administered during cardiopulmonary bypass also prevented the development of atelectasis and shunt [215]. This group also evaluated the effects of prenalterol, a selective β1-adrenoceptor agonist, on pulmonary gas exchange measured with MIGET in patients recovering from open-heart surgery for aortic or mitral valve replacement [14]. Prenalterol increased cardiac output without an effect on pulmonary arterial or wedge pressure, and gas exchange was worsened because of an increase in the amount of shunt measured by MIGET.

Hedenstierna et al., [138] evaluated pulmonary gas exchange in patients undergoing vascular reconstructive surgery for peripheral vascular disease. Unsurprisingly, because of the nature of this disease process, nine of the ten patients were either current or former smokers, and four patients had findings on spirometry consistent with COPD. This was also reflected in the $\dot{V}_A/\dot{Q}$ distributions recovered from these subjects at baseline before anesthesia, and several had bimodal distributions similar to that reported for patients with COPD. By 20 min of anesthesia, there was an increase in $\dot{V}_A/\dot{Q}$ mismatch as measured by the LogSD,Q and the development of both high and low $\dot{V}_A/\dot{Q}$ modes. This pattern of gas exchange generally persisted for the duration of anesthesia; however, two patients developed progressive shunt with increasing duration. Valentine et al., in 1994 [379], compared the gas exchange consequences of several modes of ventilator support (synchronized intermittent mandatory ventilation, pressure support ventilation, and airway pressure release ventilation) in patients just before weaning from the ventilator after cardiac surgery. While there were specific differences in gas exchange, they were relatively small, and the authors concluded that all three methods were adequate for gas exchange under these specific conditions.

7.6.6 *Effects Associated with Laparoscopic Surgery*

As part of laparoscopic surgery, the abdomen is inflated with carbon dioxide. Besides predictable effects on pH and arterial CO_2 because of absorption across the peritoneum, pneumoperitoneum also has been suggested to transiently increase P_{O_2}. To investigate the mechanism of this, Andersson et al., [9] evaluated pulmonary gas exchange in patients with normal cardiopulmonary function undergoing laparoscopic cholecystectomy. Pneumoperitoneum decreased shunt a small amount from ~6% to ~4% but this effect was transient. In pigs [347], pneumoperitoneum with CO_2 resulted in redistribution of blood flow away from collapsed regions and improvement of gas exchange because of reduction of shunt. The amount of atelectasis induced by pneumoperitoneum was increased as the inflating pressure increased; however because pulmonary blood flow was redistributed away from the collapsed regions, gas exchange (reduction of shunt) was also improved at higher inflating pressures [346]. Strang et al., [345] also showed that pneumoperitoneum with CO_2 but not air improved gas exchange by reducing shunt and shifting blood flow to better ventilated lung regions. This effect was blunted by nitroprusside suggesting that CO_2 was enhancing the hypoxic pulmonary vasoconstrictive response.

Loeckinger et al., [204] also studied, in pigs under general anesthesia, the effects of CO_2 pneumoperitoneum on gas exchange because of its use in laparoscopic surgery. They found that gas exchange defects associated with pneumoperitoneum could be mitigated by use of PEEP. In 2001, Kleinsasser et al., [190] compared sevoflurane to isoflurane and propofol anesthesia in pigs with an induced pneumoperitoneum and found that sevoflurane caused greater gas exchange disturbances than either of the other two agents, likely because of the high alveolar sevoflurane concentrations needed.

7.7 In Critical Care and Mechanical Ventilation

7.7.1 Acute Lung Injury

If ever there was a situation in which MIGET could play a meaningful clinical role, it is in the intensive care unit. Lung function is often greatly disturbed; extrapulmonary factors (metabolic rate, temperature, acid/base status, Hb concentration, cardiac output, inspired O_2 concentration) can play major, independent roles in arterial oxygenation, and especially when any of these factors change (quickly)—as is common in the ICU—it becomes difficult to determine if changes in oxygenation in a patient reflect changes in lung function per se or effects of changes in one or more these extrapulmonary factors. Within the lungs and chest cavity, rapid events are not unusual—pneumothorax, bleeding, pleural effusion, edema, consolidation, and secretions—and these may also cause sudden changes in arterial oxygenation. MIGET has the capacity to identify intrapulmonary changes in terms of altered $\dot{V}_A/\dot{Q}$ distribution and shunt and clearly separate them from extrapulmonary influences on arterial oxygenation. Thus, if arterial P_{O_2} in a patient changes, MIGET can answer the question of why the P_{O_2} has changed and quantitatively apportion the contributions of any factors to such changes. The problem of course with MIGET in the ICU is that it is a complex procedure and, at best, results would take about an hour and a half to produce. Rodriguez-Roisin in 1994 [302] published a nice summary of MIGET findings in this setting, to which the reader is referred. Another review of gas exchange in ARDS (in Spanish) was published in 2003 [284]. An important technical issue was addressed by Ferrer et al., [92] in 1998, namely, evaluating the gas compression effect of including the standard 10 l expired gas mixing box used in MIGET in the ventilator circuit, something that could greatly reduce delivered tidal volume. Replacing this box with a smaller 1 l mixing chamber essentially eliminated the problem and was shown not to affect the MIGET methodology.

Not surprisingly, a large number of research studies have been done in critically ill patients. Some were directed at the basic pathophysiology of acute lung injury, others at evaluating the effects of proposed therapeutic interventions. For the purposes of this section, we will make no distinction between the various terms used to describe critically ill patients in the ICU—terms including acute lung injury, acute respiratory failure, and ARDS—with or without additional sepsis and without subdivision by cause (surgery, trauma, infection, etc.). Doing so would substantially expand the size of this section, and we would refer the reader to the original publications for patient description. A review by Rodriguez-Roisin [302] may also be helpful to the reader. As it is and despite very short summaries of each study in the following, the sheer amount of information presented is hard to grasp. We have therefore made an attempt to synthesize and summarize common findings across these many studies at the end of this section.

7.7.2 *Basic Findings in the Critically Ill: Role of PEEP and Other Ventilator Strategies*

In 1974 the, at that time new, MIGET technique was used to study respiratory failure in a canine model [410] with three gas exchange lesions, pulmonary embolism, oleic acid pulmonary edema, and pneumococcal pneumonia, which demonstrated that MIGET was capable of distinguishing different gas exchange patterns from these three different lung insults. The first study in humans was in 1979. Here, Dantzker et al., [53] found about 50% of the cardiac output flowing through regions of zero or near-zero $\dot{V}_A/\dot{Q}$ ratio, without other differences from normal subjects (Fig. 7.12), and that the measured arterial P_{O_2} was fully explained by these findings, showing that diffusion limitation of O_2 uptake was not occurring. The addition of positive end-expiratory pressure (PEEP) reduced shunt and increased ventilation of unperfused regions (deadspace). Whether shunt reduction was due to gas volume recruitment restoring ventilation of previously unventilated alveoli or due to reduction in perfusion of still unventilated alveoli could not be determined. Logic would dictate the former, as PEEP would first expand the well-ventilated alveoli, thus squeezing capillaries (Zone 1) and reducing their perfusion, which would be redirected to unventilated alveoli that did not "see" the PEEP. This would, if anything, increase the shunt. However, in rabbit lungs, Lamm et al., [199] showed that regions of the lung in Zone I could have perfusion through alveolar corner vessels and that these regions participated somewhat in gas exchange and

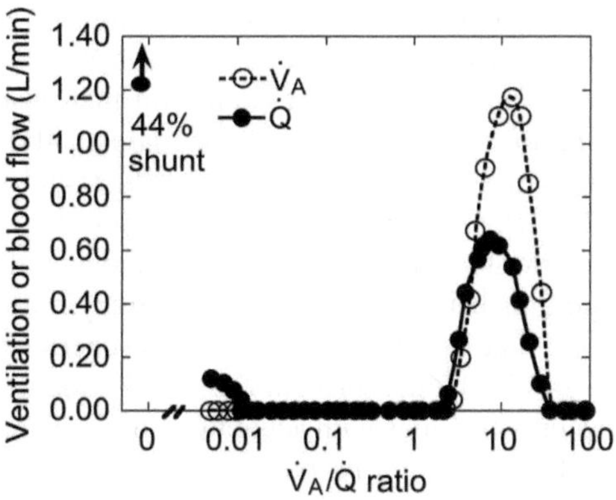

Fig. 7.12 Ventilation-perfusion relationships in a patient with ARDS. There are both large regions of shunt and regions of the lung that have a normal $\dot{V}_A/\dot{Q}$ ratio. Redrawn from [53]

thus may contribute a high $\dot{V}_A/\dot{Q}$ mode. Another possible mechanism may have been the reduction in cardiac output caused by PEEP, which in itself could alter the perfusion distribution so as to reduce flow through shunting regions. Matamis et al., [223] addressed Dantzker's findings by preventing the usual PEEP-induced fall in cardiac output with dopamine and found that PEEP still reduced shunt in patients with respiratory failure.

Several years later, Bein et al., [29] showed similar, complex $\dot{V}_A/\dot{Q}$ relationships with large shunts and areas of low and of high $\dot{V}_A/\dot{Q}$ ratio in lung injury. They found that the extent of $\dot{V}_A/\dot{Q}$ mismatch correlated with the Murray score of clinical severity. Studies in pigs have shown that the extent of $\dot{V}_A/\dot{Q}$ mismatch and shunts correlated with measures of extravascular lung water measured by thermodilution [125]. In 1985, Ralph et al., [286] studied the effects of PEEP in ARDS patients with preserved cardiac function and found, as did Dantzker et al., $\dot{V}_A/\dot{Q}$ patterns characterized by large shunts and areas of low $\dot{V}_A/\dot{Q}$ ratio that were improved with PEEP with little change in cardiac output. They also found that PEEP induced the development of high $\dot{V}_A/\dot{Q}$ areas and/or increased deadspace, suggestive of the creation of Zone 1 conditions in the most expanded ("normal") alveoli. If this mechanism is true, it may portend impending barotrauma.

Coffey et al., [46] induced permeability edema with oleic acid infusion in dogs and found essentially the same outcomes as Dantzker showed above. Batchinsky et al., [23] studied chlorine gas inhalation in sheep and found substantial, dose-related $\dot{V}_A/\dot{Q}$ inequality characterized by transient appearance of low $\dot{V}_A/\dot{Q}$ regions followed by shunt development. The same group used a lung contusion approach by blunt force trauma in pigs and found the same disturbances [24]. Bickenbach et al., [31] used saline lavage in pigs to induce lung injury and also note low $\dot{V}_A/\dot{Q}$ regions and shunt, responsive to PEEP. Rossaint et al., [311] studied patients with severe ARDS receiving extracorporeal circulatory support in order to vary mixed venous P_{O_2} independently of other factors, especially cardiac output. They found unpredictable changes in shunt, attesting to, if nothing else, the complexity of gas exchange in the critically ill. Breen et al., [38] showed in dog oleic acid lung injury and pulmonary edema that worsening P_{O_2} with increasing cardiac outputs was not due to diffusion limitation. Neumann et al., [243] compared PEEP with inverse ratio ventilation (i.e., shorter expiration than inspiration times) and found PEEP to be superior in mitigating $\dot{V}_A/\dot{Q}$ inequality and shunt in oleic acid-induced lung injury in pigs. In 2004, Henzler et al., [145] found that biphasic positive airway pressure (BIPAP) ventilation in pigs with acute lung injury produced by saline lavage was superior to conventional ventilation in that there was less cardiovascular depression and thus better maintained O_2 delivery. Subsequently, they reported that pressure-controlled assisted ventilation was superior to BIPAP in the same pig model [146].

In 1994, Putensen et al., [278, 280] found that allowing spontaneous breathing to occur while dogs with acute lung injury were mechanically ventilated improved gas exchange. Similarly, Neumann et al., [246] in 2005 found that in pigs with acute lung injury, addition of spontaneous breathing to mechanical ventilation (using airway pressure release ventilation) improved both shunt and low $\dot{V}_A/\dot{Q}$ regions. Vimlati et al., [381] reported similar findings in a piglet model of atelectasis. Gas exchange was improved further over intermittent airway pressure release if continuous positive airway pressure (CPAP) was applied, likely because of rapid lung collapse during the phases of airway pressure release below the level of CPAP [245].

Hartmann et al., [124] in 2012 found that in acute lung injury in pigs, maintaining a high respiratory frequency provided gas exchange benefits similar to those of PEEP, at least for a while. However, with time, gas exchange deteriorated compared to PEEP. In isolated perfused rabbit lungs, Hermle et al., [147] showed that the lower the PEEP level, the greater the shunt, but the higher the PEEP level, the greater the extent of high $\dot{V}_A/\dot{Q}$ areas reflecting reduced perfusion from alveolar hyperinflation. Prior to this, Kallas et al., [185] had shown that PEEP redistributed blood flow from nondependent to dependent lung regions, which would account for both the development and location of the high $\dot{V}_A/\dot{Q}$ areas in the nondependent regions. Schweiger et al., [333] combined chest wall injury with oleic acid lung injury in pigs and found lung injury raised shunt more than just chest wall injury, with positive pressure ventilation improving gas exchange in all settings.

Meconium aspiration in newborns causes hypoxemia, an acute rise in pulmonary arterial pressure, and respiratory failure. In a neonatal animal model, Truog and colleagues [372] evaluated the effect of PEEP compared to tolazoline, which is a pulmonary vasodilator, in experimental meconium aspiration. Early application of PEEP reduced hypoxemia by decreasing shunt, without depressing pulmonary blood flow, whereas tolazoline had no effect on gas exchange but reduced systemic blood pressure in some animals. This led the authors to suggest that consideration be given to the early application of PEEP in this condition.

Reducing tidal volume to avert barotrauma has become a popular ventilator strategy, although arterial P_{O_2} will fall and P_{CO_2} rise. Pfeiffer et al., [262] studied permissive hypercapnia in patients with ARDS and found an increase in shunt and a higher cardiac output, thus raising mixed venous P_{O_2}, such that arterial P_{O_2} was essentially unchanged. Feihl et al., [91] compared this approach to conventional ventilation and found that shunt increased dramatically. Thus the gas exchange consequences of permissive hypercapnia are far more than just those expected from hypoventilation. Despite a substantial rise in cardiac output, arterial P_{O_2} fell. By using dobutamine to control for this rise in cardiac output, they further showed that the changes in gas exchange were the result of both reduced tidal volume and higher cardiac output. Reducing tidal volume not only increases P_{CO_2} but also reduces lung expansion. Sinclair et al., [339] sought to determine which if either consequence of low tidal volume was more beneficial. In rabbits with acute lung injury from saline lavage, similar hypercapnia was produced by either low tidal volume or inhaled CO_2. There appeared to be little difference in gas exchange between these two approaches. Ziebart et al., [427] compared pressure support ventilation to volume-controlled ventilation in pigs given lipopolysaccharide. In both cases, a low tidal volume strategy was used and found no benefit to pressure-controlled ventilation.

Finally, it should be remembered that mechanical ventilation alone in normal animals is not without deleterious effects on gas exchange. Simonson et al., [337] assessed gas exchange serially over 4 days in normal, ventilated baboons given room air to breathe. They found both shunt and $\dot{V}_A/\dot{Q}$ inequality developing over time, without substantial lung structural or inflammatory changes.

7.7.3 *Early Changes in Gas Exchange*

Sorensen et al., [341] investigated the basis for hypoxemia in neonatal streptococcal sepsis in pigs and found $\dot{V}_A/\dot{Q}$ inequality developed without shunting; moreover, $\dot{V}_A/\dot{Q}$ relationships reverted toward normal after streptococcal administration was discontinued. Similarly, Forsgren et al., [93] found reversible $\dot{V}_A/\dot{Q}$ inequality without shunt in pigs given endotoxin, and Huttemeier et al. also confirmed this finding in sheep [179]. This suggests an initial mechanism based on vasoactivity rather than structural damage such as edema. Truog et al., [370] then pharmacologically inhibited thromboxane synthase in neonatal pigs given streptococcus infusion. This mitigated pulmonary arterial hypertension, but had no effect on arterial P_{O_2} or the abnormal $\dot{V}_A/\dot{Q}$ relationships induced by streptococcus. Sometime later, Huang et al., [175] induced acute lung injury in baboons and followed $\dot{V}_A/\dot{Q}$ relationships over time. The first changes were of extensive $\dot{V}_A/\dot{Q}$ inequality, with shunts developing only later, also consistent with the idea that initially, $\dot{V}_A/\dot{Q}$ changes may reflect functional alterations in airway or vascular tone before structural factors develop to shape the course of lung damage. Batchinsky et al., [22] used a blunt force trauma model of lung contusion in pigs. Early after injury, shunt was increased and areas of low $\dot{V}_A/\dot{Q}$ were noted. Over the next 2 h, shunt increased as $\dot{V}_A/\dot{Q}$ inequality subsided.

7.7.4 *Body Position and Gas Exchange*

Albert et al., [6] found in dogs that in acute lung injury produced by oleic acid infusion, the prone position led to better $\dot{V}_A/\dot{Q}$ relationships than when supine, starting a therapeutic movement in ICU patient positioning. This was followed up by Pappert et al., [260] in 12 patients with ARDS. Eight of them improved oxygenation due to reduced shunting while in the prone position, but four did not. Presumably, whether a patient benefits from prone positioning to some extent depends on the location of the lung damage with respect to gravitational axis, but this is unexplored. Bein et al., [28] used continual positional rotation and found that only those patients with mild to moderate lung injury benefited in terms of $\dot{V}_A/\dot{Q}$ relationships.

7.7.5 *Alternative Ventilator Strategies*

7.7.5.1 Constant Flow Ventilation

Between 1987 and 1989, Schumacker and colleagues studied gas exchange during constant flow ventilation in apneic, anesthetized, normal dogs [329–331]. They found that this modality caused significant but reversible $\dot{V}_A/\dot{Q}$ inequality (appearance of low $\dot{V}_A/\dot{Q}$ regions) and reduction in arterial P_{O_2}, but did not result in increased shunt compared to conventional positive pressure ventilation. They found evidence that individual lobes behaved similarly to the whole lung.

7.7.5.2 High-Frequency Ventilation

The physiology of high-frequency ventilation has been reviewed in [389]. McEvoy et al., [226] studied high-frequency oscillatory ventilation (HFOV), also in normal dogs, and found what at first sight appeared to be substantial ventilation associated with very high $\dot{V}_A/\dot{Q}$ ratio regions. This was however found to represent enhanced excretion of the most soluble inert gas used in MIGET (acetone) relative to that of the next most soluble gas (ether), explained by dissolution/evaporation kinetics in the conducting airways. Low $\dot{V}_A/\dot{Q}$ regions were not different between HFOV and conventional ventilation. In 1985, Truog and Standaert [373] repeated this work in normal neonatal lambs and also found no significant difference in gas exchange between HFOV and conventional ventilation in terms of low $\dot{V}_A/\dot{Q}$ regions and shunt, but did not address high $\dot{V}_A/\dot{Q}$ ratio development. Then Dembinski et al., in 2002 [64] studied pigs with acute lung injury induced by oleic acid infusion and also found similar gas exchange patterns comparing conventional ventilation to HFOV. However, they noted that to achieve these outcomes, mean airway pressure had to be raised considerably during HFOV. This was also confirmed by Kaiser et al., in a canine model [184].

Hastings and Powell [127] used MIGET during high-frequency ventilation to show that gas exchange can occur by mechanisms other than simple bulk convection in the parabronchial lung-air sac system of birds as well as in the alveolar lungs of mammals. Also by evaluating lung structures other than the mammalian lung, they were able to provide additional information clarifying the proposed mechanism of gas exchange during high-frequency ventilation. The data suggested that because of differences in lung structure, ventilation of proximal lung structures was not required, nor was asynchronous ventilation of lung regions, resulting in pendelluft.

7.7.5.3 Partial Liquid Ventilation

Partial liquid ventilation (PLV) with perfluorocarbon emulsion has been considered as a unique therapeutic strategy. Lim et al., in 2001 [202], created acute lung injury by saline lavage in rabbits and used MIGET to assess gas exchange using PLV. They found considerably reduced shunt with correspondingly increased perfusion of low $\dot{V}_A/\dot{Q}$ regions, suggesting opening of alveoli as the air/liquid interface and thus surface tension, was eliminated, but restoration of ventilation must have been minimal. Whether MIGET operates as usual in the presence of large volumes of perfluorocarbon liquid in the airways is open for discussion. Certainly, the solubility of some, maybe all, of the gases used in MIGET might have been materially changed. A variant of PLV is to deliver vaporized perfluorocarbon. Hubler et al., [177] did this in sheep with oleic acid-induced acute lung injury and found improvement in shunt and arterial P_{O_2}. However, $\dot{V}_A/\dot{Q}$ relationships per se worsened.

7.7.5.4 Other Considerations

In dogs, Barie et al., [21] asked whether there was an interaction between toxic smoke components (specifically HCl and CO) such that CO might exert an effect separate from that of acid alone. Indeed, the two agents together accelerated development of $\dot{V}_A/\dot{Q}$ inequality, although the mechanism of their interaction is not clear. In addition, the rate of CO uptake and its subsequent elimination were reduced due to the $\dot{V}_A/\dot{Q}$ mismatch, as expected on basic principles of gas exchange.

Tornabene et al., [364] evaluated the effect of fat embolism on pulmonary gas exchange in dogs. Fat injection increases pulmonary arterial pressure and pulmonary vascular resistance and causes the development of regions of high $\dot{V}_A/\dot{Q}$ ratio with gradual resolution over the 2-h period of the study. There was no development of shunt or regions of low $\dot{V}_A/\dot{Q}$ ratio.

Yamaya et al., [424] used MIGET to study the effects of IV injection of Optison, a contrast agent used in myocardial echocardiography in order to evaluate the potential for the development of a gas exchange pattern consistent with microembolism. There were no effects on gas exchange seen at any dose, but there was evidence or a transient increase in pulmonary arterial pressure in animals that were otherwise healthy but not in those that were given IV injection of polystyrene beads creating pulmonary hypertension.

Neumann et al., [244] asked if lung injury induced with oleic acid injection, endotoxin infusion, or repeated lung lavage produced similar gas exchange patterns in a pig model of acute lung injury. Gas exchange was less impaired in the endotoxin group, because shunt was the sole defect, whereas in oleic acid injection and lung lavage-induced injury, increased $\dot{V}_A/\dot{Q}$ mismatch was also observed. Oxygen diffusion limitation was not observed. However since a specific pattern of $\dot{V}_A/\dot{Q}$ mismatch was not seen, the authors concluded that a recovered $\dot{V}_A/\dot{Q}$ distribution could not be used to diagnose a specific type of lung injury.

Using dogs with oleic acid-induced acute lung injury, Putensen et al., [281] asked if inhaled isoflurane used for sedation in ventilated patients affected their gas exchange. They found that isoflurane reduced cardiac output and worsened $\dot{V}_A/\dot{Q}$ relationships, shunt, and oxygenation in a dose-related manner.

Tusman et al., [376] found that the slope of phase III (i.e., the alveolar plateau) of the expired CO_2 tracing was closely correlated with the amount of $\dot{V}_A/\dot{Q}$ inequality measured by MIGET in pigs with acute lung injury from saline lavage. This is as expected from basic physiological principles, but what must be taken into account is that only ventilated alveoli can contribute to the slope of phase III, and so no information about the amount of shunt can be gained.

Radermacher et al., [282] used MIGET to show that a simple with a compartment representing fractional blood flow to a perfused and open gas-filled but non-ventilated compartment (i.e., shunt) and a normal compartment calculated from the alveolar-arterial difference for nitrogen ($aADN_2$) compared favorably with data from MIGET. They suggested that the measurement of the $aADN_2$ and the calculation of the two-compartment lung with one ventilated and perfused compartment and one shunt compartment allowed the estimation of the contribution of low $\dot{V}_A/\dot{Q}$ areas and true right-to-left shunt to arterial hypoxemia.

7.7.6 Systemically Administered Vasodilators

Melot et al., [235] found that reduction in pulmonary artery tone and vascular pressures (by the drug diltiazem) *worsened* gas exchange in ARDS patients by increasing shunt. Two years later, using the vasodilator PGE_1, these authors [231] found the same outcome—reduced pulmonary artery pressure but *worsened* $\dot{V}_A/\dot{Q}$ relationships and gas exchange in ARDS patients. Radermacher et al., [283] gave ARDS patients prostacyclin (PGI_2) with the same results (reduced pulmonary artery pressure *worsened* $\dot{V}_A/\dot{Q}$ relationships, and higher shunt), but without reduction in arterial P_{O_2} (because of an increase in cardiac output raising mixed venous P_{O_2}). Reyes et al., [290] increased pulmonary vascular tone in ARDS patients with the drug almitrine and found reduction in shunt and *improved* gas exchange. Rennotte et al., [289] studied the individual effects of infused dopamine and dobutamine on gas exchange in mechanically ventilated patients and found that both drugs raised cardiac output and worsened gas exchange by raising the amounts of shunted blood and increasing perfusion of low $\dot{V}_A/\dot{Q}$ areas. Dopamine had the larger effect. Presumably, these studies show, using systemically active vasodilators, that hypoxic pulmonary vasoconstriction in shunting blood vessels is active and partially limits the size of the shunt in patients with ARDS. In contrast, Ukita et al., [378] found no effect of almitrine on gas exchange in dogs with paraquat-induced lung injury.

7.7.7 Inhaled Vasodilators

Ogura et al., [253] studied inhaled NO in pigs given lipopolysaccharide (LPS) intravenously. LPS caused severe $\dot{V}_A/\dot{Q}$ inequality and shunt, and these were *improved* by inhaled NO. This outcome therefore is opposite to those of Melot and Radermacher above, but there is a key difference, inhaled versus systemic dosing of the vasodilator, and this may well explain the opposite findings. Indeed, this is supported by the studies of Putensen et al., [277] and Ogura et al., [255] who used inhaled NO in oleic acid-induced lung injury in dogs and in pigs given LPS, respectively. In both studies, $\dot{V}_A/\dot{Q}$ relationships were improved by inhaled NO, attributed to NO reaching the alveoli of the better ventilated alveoli, increasing their perfusion, and thus reducing shunt. Putensen et al., [279] in 1994 found that in dogs with acute lung injury, inhaled NO was more beneficial to gas exchange when CPAP ventilation was used, reasoning wider distribution of NO due to alveolar recruitment by CPAP. Hermle et al., [148] found that inhaled NO attenuated pulmonary hypertension, $\dot{V}_A/\dot{Q}$ inequality, and shunt after ischemia/reperfusion injury of excised perfused rabbit lungs. The same group, using the same model, also found that pretreatment with aerosolized vasodilators (nitroprusside, PGE_1, and iloprost) each largely prevented deleterious effects of ischemia/reperfusion [207]. Similarly, in pigs with acute lung injury from lavage, Dembinski et al., [66] found greatly improved $\dot{V}_A/\dot{Q}$ relationships and reduced shunt after inhaled NO (given together with infused almitrine). They also found improvement using the combination of inhaled prostacyclin and infused almitrine in the same pig model [65]. Trachsel et al., [367] found that the response to inhaled nitric oxide was blocked by endothelin receptor antagonists. Maurenbrecher et al., [224] studied NO inhalation in pigs given endotoxin to produce acute lung injury and confirmed improvement in oxygenation, $\dot{V}_A/\dot{Q}$ inequality, and shunt acutely. However, it appeared that after some hours, NO inhalation lost its effectiveness. In 2002,

Sticher et al., [344] examined the dependence of shunt flow on inhaled NO levels during one-lung ventilation in piglets and found that lower inhaled NO concentrations of 4 ppm were associated with less shunting than higher doses of the gas up to 32 ppm. Schermuly et al., [325] studied isolated rabbit lungs with acute lung injury produced by infusion of a thromboxane mimetic and found mitigation of $\dot{V}_A/\dot{Q}$ relationships, edema, and shunt when inhaled iloprost (PGI_2) was given together with infused zardaverine (phosphodiesterase inhibitor). Walmrath et al., [416] studied the effects of aerosolized prostaglandin I_2 in mechanically ventilated patients with community-acquired pneumonia and compared those without preexisting lung disease to those with interstitial fibrosis. Ventilation-perfusion distribution was assessed by the multiple inert gas elimination technique. Low doses of aerosolized prostaglandin had beneficial effects in those with previously normal lung, decreasing pulmonary arterial pressure and improving gas exchange by decreasing shunt. In the patients with pulmonary fibrosis, the drug was also effective, but a threefold increase in dose was required.

Shermudy et al., [326] also studied the effect of stabilization of the second-messenger cAMP by phosphodiesterase inhibition on the increase of the vasodilator response to nebulized prostacyclin, a selective pulmonary vasodilator, in perfused rabbit lungs. Drugs evaluated included the phosphodiesterase 3 inhibitor motapizone, the phosphodiesterase 4 inhibitor rolipram, and the dual-selective phosphodiesterase 3/4 inhibitor tolafentrine. The drugs that were most effective in increasing and prolonging the effect of prostacyclin were rolipram plus motapizone, and they reduced lung edema formation and there were substantial decreases in shunt. This suggests that this approach could be used for selective pulmonary vasodilation in respiratory failure and pulmonary hypertension to reduce pulmonary vascular pressures and maintain gas exchange.

Recently, Grubb et al., [105, 106] studied gas exchange in isoflurane-anesthetized horses and examined the effects of inhaled nitric oxide pulses on gas exchange. Inhaled NO was found to reduce both $\dot{V}_A/\dot{Q}$ inequality and shunt and thus improve arterial oxygenation. They also assessed regional lung perfusion scintigraphically and found that inhaled NO redistributed pulmonary blood flow from dependent, atelectatic regions to nondependent aerated regions to explain these findings.

7.7.8 Mediators and Gas Exchange

Saba et al., in 1983 [318], showed that fibronectin replacement by plasma cryoprecipitate infusion improved $\dot{V}_A/\dot{Q}$ relationships both in patients and sheep with sepsis. Wu et al., [421] studied tracheal acid aspiration in dogs and found that inhibition of cyclooxygenase, but not of thromboxane synthetase, mitigated the development of post-aspiration $\dot{V}_A/\dot{Q}$ inequality and arterial oxygenation. In blood-free perfused rabbit lungs, Walmrath et al., [414] studied intravenously infused staphylococcal alpha-toxin and found extensive $\dot{V}_A/\dot{Q}$ inequality and pulmonary hypertension mitigated by either aspirin or thromboxane block. These authors repeated the study using LPS with similar findings [413]. Hartmann et al., [123] studied acute lung injury using oleic acid in pigs asking whether an inhaled TIP peptide (related to the lectin-like domain of TNF-alpha) would mitigate the gas exchange consequences of the injury. The answer was no. Treml et al., [368] gave angiotensin-converting enzyme 2 to piglets with acute lung injury from lipopolysaccharide infusion and found considerable mitigation of the lung gas exchange disturbances.

7.7.9 Summary

The preceding recounts a huge amount of information, but there are some overarching points that might be made.

Studies vary in many ways no matter what primary question is being asked:

- By the animal species chosen or human disease phenotype
- By the experimental model of lung injury (animal) or etiology (human)
- By when in the course of disease studies were done
- By supportive management choices (ventilator parameters, F_{IO_2}, drugs, etc.)

Acute lung injury in humans and experimental animals, no matter how the injury was generated, shares common pathophysiological features evident when MIGET is used. These include the universal development of $\dot{V}_A/\dot{Q}$ inequality with low $\dot{V}_A/\dot{Q}$ regions dominating, especially early in the course of disease. In many studies, early $\dot{V}_A/\dot{Q}$ inequality is reversible if the injury is discontinued, but if the insult is maintained, shunts develop and increase over time, while low $\dot{V}_A/\dot{Q}$ regions may diminish (likely representing conversion of low $\dot{V}_A/\dot{Q}$ alveoli to unventilated alveoli causing shunting). Overall, this suggests that the earliest changes are functional and related to airway and/or vascular tone changes, but over time there are structural changes that worsen gas exchange and resist treatment. These reports also clearly show that when in the course of disease studies are done is important and may affect outcome of specific interventions.

The dominant gas exchange abnormality in both humans and animal models of established acute lung injury is clearly agreed to be shunt. In some studies, areas of high $\dot{V}_A/\dot{Q}$ are seen as well. This may reflect high alveolar gas pressures in the least damaged and thus best ventilated regions, high pressures caused by the ventilator strategy. We suggest this may be a useful early indicator of possible barotrauma. This also implies redistribution of blood flow from the healthier alveoli to those that are poorly ventilated or unventilated. Thus, titrating ventilator protocols to minimize development of high $\dot{V}_A/\dot{Q}$ regions is worthy of consideration.

That said, raising alveolar gas pressures by any of several ventilator strategies including PEEP clearly improves gas exchange, likely by expanding collapsed alveoli, thus reducing shunts. Other strategies such as blocking various inflammatory mediators, providing exogenous surfactant, and minimizing oxidative injury all show experimental benefits short term. How these may affect morbidity long term and mortality is not known.

Strikingly, vasodilators have profound effects on gas exchange in acute lung injury, and these are in fact opposite depending on whether the vasodilator is infused intravenously (bad) or inhaled (good). Note that given by either route, they reduce pulmonary arterial pressures. However, infusion directs the vasodilator to all perfused vessels and may interfere with hypoxic pulmonary vasoconstriction that exists in low $\dot{V}_A/\dot{Q}$ regions, thus increasing their blood flow and worsening gas exchange. Inhaled vasodilators on the other hand best reach the better ventilated parts of the lung, increasing their blood flow and thus reducing flow through areas of low or zero $\dot{V}_A/\dot{Q}$ ratio.

Strategies that alter cardiac output (whether ventilator induced or pharmacologically based) affect not only the $\dot{V}_A/\dot{Q}$ distribution and shunt through direct mechanical and vasoactive means as above, but they additionally affect arterial oxygenation through the effects of altered cardiac output on mixed venous P_{O_2}. Thus, a fall in cardiac output with no change in $\dot{V}_A/\dot{Q}$ distribution will result in a fall in arterial P_{O_2} caused in turn by a fall in mixed venous P_{O_2}. This leads to complex interactions affecting gas exchange.

Examples include PEEP where (a) shunts are reduced by alveolar expansion (good) but (b) cardiac output is reduced by a fall in venous return (bad). These two effects are opposite in how they affect arterial P_{O_2}, so that the net result may be no significant change in P_{O_2}. Another example is with vasodilators where (a) perfusion of low $\dot{V}_A/\dot{Q}$ regions or shunt pathways is increased (bad) but (b) cardiac output is increased (good). Again, arterial oxygenation may in the end show little if any change despite large changes in total blood flow and in $\dot{V}_A/\dot{Q}$ distribution.

These examples underlie a major principle elucidated by studies using MIGET: arterial oxygenation depends not only on the nature and severity of intrapulmonary damage, but to a large extent on mixed venous and inspired P_{O_2}, as well as on other factors such as body temperature, acid/base status, and Hb concentration.

Finally, avoidance of barotrauma by use of permissive hypercapnia resulting from low tidal volume ventilation may improve mortality, but research with MIGET has shown that gas exchange is affected though worsening of $\dot{V}_A/\dot{Q}$ relationships and shunt.

7.7.10 In Lung Injury Arising from Smoke Inhalation

Smoke inhalation causes airway injury followed by pulmonary edema and infection. In sheep, Ogura et al., [254] showed that smoke inhalation resulted in $\dot{V}_A/\dot{Q}$ mismatch largely due to the development of regions of low $\dot{V}_A/\dot{Q}$ ratio. Shimatzu et al., [336] showed that the lung injury was such that the pattern of gas exchange worsened with the severity of exposure and the time of measurement after exposure up to 72 h. They also confirmed that the pattern of abnormality was the development of large regions of low $\dot{V}_A/\dot{Q}$ ratio, with the development of regions of deadspace and shunt not being consistent findings. Tasaki et al., [352] also showed in a sheep model that the extent of the smoke inhalation-induced lung injury was not increased in animals that also suffered from full-thickness burns despite lower plasma proteins and increased fluid retention, which might be expected in increased lung edema.

MIGET showed that Sulfo Lewis C, a blocker of selectins administered after smoke inhalation, reduced the severity of the injury improving gas exchange and implicated neutrophils in the development of the lung injury [354]. In animals treated with pentoxifylline, a drug though to increase microvascular flow, Ogura et al., [254] also showed that the gas exchange abnormalities were lessened and $\dot{V}_A/\dot{Q}$ mismatch was less, suggesting its use as therapeutic agent. Tasaki et al., [353] demonstrated that aerosolized inhaled heparin did not improve gas exchange; however heparin in combination with lisofylline (which is converted to pentoxifylline in the liver) was beneficial in reducing blood flow to low $\dot{V}_A/\dot{Q}$ regions and improving gas exchange.

7.7.11 Hyperoxia and Lung Damage: Surfactant Metabolism

Lemaire et al., [201] asked whether 100% O_2 led to alveolar collapse in patients with acute respiratory failure (as had been previously predicted theoretically and shown experimentally). They did not find significant effects of hyperoxia on shunt, but it is possible that the high tidal volumes used (12 ml/kg) prevented this and that at more normal tidal volumes, absorption atelectasis might have occurred. Huang et al., [174] sought to determine if freshly extracted porcine surfactant replacement in primate lungs subjected to long-term 100% O_2 breathing would prevent hyperoxic lung damage. The answer was that the severe $\dot{V}_A/\dot{Q}$ inequality that developed with 100% O_2 was not improved by that surfactant, but that the combination of PEEP and surfactant did mitigate damage. They then used an artificial (commercial) protein-free surfactant [176] and found improvement in shunt and $\dot{V}_A/\dot{Q}$ relationships. Next they used aerosolized recombinant human manganese superoxide dismutase in hyperoxic baboons and found considerable protection of gas exchange [338]. Yamaguchi et al., [422] evaluated the effects of short-term hyperoxia on $\dot{V}_A/\dot{Q}$ relationships in dogs given oleic acid and found that while $\dot{V}_A/\dot{Q}$ relationships worsened with hyperoxia in control animals, in oleic acid-treated dogs, $\dot{V}_A/\dot{Q}$ inequality actually improved. This may reflect conversion of low $\dot{V}_A/\dot{Q}$ alveoli into unventilated alveoli due to absorption atelectasis. Schermuly et al., [323] compared tracheal instillation versus nebulization of surfactant in a Tween 20 model of acute lung injury in isolated perfused rabbit lungs and found rescue only using the nebulized administration. The same group [324] created injury in rabbit lungs using a combination of aerosolized fibrinogen and thrombin and found the $\dot{V}_A/\dot{Q}$ inequality could be rescued by nebulized urokinase and bovine surfactant given together.

7.7.12 Following Head Injury

One study has evaluated the mechanism of gas exchange abnormalities in the delayed pulmonary dysfunction that occurs in some patients following head injury. Popp et al., [266] showed that there was a heterogeneous response. In some patients, pulmonary vascular resistance and wedge pressure were elevated, and these individuals had large shunts of over 15% of cardiac output measured by the oxygen technique. MIGET studies in two of these patients showed a bimodal $\dot{V}_A/\dot{Q}$ distribution with both a low $\dot{V}_A/\dot{Q}$ mode and a normal mode and increased shunt of ~6%, suggesting that either pulmonary edema, retained secretions, or atelectasis was responsible. Five other patients with no detectable shunt measured by the oxygen technique had normal $\dot{V}_A/\dot{Q}$ distributions.

7.7.13 In Sepsis

Several studies have used MIGET to evaluate the effects of sepsis on pulmonary gas exchange. Sorensen et al., [341] demonstrated the presence of regions of high $\dot{V}_A/\dot{Q}$ ratio leading to marked $\dot{V}_A/\dot{Q}$ mismatch in newborn piglets after the administration of group b streptococcus used to induce sepsis. This is not a universal finding, however, and some authors report the development of regions of low $\dot{V}_A/\dot{Q}$ ratio in a similar animal model [370]. Neither study reported shunting perhaps because of the short-term nature of the follow-up measurements. In 1996, Huang et al., [175] investigated the mechanism of hypoxemia that occurs during bacterial sepsis using MIGET. Using infusion of *E. coli* in baboons, it was shown that early gas exchange findings included marked $\dot{V}_A/\dot{Q}$ mismatch and a transient increase in deadspace, without evidence for diffusion limitation. Intrapulmonary shunt developed later as pulmonary edema and alveolar flooding occurred. Ogura et al., [253] showed similar findings in pigs treated with *E. coli* lipopolysaccharide with the development of shunt, regions of high $\dot{V}_A/\dot{Q}$ ratio, and marked $\dot{V}_A/\dot{Q}$ mismatch.

Truog et al., [370] demonstrated that the selective thromboxane A_2 synthase inhibitor, dazmegrel, improved pulmonary arterial pressures but did not improve the pattern of gas exchange in *E. coli* sepsis. This suggested that either the edema was too far advanced or that inhibition of thromboxane A_2 synthase resulted in an overactivity of prostacyclin and blocking hypoxic pulmonary vasoconstriction. The administration of inhaled nitric oxide [253] improved $\dot{V}_A/\dot{Q}$ matching by reducing the extent of pulmonary edema and shunt with redistribution of blood flow to high $\dot{V}_A/\dot{Q}$ regions resulting in more uniform matching. Also in the same animal model, these authors investigated the role of failure of hypoxic pulmonary vasoconstriction to mediate gas exchange in sepsis by using N-nitro-L-arginine methyl ester to block NO synthesis. They showed that treatment with N-nitro-L-arginine methyl ester did not improve gas exchange [255]. However, addition of inhaled NO in addition to the N-nitro-L-arginine methyl ester did improve gas exchange, suggesting that overproduction of NO inhibiting HPV was not the basis of the $\dot{V}_A/\dot{Q}$ mismatch in sepsis.

Walmrath et al., used staphylococcal alpha-toxin in isolated perfused rabbit lung and showed the development of a rapid pulmonary vasoconstrictor response, lung edema, and later development of regions of low $\dot{V}_A/\dot{Q}$ ratio. Pretreatment with acetylsalicylic acid (ASA) or the thromboxane inhibitor BM-13.505 or treatment with BM-13.505 after edema formation normalized the gas exchange abnormalities. In another study, Walmrath et al., [412] used low-dose *E. coli* hemolysin and lipopolysaccharides, alone and in combination in the same type of isolated rabbit lung model. Lipopolysaccharides alone did not change the pattern of gas exchange from control lungs. *E. coli* hemolysin alone caused a modest increase in pulmonary arterial pressure and $\dot{V}_A/\dot{Q}$ mismatch. However, the addition of *E. coli* hemolysin to lungs that previously received lipopolysaccharides resulted in a rapid increase in pulmonary arterial pressure, edema formation, and $\dot{V}_A/\dot{Q}$ mismatch with large regions of shunt. As above, the pulmonary arterial pressure response and $\dot{V}_A/\dot{Q}$ mismatch were ameliorated by pretreatment with ASA or BM-13.505, and administration of BM-13.505 after edema was established reversed the rise in pulmonary arterial pressure and shunt. The net result is that these studies suggest that both the pulmonary vasoconstrictor response and shunt are related to thromboxane generation induced by the bacterial toxins.

7.7.14 In CPR and Resuscitation from Shock

Four studies have explored factors relevant to cardiopulmonary resuscitation with MIGET. Thrush et al., [358] working with a swine model showed that CPR and ventricular fibrillation caused an increase in both shunt and deadspace. Shunt increased from 6% to 10% with CPR and was significantly worsened to almost 30% by the use of epinephrine during CPR. The use of methoxamine instead of epinephrine during CPR cut the increase in shunt in half. Deadspace was increased by about 5% from baseline with CPR and was similar with administration of methoxamine. However, epinephrine increased both deadspace (to about 50%) and regions of high $\dot{V}_A/\dot{Q}$ ratio (10–100). More recently Hartmann et al., [126] evaluated the effect of differing F_{IO_2} on the pattern of gas exchange during CPR. These animals were given a continuous infusion of epinephrine throughout the study. Similar to the previous study, CPR increased shunt over baseline conditions, and the increase in shunt was greater with increasing inspired oxygen concentrations averaging ~5% of cardiac output in normoxia, 10% with 70% O_2, and reaching almost 20% with 100% O_2 ventilation. Significant regions of low $\dot{V}_A/\dot{Q}$ ratio did not develop in any of the groups. There was also an increase in regions of high $\dot{V}_A/\dot{Q}$ ratio with CPR, but the extent of high $\dot{V}_A/\dot{Q}$ regions did not show a consistent pattern with F_{IO_2}. Electrical impedance tomography in these animals showed that there were changes in the distribution of ventilation in the ventral, presumably nondependent lung (although posture is not explicitly stated), consistent with hyper-inflation and atelectasis formation in the dorsal (dependent) lung regions. Loeckinger et al., [206] showed in a swine model that when spontaneous circulation returned after CPR with either epinephrine or vasopressin used as a pressor agent, pulmonary gas exchange was more efficient in the animals given vasopressin. This was manifest as more perfusion to regions of normal $\dot{V}_A/\dot{Q}$ ratio and a reduction in the extent of perfusion of regions of low $\dot{V}_A/\dot{Q}$ ratio. Finally Robinson et al., [296] showed that in hemorrhagic shock in dogs, the mean $\dot{V}_A/\dot{Q}$ ratio was increased consistent with a decrease in overall blood flow, but that heterogeneity about this new mean was unchanged. Following resuscitation $\dot{V}_A/\dot{Q}$ relationships were unchanged from the baseline pre-shock measurements.

7.8 In Comparative Physiology

One of the more interesting uses of MIGET has been in comparing gas exchange across different species. Generally speaking, animals with higher metabolic rates have more complex lungs with greater surface area, facilitating diffusion of oxygen [268]. However, the structure of these lungs varies widely, and the alveolar mammalian lung is only one of a variety of structures that enable the exchange of respiratory gases. For example, the lungs of amphibians and reptiles may include simple unicameral saclike structures, whereas in the bird the processes of ventilation and gas exchange are served by different structures.

7.8.1 Birds

The mammalian lung is more structurally uniform compared to other vertebrates [84], and the functions of ventilation and gas exchange are shared by the same structures: fresh air traverses the respiratory bronchioles before reaching the alveolar ducts and passing into the alveoli space where gas exchange takes place. Approximately 80% of the human lung is devoted to gas exchange, and only ~20% is devoted to solely to ventilation alone (anatomical deadspace). On the other hand, the bird respiratory system completely separates the functions of ventilation and gas exchange [84]. In birds, air sacs ventilate the small, constant volume gas-exchanging region called the parabronchi. These air sacs occupy approximately 90% of the total respiratory system volume with the gas-exchanging parabronchi comprising only 10% of the lung.

As might be expected from such different lung structures, models of gas exchange in birds and mammal also differ. The parabronchi in birds where gas exchange takes place are small parallel structures. These are ventilated with air from the trachea or air sacs and perfused along their entire length. This means that ventilation and perfusion can be considered as occurring at right angles and is modeled by a cross-current model [269, 270], which is thought to be more efficient than alveolar exchange.

MIGET has been used to evaluate pulmonary gas exchange in birds both at rest in ducks [32], geese [127, 270], and emus [328] and during exercise in emus [328]. Figure 7.13 shows $\dot{V}_A/\dot{Q}$ distributions in birds, mammals, and reptiles. Despite the differences in lung structure, the distributions in birds and mammals are remarkably similar, although in anesthetized geese there was a tendency toward a bimodal distribution with a separate high $\dot{V}_A/\dot{Q}$ mode in some animals for reasons that are unclear [271]. This high $\dot{V}_A/\dot{Q}$ mode was not seen in the emu studies [328]. In birds the values for LogSD,Q and LogSD,V_A are at the upper limit of the range for healthy humans. Also (as discussed below) reptiles with high metabolic scope also exhibit a similar degree of $\dot{V}_A/\dot{Q}$ matching. The logical conclusion is that most vertebrate lungs, particularly in animals capable of relatively high metabolic rates, have evolved with similar amounts of $\dot{V}_A/\dot{Q}$ mismatch despite structural differences. However by comparing MIGET which measures intrapulmonary shunt to oxygen measures of shunt, Bicker et al., [32] showed that birds have ~5% extrapulmonary shunt.

In contrast to mammals and reptiles, the sole study in birds showed that exercise does not increase $\dot{V}_A/\dot{Q}$ mismatch in exercising birds [328]. Interestingly the extent of $\dot{V}_A/\dot{Q}$ mismatch is also unaffected by hypoxia or by hypoxic exercise, perhaps explaining some of the bird's advantage during hypoxic exercise. In addition, birds have not been shown to have significant amounts of intrapulmonary shunt [32, 270, 271, 328] under any of the conditions studied.

7.8.2 Reptiles

Reptilian lung structure is very diverse, and lung structure in these animals is an interesting intermediate between birds and mammals. Reptilian lungs vary from simple saclike organs called unicameral lungs to those with a few chambers (paucicameral) and more complex multi-chambered multicameral lungs [261]. Generally speaking the more metabolically active animals such as varanid lizards have the relatively complex multicameral lungs, whereas animals such as the tegu lizard which have lower metabolic demands have less complex lungs such as a unicameral lung. In the tegu this lung has a large central lumen lined with gas-exchanging "falveoli," which are honeycomb structures in the wall of the organ [153].

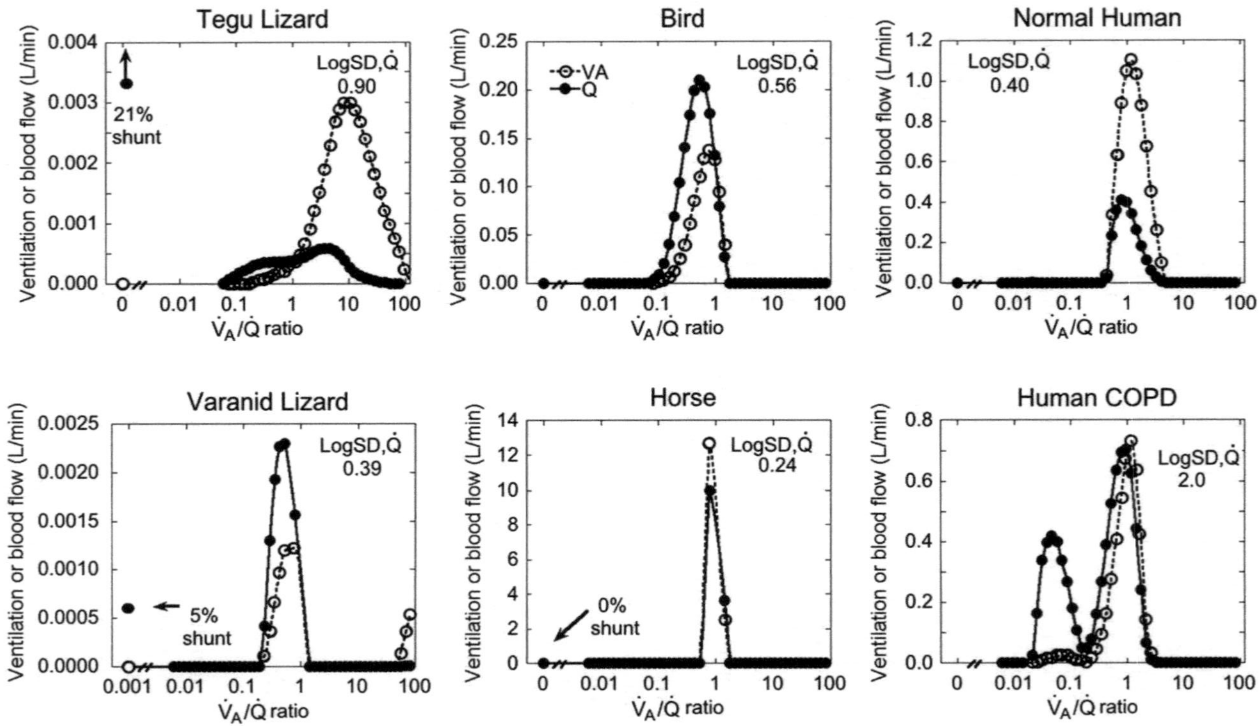

Fig. 7.13 Ventilation-perfusion relationships in two species of reptiles, a bird (emu), a horse, a healthy normal human and a patient with severe COPD. Across species the recovered $\dot{V}_A/\dot{Q}$ distributions are remarkably similar, albeit with varying shunt in birds and reptiles. Compare these to the patient with COPD. Data from [153, 156, 158, 160, 328, 397]

Given this variation in lung structure, it is perhaps surprising that the extent of $\dot{V}_A/\dot{Q}$ mismatch in these lungs of differing structure is rather similar. In Fig. 7.13 it can be seen that reptilian $\dot{V}_A/\dot{Q}$ distributions are, in general, unimodal and can be relatively narrow although $\dot{V}_A/\dot{Q}$ heterogeneity tends to be larger in the reptiles with less complicated lung structure such as the tegu and turtle. In the turtle, Hopkins et al., [166] showed that $\dot{V}_A/\dot{Q}$ mismatch is decreased by increasing pulmonary blood flow and awake alligators [268] show less $\dot{V}_A/\dot{Q}$ mismatch than anesthetized animals [267]. Since cardiac output was depressed in the anesthetized alligators, this may provide additional support for the effect of increased cardiac output improving $\dot{V}_A/\dot{Q}$ matching in some species. However, varanid lizards show an increase in $\dot{V}_A/\dot{Q}$ mismatch with exercise to a similar extent as humans.

Reptiles differ from birds and mammals in that there is a large variation in cardiac anatomy and there are significant shunts related to this structural variation. In the case of turtles, tegus, and varanid lizards, the shunts result from three chambered hearts whereby the animal has a common ventricle allowing mixing of oxygenated and deoxygenated blood. Crocodilians have four-chambered hearts but a double aortic arch and connections between left and right aortas that result in shunting [149]. In these animals, shunts reduce P_{aO_2} significantly, and in order to assess the effects of $\dot{V}_A/\dot{Q}$ mismatch on gas exchange, it is necessary to sample pulmonary venous blood from the left atrium instead of systemic arterial blood to assess pulmonary gas exchange.

In varanids, like humans, at rest the expired-left atrial P_{O_2} difference is explained almost completely by $\dot{V}_A/\dot{Q}$ mismatch (with an average LogSD,Q of 0.39) and an ~5% intrapulmonary shunt [160]. $\dot{V}_A/\dot{Q}$ mismatch is increased during exercise in varanids and the LogSD,Q averages 0.78, but the extent of intrapulmonary shunting was reduced to ~2% of cardiac output. There is no evidence for O_2 diffusion limitation in resting varanids, but during exercise approximately 40% of the expired-left atrial P_{O_2} difference is not explained by $\dot{V}_A/\dot{Q}$ inequality and therefore attributable to diffusion limitation. Interestingly there was evidence of incomplete intrapulmonary gas mixing at rest in varanid lizards, and the error between the measured retention and the best-fit retention for enflurane, the heaviest gas, and cyclopropane, the lightest gas, increased the residual sum of squares.

Turtles experience long periods of apnea, and during times when the lung is not ventilated, perfusion to the lungs is reduced by marked increases in pulmonary vascular resistance. Compared to varanids, turtles have a larger expired-arterial P_{O_2} difference, and this difference varies with ventilatory state and pulmonary blood flow. The expired-arterial P_{O_2} difference is explained by approximately equal contributions from $\dot{V}_A/\dot{Q}$ mismatch and a 17% intrapulmonary shunt [166]. Also, turtles may have some diffusion limitation for O_2 at rest, but this is difficult to model accurately because of the presence of multiple hemoglobins. Data to partition the expired-arterial P_{O_2} difference for alligators and tegu lizards are not available.

7.8.3 Mammals

For all the similarity in overall lung structure, MIGET studies have shown that there are some important differences in pulmonary gas exchange between mammalian species. At rest, studies in dogs [27, 73, 161, 349, 374, 375, 408], pigs [137, 165], and sheep [23, 264, 336] generally show smooth unimodal distributions with LogSD,Q and LogSD,V_A that are in the normal range for resting humans with minimal shunt (<1.0%). MIGET studies in anesthetized rabbits also showed relatively high LogSD,Q and LogSD,V_A (~ 0.80–1.05) [25, 57, 194, 195, 259, 339], shunt ranging from 0% to 3% of cardiac output. MIGET studies in resting anesthetized rats also show that these animals have greater $\dot{V}_A/\dot{Q}$ heterogeneity than other mammalian species with the LogSD,Q ~1.03 [7] and slightly greater intrapulmonary shunting (2–3%). However because of their small body size and the relatively large volume of blood required for MIGET sampling, these studies really push the limits of the technique and may induce physiological disturbances to the animal. Alternately it has been suggested that stratified inhomogeneity affecting high molecular weight gases may be distorting the $\dot{V}_A/\dot{Q}$ distributions recovered from MIGET in rats [152]. There was no evidence for diffusion limitation at rest in the normal mammalian lung in any of the studies.

There are several MIGET studies conducted in horses. The majority of these show values of LogSD,Q and LogSD,V_A that are the lowest of any species studied [140, 156, 220, 250, 334, 402] (range 0.24–0.40) with most reporting values at the low end of this range. There is no evidence of significant shunt in conscious adult animals, but large shunts are often seen during anesthesia in these animals [221, 248]. Work in foals by Stewart and colleagues [342, 343] showed that there were large (>30%) shunts of both intrapulmonary and extrapulmonary origin in neonatal animals that explained neonatal hypoxemia without evidence of regions of low $\dot{V}_A/\dot{Q}$ ratio. These were reduced with increasing age of the animal. As mentioned earlier, a case report in a giraffe [251] suggests that $\dot{V}_A/\dot{Q}$ matching under anesthesia in this species shows $\dot{V}_A/\dot{Q}$ distributions comparable to that of the horse (and other animals such as the rabbit and the rat) under anesthesia, but absence of shunt for reasons that are unclear.

Many of the differences in pulmonary gas exchange between species become more pronounced during exercise. Pulmonary gas exchange in exercising humans is impacted by both $\dot{V}_A/\dot{Q}$ heterogeneity and diffusion limitation, although the relative contribution varies between individuals and with aerobic capacity [154]. There is no gas exchange evidence for intrapulmonary shunting [101, 154, 155, 162, 258, 291]. In elite athletes, $\dot{V}_A/\dot{Q}$ heterogeneity may be responsible for 60% of the alveolar-arterial P_{O_2} difference, which can become as large as 40 Torr [154]. However, the individuals with the most marked hypoxemia typically have greater diffusion limitation [154, 291]. The horse is remarkable in that they are capable of a very high mass-specific $\dot{V}_{O_2max}$ but also experience substantial exercise-induced pulmonary limitations [334]. During exercise in the horse pulmonary diffusion limitation, combined with mechanical constraints on ventilation, causes marked arterial hypoxemia [402]. P_{aO_2} may fall to very low levels in these animals during heavy exercise [86, 87, 156, 334, 402]. Compared to other mammalian species, horses show little increase, if any, in $\dot{V}_A/\dot{Q}$ mismatch with exercise [87, 100, 156, 334, 402], and the slope of the relationship between mass-specific $\dot{V}_{O_2}$ and LogSD, Q is significantly less than humans [334]. However, horses have evidence for substantial contribution of diffusion limitation [87, 402] to the $AaDO_2$ which combined with alveolar hypoventilation explains the extent of hypoxemia in these animals [402]. The P_{aO_2} was improved and $AaDO_2$ was reduced in horses by breathing helium-oxygen mixtures [87] and this was explained by increased alveolar ventilation and less diffusion limitation, the latter presumably on the basis on increased P_{AO_2} induced by the former. However, the converse was not the case in hypoxic helium mixtures [86] largely because in hypoxia lung units were on the more linear portion of the dissociation curve. The one situation where horses developed marked increase in $\dot{V}_A/\dot{Q}$ mismatch with exercise was in animals that were suffering from red cell hypervolemia [100], where it was thought that the increased $\dot{V}_A/\dot{Q}$ mismatch might be related to bleeding into the alveolar space as a result of stress failure.

Dogs despite also having a high aerobic capacity, similar to that of horses, show only small increases in $\dot{V}_A/\dot{Q}$ inequality with minimal shunt and no diffusion limitation during exercise [169, 351]. On the other hand, pigs significantly increase $\dot{V}_A/\dot{Q}$ heterogeneity with exercise but show no appreciable diffusion limitation [165].

MIGET has been used to evaluate horses with lung disease. In horses with chronic bronchiolitis [249], minute ventilation was elevated, P_{aO_2} was reduced, and the $\dot{V}_A/\dot{Q}$ distribution showed marked regions of high $\dot{V}_A/\dot{Q}$ resulting in increased heterogeneity. Funkquist et al., [99] studied 3-year-old standard-bred trotters with a history of *Rhodococcus equi* pneumonia to assess the possibility of long-term pulmonary dysfunction after infection. However, pulmonary gas exchange during exercise in these animals was not different from that in horses that had never experienced infection. Similarly Nyman et al., [247] showed that pulmonary gas exchange was not different during exercise in standardbreds compared to healthy animals although both pulmonary arterial pressures and systemic blood pressure were higher.

References

1. Agusti AG, Barbera JA. Contribution of multiple inert gas elimination technique to pulmonary medicine. 2. Chronic pulmonary diseases: chronic obstructive pulmonary disease and idiopathic pulmonary fibrosis. Thorax. 1994;49:924–32.
2. Agusti AG, Barbera JA, Roca J, Wagner PD, Guitart R, Rodriguez-Roisin R. Hypoxic pulmonary vasoconstriction and gas exchange during exercise in chronic obstructive pulmonary disease. Chest. 1990;97:268–75.

3. Agusti AG, Roca J, Bosch J, Garcia-Pagan JC, Wagner PD, Rodriguez-Roisin R. Effects of propranolol on arterial oxygenation and oxygen transport to tissues in patients with cirrhosis. Am Rev Respir Dis. 1990;142:306–10.
4. Agusti AG, Roca J, Gea J, Wagner PD, Xaubet A, Rodriguez-Roisin R. Mechanisms of gas-exchange impairment in idiopathic pulmonary fibrosis. Am Rev Respir Dis. 1991;143:219–25.
5. Agusti AG, Roca J, Rodriguez-Roisin R, Mastai R, Wagner PD, Bosch J. Pulmonary hemodynamics and gas exchange during exercise in liver cirrhosis. Am Rev Respir Dis. 1989;139:485–91.
6. Albert RK, Leasa D, Sanderson M, Robertson HT, Hlastala MP. The prone position improves arterial oxygenation and reduces shunt in oleic-acid-induced acute lung injury. Am Rev Respir Dis. 1987;135:628–33.
7. Alfaro V, Roca-Acin J, Palacios L, Guitart R. Multiple inert gas elimination technique for determining ventilation/perfusion distributions in rat during normoxia, hypoxia and hyperoxia. Clin Exp Pharmacol Physiol. 2001;28:419–24.
8. Altemeier WA, Robertson HT, Glenny RW. Pulmonary gas-exchange analysis by using simultaneous deposition of aerosolized and injected microspheres. J Appl Physiol (1985). 1998;85:2344–51.
9. Andersson L, Lagerstrand L, Thorne A, Sollevi A, Brodin LA, Odeberg-Wernerman S. Effect of CO(2) pneumoperitoneum on ventilation-perfusion relationships during laparoscopic cholecystectomy. Acta Anaesthesiol Scand. 2002;46:552–60.
10. Andrivet P, Lofaso F, Carette MF, Allegrini J, Adnot S. Haemodynamics and gas exchange before and after coil embolization of pulmonary arteriovenous malformations. Eur Respir J. 1995;8:1228–30.
11. Anjou-Lindskog E, Broman L, Broman M, Holmgren A. Effects of nitroglycerin on central haemodynamics and VA/Q distribution during ventilation with $FIO_2 = 1.0$ in patients after coronary bypass surgery. Acta Anaesthesiol Scand. 1984;28:27–33.
12. Anjou-Lindskog E, Broman L, Broman M, Holmgren A. Effects of oxygen on central haemodynamics and VA/Q distribution after coronary bypass surgery. Acta Anaesthesiol Scand. 1983;27:378–84.
13. Anjou-Lindskog E, Broman L, Holmgren A. Effects of nitroglycerin on central haemodynamics and VA/Q distribution early after coronary bypass surgery. Acta Anaesthesiol Scand. 1982;26:489–97.
14. Anjou-Lindskog E, Broman L, Holmgren A, Wedelin B. Effects of prenalterol on central haemodynamics and VA/Q distribution early after open-heart surgery. Acta Anaesthesiol Scand. 1983;27:142–8.
15. Balgos AA, Willford DC, West JB. Ventilation-perfusion relationships during induced normovolemic polycythemia in dogs. J Appl Physiol (1985). 1988;65:1686–92.
16. Ballester E, Reyes A, Roca J, Guitart R, Wagner PD, Rodriguez-Roisin R. Ventilation-perfusion mismatching in acute severe asthma: effects of salbutamol and 100% oxygen. Thorax. 1989;44:258–67.
17. Barbera JA, Ramirez J, Roca J, Wagner PD, Sanchez-Lloret J, Rodriguez-Roisin R. Lung structure and gas exchange in mild chronic obstructive pulmonary disease. Am Rev Respir Dis. 1990;141:895–901.
18. Barbera JA, Reyes A, Roca J, Montserrat JM, Wagner PD, Rodriguez-Roisin R. Effect of intravenously administered aminophylline on ventilation/perfusion inequality during recovery from exacerbations of chronic obstructive pulmonary disease. Am Rev Respir Dis. 1992;145:1328–33.
19. Barbera JA, Riverola A, Roca J, Ramirez J, Wagner PD, Ros D, Wiggs BR, Rodriguez-Roisin R. Pulmonary vascular abnormalities and ventilation-perfusion relationships in mild chronic obstructive pulmonary disease. Am J Respir Crit Care Med. 1994;149:423–9.

20. Barbera JA, Roca J, Ramirez J, Wagner PD, Ussetti P, Rodriguez-Roisin R. Gas exchange during exercise in mild chronic obstructive pulmonary disease. Correlation with lung structure. Am Rev Respir Dis. 1991;144:520–5.
21. Barie PS, Wu W, Hariri RJ, Halebian PH, Shires GT. Alterations of pulmonary gas exchange after superimposed carbon monoxide poisoning in acute lung injury. Surgery. 1994;115:678–86.
22. Batchinsky AI, Jordan BS, Necsoiu C, Dubick MA, Cancio LC. Dynamic changes in shunt and ventilation-perfusion mismatch following experimental pulmonary contusion. Shock. 2010;33:419–25.
23. Batchinsky AI, Martini DK, Jordan BS, Dick EJ, Fudge J, Baird CA, Hardin DE, Cancio LC. Acute respiratory distress syndrome secondary to inhalation of chlorine gas in sheep. J Trauma. 2006;60:944–956.; discussion 956–957.
24. Batchinsky AI, Weiss WB, Jordan BS, Dick EJ Jr, Cancelada DA, Cancio LC. Ventilation-perfusion relationships following experimental pulmonary contusion. J Appl Physiol (1985). 2007;103:895–902.
25. Baumgardner JE, Choi IC, Vonk-Noordegraaf A, Frasch HF, Neufeld GR, Marshall BE. Sequential V(A)/Q distributions in the normal rabbit by micropore membrane inlet mass spectrometry. J Appl Physiol (1985). 2000;89:1699–708.
26. Bebout DE, Story D, Roca J, Hogan MC, Poole DC, Gonzalez-Camarena R, Ueno O, Haab P, Wagner PD. Effects of altitude acclimatization on pulmonary gas exchange during exercise. J Appl Physiol (1985). 1989;67:2286–95.
27. Beck KC, Vettermann J, Rehder K. Gas exchange in dogs in the prone and supine positions. J Appl Physiol (1985). 1992;72:2292–7.
28. Bein T, Reber A, Metz C, Jauch KW, Hedenstierna G. Acute effects of continuous rotational therapy on ventilation-perfusion inequality in lung injury. Intensive Care Med. 1998;24:132–7.
29. Bein T, Reber A, Stjernstrom H, Metz C, Taeger K, Hedenstierna G. Ventilation-perfusion ratio in patients with acute respiratory insufficiency. Anaesthesist. 1996;45:337–42.
30. Bencowitz HZ, LeWinter MM, Wagner PD. Effect of sodium nitroprusside on ventilation-perfusion mismatching in heart failure. J Am Coll Cardiol. 1984;4:918–22.
31. Bickenbach J, Czaplik M, Dembinski R, Pelosi P, Schroeder W, Marx G, Rossaint R. In vivo microscopy in a porcine model of acute lung injury. Respir Physiol Neurobiol. 2010;172:192–200.
32. Bickler PE, Maginniss LA, Powell FL. Intrapulmonary and extrapulmonary shunt in ducks. Respir Physiol. 1986;63:151–60.
33. Bindslev L, Hedenstierna G, Santesson J, Gottlieb I, Carvallhas A. Ventilation-perfusion distribution during inhalation anaesthesia. Effects of spontaneous breathing, mechanical ventilation and positive end-expiratory pressure. Acta Anaesthesiol Scand. 1981;25:360–71.
34. Bratel T, Hedenstierna G, Nyquist O, Ripe E. The effect of a new calcium antagonist, felodipine, on pulmonary hypertension and gas exchange in chronic obstructive lung disease. Eur J Respir Dis. 1985;67:244–53.
35. Bratel T, Hedenstierna G, Nyquist O, Ripe E. Long-term treatment with a new calcium antagonist, felodipine, in chronic obstructive lung disease. Eur J Respir Dis. 1986;68:351–61.
36. Bratel T, Hedenstierna G, Nyquist O, Ripe E. The use of a vasodilator, felodipine, as an adjuvant to long-term oxygen treatment in COLD patients. Eur Respir J. 1990;3:46–54.
37. Bratel T, Lagerstrand L, Brodin LA, Nowak J, Randmaa I. Ventilation-perfusion relationships in pulmonary arterial hypertension: effect of intravenous and inhaled prostacyclin treatment. Respir Physiol Neurobiol. 2007;158:59–69.
38. Breen PH, Schumacker PT, Hedenstierna G, Ali J, Wagner PD, Wood LD. How does increased cardiac output increase shunt in pulmonary edema? J Appl Physiol Respir Environ Exerc Physiol. 1982;53:1273–80.

39. Brogan TV, Hedges RG, McKinney S, Robertson HT, Hlastala MP, Swenson ER. Pulmonary NO synthase inhibition and inspired CO_2: effects on V'/Q' and pulmonary blood flow distribution. Eur Respir J. 2000;16:288–95.
40. Brogan TV, Robertson HT, Lamm WJ, Souders JE, Swenson ER. Carbon dioxide added late in inspiration reduces ventilation-perfusion heterogeneity without causing respiratory acidosis. J Appl Physiol (1985). 2004;96:1894–8.
41. Burnham KJ, Arai TJ, Dubowitz DJ, Henderson AC, Holverda S, Buxton RB, Prisk GK, Hopkins SR. Pulmonary perfusion heterogeneity is increased by sustained, heavy exercise in humans. J Appl Physiol (1985). 2009;107:1559–68.
42. Bylin G, Hedenstierna G, Lagerstrand L, Wagner PD. No influence of acetylcysteine on gas exchange and spirometry in chronic asthma. Eur J Respir Dis. 1987;71:102–7.
43. Cano I, Roca J, Wagner PD. Effects of lung ventilation-perfusion and muscle metabolism-perfusion heterogeneities on maximal O_2 transport and utilization. J Physiol. 2015;593:1841–56.
44. Cardus J, Burgos F, Diaz O, Roca J, Barbera JA, Marrades RM, Rodriguez-Roisin R, Wagner PD. Increase in pulmonary ventilation-perfusion inequality with age in healthy individuals. Am J Respir Crit Care Med. 1997;156:648–53.
45. Castaing Y, Manier G, Guenard H. Effect of 26% oxygen breathing on ventilation and perfusion distribution in patients with cold. Bull Eur Physiopathol Respir. 1985;21:17–23.
46. Coffey RL, Albert RK, Robertson HT. Mechanisms of physiological dead space response to PEEP after acute oleic acid lung injury. J Appl Physiol Respir Environ Exerc Physiol. 1983;55:1550–7.
47. Crapo RO, Morris AH, Clayton PD, Nixon CR. Lung volumes in healthy nonsmoking adults. Bull Eur Physiopathol Respir. 1982;18:419–25.
48. Crapo RO, Morris AH, Gardner RM. Reference values for pulmonary tissue volume, membrane diffusing capacity, and pulmonary capillary blood volume. Bull Eur Physiopathol Respir. 1982;18:893–9.
49. Crawford AB, Regnis J, Laks L, Donnelly P, Engel LA, Young IH. Pulmonary vascular dilatation and diffusion-dependent impairment of gas exchange in liver cirrhosis. Eur Respir J. 1995;8:2015–21.
50. Cremona G, Barbera JA, Melgosa T, Appendini L, Roca J, Casadio C, Donner CF, Rodriguez-Roisin R, Wagner PD. Mechanisms of gas exchange response to lung volume reduction surgery in severe emphysema. J Appl Physiol (1985). 2011;110:1036–45.
51. D'Alonzo GE, Bower JS, DeHart P, Dantzker DR. The mechanisms of abnormal gas exchange in acute massive pulmonary embolism. Am Rev Respir Dis. 1983;128:170–2.
52. Dantzker DR, Bower JS. Mechanisms of gas exchange abnormality in patients with chronic obliterative pulmonary vascular disease. J Clin Invest. 1979;64:1050–5.
53. Dantzker DR, Brook CJ, Dehart P, Lynch JP, Weg JG. Ventilation-perfusion distributions in the adult respiratory distress syndrome. Am Rev Respir Dis. 1979;120:1039–52.
54. Dantzker DR, D'Alonzo GE. The effect of exercise on pulmonary gas exchange in patients with severe chronic obstructive pulmonary disease 1–3. Am Rev Respir Dis. 1986;134:1135–9.
55. Dantzker DR, Patten GA, Bower JS. Gas exchange at rest and during exercise in adults with cystic fibrosis. Am Rev Respir Dis. 1982;125:400–5.
56. Dantzker DR, Wagner PD, Tornabene VW, Alazraki NP, West JB. Gas exchange after pulmonary thromboembolization in dogs. Circ Res. 1978;42:92–103.
57. Deem S, Hedges RG, McKinney S, Polissar NL, Alberts MK, Swenson ER. Mechanisms of improvement in pulmonary gas exchange during isovolemic hemodilution. J Appl Physiol (1985). 1999;87:132–41.
58. Deem S, McKinney S, Polissar NL, Hedges RG, Swenson ER. Hemodilution during venous gas embolization improves gas exchange, without altering V(A)/Q or pulmonary blood flow distributions. Anesthesiology. 1999;91:1861–72.

59. Delcroix M, Melot C, Lejeune P, Leeman M, Naeije R. Cyclooxygenase inhibition aggravates pulmonary hypertension and deteriorates gas exchange in canine pulmonary embolism. Am Rev Respir Dis. 1992;145:806–10.
60. Delcroix M, Melot C, Lejeune P, Leeman M, Naeije R. Effects of vasodilators on gas exchange in acute canine embolic pulmonary hypertension. Anesthesiology. 1990;72:77–84.
61. Delcroix M, Melot C, Vachiery JL, Lejeune P, Leeman M, Vanderhoeft P, Naeije R. Effects of embolus size on hemodynamics and gas exchange in canine embolic pulmonary hypertension. J Appl Physiol (1985). 1990;69:2254–61.
62. Delcroix M, Melot C, Vanderhoeft P, Naeije R. Embolus size affects gas exchange in canine autologous blood clot pulmonary embolism. J Appl Physiol (1985). 1993;74:1140–8.
63. Delcroix M, Melot C, Vermeulen F, Naeije R. Hypoxic pulmonary vasoconstriction and gas exchange in acute canine pulmonary embolism. J Appl Physiol (1985). 1996;80:1240–8.
64. Dembinski R, Max M, Bensberg R, Bickenbach J, Kuhlen R, Rossaint R. High-frequency oscillatory ventilation in experimental lung injury: effects on gas exchange. Intensive Care Med. 2002;28:768–74.
65. Dembinski R, Max M, Lopez F, Kuhlen R, Kurth R, Rossaint R. Effect of inhaled prostacyclin in combination with almitrine on ventilation-perfusion distributions in experimental lung injury. Anesthesiology. 2001;94:461–7.
66. Dembinski R, Max M, Lopez F, Kuhlen R, Sunner M, Rossaint R. Effect of inhaled nitric oxide in combination with almitrine on ventilation-perfusion distributions in experimental lung injury. Intensive Care Med. 2000;26:221–8.
67. Dempsey JA, Hanson PG, Henderson KS. Exercised-induced arterial hypoxaemia in healthy human subjects at sea level. J Physiol (Lond). 1984;355:161–75.
68. Dempsey JA, Wagner PD. Exercise-induced arterial hypoxemia. J Appl Physiol (1985). 1999;87:1997–2006.
69. Derion T, Guy HJ, Tsukimoto K, Schaffartzik W, Prediletto R, Poole DC, Knight DR, Wagner PD. Ventilation-perfusion relationships in the lung during head-out water immersion. J Appl Physiol (1985). 1992;72:64–72.
70. Diaz O, Iglesia R, Ferrer M, Zavala E, Santos C, Wagner PD, Roca J, Rodriguez-Roisin R. Effects of noninvasive ventilation on pulmonary gas exchange and hemodynamics during acute hypercapnic exacerbations of chronic obstructive pulmonary disease. Am J Respir Crit Care Med. 1997;156:1840–5.
71. Dodam JR, Moon RE, Olson NC, Exposito AJ, Fawcett TA, Huang YC, Theil DR, Camporesi E, Swanson CR. Effects of clenbuterol hydrochloride on pulmonary gas exchange and hemodynamics in anesthetized horses. Am J Vet Res. 1993;54:776–82.
72. Domino KB, Cheney FW, Eisenstein BL, Hlastala MP. Effect of regional alveolar hypoxia on gas exchange in pulmonary edema. Am Rev Respir Dis. 1992;145:340–7.
73. Domino KB, Eisenstein BL, Cheney FW, Hlastala MP. Pulmonary blood flow and ventilation-perfusion heterogeneity. J Appl Physiol (1985). 1991;71:252–8.
74. Domino KB, Eisenstein BL, Tran T, Hlastala MP. Increased pulmonary perfusion worsens ventilation-perfusion matching. Anesthesiology. 1993;79:817–26.
75. Domino KB, Emery MJ, Swenson ER, Hlastala MP. Ventilation heterogeneity is increased in hypocapnic dogs but not pigs. Respir Physiol. 1998;111:89–100.
76. Domino KB, Hlastala MP. Hyperventilation in the treatment of metabolic acidosis does not adversely affect pulmonary gas exchange. Anesthesiology. 1994;81:1445–53.
77. Domino KB, Hlastala MP, Eisenstein BL, Cheney FW. Effect of regional alveolar hypoxia on gas exchange in dogs. J Appl Physiol (1985). 1989;67:730–5.
78. Domino KB, Lu Y, Eisenstein BL, Hlastala MP. Hypocapnia worsens arterial blood oxygenation and increases VA/Q heterogeneity in canine pulmonary edema. Anesthesiology. 1993;78:91–9.
79. Domino KB, Swenson ER, Hlastala MP. Hypocapnia-induced ventilation/perfusion mismatch: a direct CO_2 or pH-mediated effect? Am J Respir Crit Care Med. 1995;152:1534–9.

80. Domino KB, Swenson ER, Polissar NL, Lu Y, Eisenstein BL, Hlastala MP. Effect of inspired CO_2 on ventilation and perfusion heterogeneity in hyperventilated dogs. J Appl Physiol (1985). 1993;75:1306–14.
81. Dueck R, Prutow RJ, Davies NJ, Clausen JL, Davidson TM. The lung volume at which shunting occurs with inhalation anesthesia. Anesthesiology. 1988;69:854–61.
82. Dueck R, Rathbun M, Wagner PD. Chromatographic analysis of multiple tracer inert gases in the presence of anesthetic gases. Anesthesiology. 1978;49:31–6.
83. Dueck R, Young I, Clausen J, Wagner PD. Altered distribution of pulmonary ventilation and blood flow following induction of inhalation anesthesia. Anesthesiology. 1980;52:113–25.
84. Dunker H-R. Functional morphology of the respiratory system and coelomic subdivisions in reptiles, birds and mammals. Vehr Dtsch Zool Ges. 1978;1978:99–132.
85. Edell ES, Cortese DA, Krowka MJ, Rehder K. Severe hypoxemia and liver disease. Am Rev Respir Dis. 1989;140:1631–5.
86. Erickson BK, Seaman J, Kubo K, Hiraga A, Kai M, Yamaya Y, Wagner PD. Hypoxic helium breathing does not reduce alveolar-arterial PO_2 difference in the horse. Respir Physiol. 1995;100:253–60.
87. Erickson BK, Seaman J, Kubo K, Hiraga A, Kai M, Yamaya Y, Wagner PD. Mechanism of reduction in alveolar-arterial PO_2 difference by helium breathing in the exercising horse. J Appl Physiol (1985). 1994;76:2794–801.
88. Eriksson LS, Soderman C, Ericzon BG, Eleborg L, Wahren J, Hedenstierna G. Normalization of ventilation/perfusion relationships after liver transplantation in patients with decompensated cirrhosis: evidence for a hepatopulmonary syndrome. Hepatology. 1990;12:1350–7.
89. Escourrou PJ, Teisseire BP, Herigault RA, Vallez MO, Dupeyrat AJ. Improvement of gas exchange during inhibition of hypoxic pulmonary vasoconstriction by nitrendipine in piglets. Bull Eur Physiopathol Respir. 1986;22:531–7.
90. Escourrou PJ, Teisseire BP, Herigault RA, Vallez MO, Dupeyrat AJ, Gaultier C. Mechanism of improvement in pulmonary gas exchange during growth in awake piglets. J Appl Physiol (1985). 1988;65:1055–61.
91. Feihl F, Eckert P, Brimioulle S, Jacobs O, Schaller MD, Melot C, Naeije R. Permissive hypercapnia impairs pulmonary gas exchange in the acute respiratory distress syndrome. Am J Respir Crit Care Med. 2000;162:209–15.
92. Ferrer M, Zavala E, Diaz O, Roca J, Wagner PD, Rodriguez-Roisin R. Assessment of ventilation-perfusion mismatching in mechanically ventilated patients. Eur Respir J. 1998;12:1172–6.
93. Forsgren P, Jakobson S, Modig J. True shunt in relation to venous admixture in an experimental porcine model of early ARDS. Acta Anaesthesiol Scand. 1989;33:621–8.
94. Fortune JB, Mazzone RW, Wagner PD. Ventilation-perfusion relationships during hemorrhagic hypotension and reinfusion in the dog. J Appl Physiol Respir Environ Exerc Physiol. 1983;54:1071–82.
95. Frans A, Clerbaux T, Willems E, Kreuzer F. Effect of metabolic acidosis on pulmonary gas exchange of artificially ventilated dogs. J Appl Physiol (1985). 1993;74:2301–8.
96. Freiberg DB, Young IH, Laks L, Regnis JA, Lehrhaft B, Sullivan CE. Improvement in gas exchange with nasal continuous positive airway pressure in pulmonary alveolar microlithiasis. Am Rev Respir Dis. 1992;145:1215–6.
97. Freyschuss U, Hedlin G, Hedenstierna G. Ventilation-perfusion relationships during exercise-induced asthma in children. Am Rev Respir Dis. 1984;130:888–94.
98. Frostell C, Pande JN, Hedenstierna G. Effects of high-frequency breathing on pulmonary ventilation and gas exchange. J Appl Physiol Respir Environ Exerc Physiol. 1983;55:1854–61.
99. Funkquist P, Demmers S, Hedenstierna G, Jensen Waern M, Nyman G. Gas exchange during intense exercise in Standardbreds with earlier Rhodococcus equi pneumonia. Equine Vet J Suppl. 2002;34:434–41.

100. Funkquist P, Wagner PD, Hedenstierna G, Persson SG, Nyman G. Ventilation-perfusion relationships during exercise in standardbred trotters with red cell hypervolaemia. Equine Vet J. 1999;31:107–13.
101. Gale GE, Torre-Bueno JR, Moon RE, Saltzman HA, Wagner PD. Ventilation-perfusion inequality in normal humans during exercise at sea level and simulated altitude. J Appl Physiol (1985). 1985;58:978–88.
102. Ghofrani HA, Wiedemann R, Rose F, Schermuly RT, Olschewski H, Weissmann N, Gunther A, Walmrath D, Seeger W, Grimminger F. Sildenafil for treatment of lung fibrosis and pulmonary hypertension: a randomised controlled trial. Lancet. 2002;360:895–900.
103. Gibson RL, Davis CB, Standaert TA, Truog WE, Redding GJ. Lung vascular responses and VA/Q matching after chronic hypoxia in neonatal piglets. J Dev Physiol. 1993;19:157–63.
104. Giuntini C. Ventilation/perfusion scan and dead space in pulmonary embolism: are they useful for the diagnosis? Q J Nucl Med. 2001;45:281–6.
105. Grubb T, Frendin JH, Edner A, Funkquist P, Hedenstierna G, Nyman G. The effects of pulse-delivered inhaled nitric oxide on arterial oxygenation, ventilation-perfusion distribution and plasma endothelin-1 concentration in laterally recumbent isoflurane-anaesthetized horses. Vet Anaesth Analg. 2013;40:e19–30.
106. Grubb TL, Lord PF, Berger M, Larsson C, Ryden A, Frendin J, Funkquist P, Edner A, Nyman G. Effects of pulse-delivered inhaled nitric oxide administration on pulmonary perfusion and arterial oxygenation in dorsally recumbent isoflurane-anesthetized horses. Am J Vet Res. 2014;75:949–55.
107. Gunnarsson L, Strandberg A, Brismar B, Tokics L, Lundquist H, Hedenstierna G. Atelectasis and gas exchange impairment during enflurane/nitrous oxide anaesthesia. Acta Anaesthesiol Scand. 1989;33:629–37.
108. Gunnarsson L, Tokics L, Gustavsson H, Hedenstierna G. Influence of age on atelectasis formation and gas exchange impairment during general anaesthesia. Br J Anaesth. 1991;66:423–32.
109. Gunnarsson L, Tokics L, Lundquist H, Brismar B, Strandberg A, Berg B, Hedenstierna G. Chronic obstructive pulmonary disease and anaesthesia: formation of atelectasis and gas exchange impairment. Eur Respir J. 1991;4:1106–16.
110. Gunther A, Enke B, Markart P, Hammerl P, Morr H, Behr J, Stahler G, Seeger W, Grimminger F, Leconte I, Roux S, Ghofrani HA. Safety and tolerability of bosentan in idiopathic pulmonary fibrosis: an open label study. Eur Respir J. 2007;29:713–9.
111. Hachenberg T, Holst D, Ebel C, Pfeiffer B, Thomas H, Wendt M, Hedenstierna G. Effect of thoracic epidural anaesthesia on ventilation-perfusion distribution and intrathoracic blood volume before and after induction of general anaesthesia. Acta Anaesthesiol Scand. 1997;41:1142–8.
112. Hachenberg T, Tenling A, Hansson HE, Tyden H, Hedenstierna G. The ventilation-perfusion relation and gas exchange in mitral valve disease and coronary artery disease. Implications for anesthesia, extracorporeal circulation, and cardiac surgery. Anesthesiology. 1997;86:809–17.
113. Hachenberg T, Tenling A, Nystrom SO, Tyden H, Hedenstierna G. Ventilation-perfusion inequality in patients undergoing cardiac surgery. Anesthesiology. 1994;80:509–19.
114. Hambraeus-Jonzon K, Bindslev L, Frostell C, Hedenstierna G. Individual lung blood flow during unilateral hypoxia: effects of inhaled nitric oxide. Eur Respir J. 1998;11:565–70.
115. Hammond MD, Gale GE, Kapitan KS, Ries A, Wagner PD. Pulmonary gas exchange in humans during exercise at sea level. J Appl Physiol (1985). 1986;60:1590–8.
116. Hammond MD, Gale GE, Kapitan KS, Ries A, Wagner PD. Pulmonary gas exchange in humans during normobaric hypoxic exercise. J Appl Physiol (1985). 1986;61:1749–57.
117. Hammond MD, Hempleman SC. Oxygen diffusing capacity estimates derived from measured VA/Q distributions in man. Respir Physiol. 1987;69:129–47.
118. Hardie JA, Vollmer WM, Buist AS, Ellingsen I, Morkve O. Reference values for arterial blood gases in the elderly. Chest. 2004;125:2053–60.

119. Harms CA, McClaran SR, Nickele GA, Pegelow DF, Nelson WB, Dempsey JA. Effect of exercise-induced arterial O_2 desaturation on VO_{2max} in women. Med Sci Sports Exerc. 2000;32:1101–8.
120. Harms CA, McClaran SR, Nickele GA, Pegelow DF, Nelson WB, Dempsey JA. Exercise-induced arterial hypoxaemia in healthy young women. J Physiol. 1998;507(Pt 2):619–28.
121. Harms CA, Rosenkranz S. Sex differences in pulmonary function during exercise. Med Sci Sports Exerc. 2008;40:664–8.
122. Harris EA, Kenyon AM, Nisbet HD, Seelye ER, Whitlock RM. The normal alveolar-arterial oxygen-tension gradient in man. Clin Sci Mol Med. 1974;46:89–104.
123. Hartmann EK, Bentley A, Duenges B, Klein KU, Boehme S, Markstaller K, David M. TIP peptide inhalation in oleic acid-induced experimental lung injury: a post-hoc comparison. BMC Res Notes. 2013;6:385.
124. Hartmann EK, Boehme S, Bentley A, Duenges B, Klein KU, Elsaesser A, Baumgardner JE, David M, Markstaller K. Influence of respiratory rate and end-expiratory pressure variation on cyclic alveolar recruitment in an experimental lung injury model. Crit Care. 2012;16:R8.
125. Hartmann EK, Duenges B, Baumgardner JE, Markstaller K, David M. Correlation of thermodilution-derived extravascular lung water and ventilation/perfusion-compartments in a porcine model. Intensive Care Med. 2013;39:1313–7.
126. Hartmann EK, Duenges B, Boehme S, Szczyrba M, Liu T, Klein KU, Baumgardner JE, Markstaller K, David M. Ventilation/perfusion ratios measured by multiple inert gas elimination during experimental cardiopulmonary resuscitation. Acta Anaesthesiol Scand. 2014;58:1032–9.
127. Hastings RH, Powell FL. High-frequency ventilation of ducks and geese. J Appl Physiol (1985). 1987;63:413–7.
128. Hedenstierna G. Airway closure, atelectasis and gas exchange during anaesthesia. Minerva Anestesiol. 2002;68:332–6.
129. Hedenstierna G. Atelectasis formation and gas exchange impairment during anaesthesia. Monaldi Arch Chest Dis. 1994;49:315–22.
130. Hedenstierna G. Contribution of multiple inert gas elimination technique to pulmonary medicine. 6. Ventilation-perfusion relationships during anaesthesia. Thorax. 1995;50:85–91.
131. Hedenstierna G. Effects of anaesthesia on ventilation/perfusion matching. Eur J Anaesthesiol. 2014;31:447–9.
132. Hedenstierna G. Gas exchange during anaesthesia. Acta Anaesthesiol Scand Suppl. 1990;94:27–31.
133. Hedenstierna G. New aspects on atelectasis formation and gas exchange impairment during anaesthesia. Clin Physiol. 1989;9:407–17.
134. Hedenstierna G. Why may hypoxaemia develop during anaesthesia. Eur J Med. 1992;1:43–8.
135. Hedenstierna G, Freyschuss U. Assessment of ventilation-perfusion distribution by multiple inert gas elimination techniques: its application in the child. Eur Respir J Suppl. 1989;4:181S–3S.
136. Hedenstierna G, Freyschuss U, Hedlin G, Thoren C, Wallgren G. Ventilation-perfusion relationships in children. Clin Physiol. 1982;2:181–8.
137. Hedenstierna G, Hammond M, Mathieu-Costello O, Wagner PD. Functional lung unit in the pig. Respir Physiol. 2000;120:139–49.
138. Hedenstierna G, Lundh R, Johansson H. Alveolar stability during anaesthesia for reconstructive vascular surgery in the leg. Acta Anaesthesiol Scand. 1983;27:26–34.
139. Hedenstierna G, Mebius C, Bygdeman S. Ventilation-perfusion relationship during hip arthroplasty. Acta Anaesthesiol Scand. 1983;27:56–61.
140. Hedenstierna G, Nyman G, Kvart C, Funkquist B. Ventilation-perfusion relationships in the standing horse: an inert gas elimination study. Equine Vet J. 1987;19:514–9.
141. Hedenstierna G, Soderman C, Eriksson LS, Wahren J. Ventilation-perfusion inequality in patients with non-alcoholic liver cirrhosis. Eur Respir J. 1991;4:711–7.

142. Hedenstierna G, Tokics L, Strandberg A, Lundquist H, Brismar B. Correlation of gas exchange impairment to development of atelectasis during anaesthesia and muscle paralysis. Acta Anaesthesiol Scand. 1986;30:183–91.
143. Hedenstierna G, White FC, Mazzone R, Wagner PD. Redistribution of pulmonary blood flow in the dog with PEEP ventilation. J Appl Physiol Respir Environ Exerc Physiol. 1979;46:278–87.
144. Hedlin G, Freyschuss U, Hedenstierna G. Histamine-induced asthma in children: effects on the ventilation-perfusion relationship. Clin Physiol. 1985;5:19–34.
145. Henzler D, Dembinski R, Bensberg R, Hochhausen N, Rossaint R, Kuhlen R. Ventilation with biphasic positive airway pressure in experimental lung injury. Influence of transpulmonary pressure on gas exchange and haemodynamics. Intensive Care Med. 2004;30:935–43.
146. Henzler D, Pelosi P, Bensberg R, Dembinski R, Quintel M, Pielen V, Rossaint R, Kuhlen R. Effects of partial ventilatory support modalities on respiratory function in severe hypoxemic lung injury. Crit Care Med. 2006;34:1738–45.
147. Hermle G, Mols G, Zugel A, Benzing A, Lichtwarck-Aschoff M, Geiger K, Guttmann J. Intratidal compliance-volume curve as an alternative basis to adjust positive end-expiratory pressure: a study in isolated perfused rabbit lungs. Crit Care Med. 2002;30:1589–97.
148. Hermle G, Schutte H, Walmrath D, Geiger K, Seeger W, Grimminger F. Ventilation-perfusion mismatch after lung ischemia-reperfusion. Protective effect of nitric oxide. Am J Respir Crit Care Med. 1999;160:1179–87.
149. Hicks JW. Cardiac shunting in reptiles: mechanisms, regulation and physiological functions. In: Gaunt A, Gans C, editors. Biology of the reptilia (Morphology G). Ithaca, NY: Society for the Study of Amphibians and Reptiles; 1998. p. 425–83.
150. Hirano M. Effect of a Ca2+ antagonist (verapamil) on Va/Q distribution. Masui. 1989;38:992–9.
151. Hlastala MP. Multiple inert gas elimination technique. J Appl Physiol Respir Environ Exerc Physiol. 1984;56:1–7.
152. Hlastala MP, Scheid P, Piiper J. Interpretation of inert gas retention and excretion in the presence of stratified inhomogeneity. Respir Physiol. 1981;46:247–59.
153. Hlastala MP, Standaert TA, Pierson DJ, Luchtel DL. The matching of ventilation and perfusion in the lung of the Tegu lizard, Tupinambis nigropunctatus. Respir Physiol. 1985;60:277–94.
154. Hopkins SR. Exercise induced arterial hypoxemia: the role of ventilation-perfusion inequality and pulmonary diffusion limitation. Adv Exp Med Biol. 2006;588:17–30.
155. Hopkins SR, Barker RC, Brutsaert TD, Gavin TP, Entin P, Olfert IM, Veisel S, Wagner PD. Pulmonary gas exchange during exercise in women: effects of exercise type and work increment. J Appl Physiol (1985). 2000;89:721–30.
156. Hopkins SR, Bayly WM, Slocombe RF, Wagner H, Wagner PD. Effect of prolonged heavy exercise on pulmonary gas exchange in horses. J Appl Physiol (1985). 1998;84:1723–30.
157. Hopkins SR, Belzberg AS, Wiggs BR, McKenzie DC. Pulmonary transit time and diffusion limitation during heavy exercise in athletes. Respir Physiol. 1996;103:67–73.
158. Hopkins SR, Gavin TP, Siafakas NM, Haseler LJ, Olfert IM, Wagner H, Wagner PD. Effect of prolonged, heavy exercise on pulmonary gas exchange in athletes. J Appl Physiol (1985). 1998;85:1523–32.
159. Hopkins SR, Harms CA. Gender and pulmonary gas exchange during exercise. Exerc Sport Sci Rev. 2004;32:50–6.
160. Hopkins SR, Hicks JW, Cooper TK, Powell FL. Ventilation and pulmonary gas exchange during exercise in the savannah monitor lizard (Varanus exanthematicus). J Exp Biol. 1995;198:1783–9.

161. Hopkins SR, Johnson EC, Richardson RS, Wagner H, De Rosa M, Wagner PD. Effects of inhaled nitric oxide on gas exchange in lungs with shunt or poorly ventilated areas. Am J Respir Crit Care Med. 1997;156:484–91.
162. Hopkins SR, McKenzie DC, Schoene RB, Glenny RW, Robertson HT. Pulmonary gas exchange during exercise in athletes. I. Ventilation-perfusion mismatch and diffusion limitation. J Appl Physiol (1985). 1994;77:912–7.
163. Hopkins SR, Olfert IM, Wagner PD. Last word on point: counterpoint: exercise-induced intrapulmonary shunting is imaginary vs. real. J Appl Physiol (1985). 2009;107:1002.
164. Hopkins SR, Olfert IM, Wagner PD. Point: exercise-induced intrapulmonary shunting is imaginary. J Appl Physiol (1985). 2009;107:993–4.
165. Hopkins SR, Stary CM, Falor E, Wagner H, Wagner PD, McKirnan MD. Pulmonary gas exchange during exercise in pigs. J Appl Physiol (1985). 1999;86:93–100.
166. Hopkins SR, Wang T, Hicks JW. The effect of altering pulmonary blood flow on pulmonary gas exchange in the turtle Trachemys (Pseudemys) scripta. J Exp Biol. 1996;199:2207–14.
167. Hsia CC, Carlin JI, Wagner PD, Cassidy SS, Johnson RL Jr. Gas exchange abnormalities after pneumonectomy in conditioned foxhounds. J Appl Physiol (1985). 1990;68:94–104.
168. Hsia CC, Dane DM, Estrera AS, Wagner HE, Wagner PD, Johnson RL Jr. Shifting sources of functional limitation following extensive (70%) lung resection. J Appl Physiol (1985). 2008;104:1069–79.
169. Hsia CC, Herazo LF, Ramanathan M, Johnson RL Jr, Wagner PD. Cardiopulmonary adaptations to pneumonectomy in dogs. II. VA/Q relationships and microvascular recruitment. J Appl Physiol (1985). 1993;74:1299–309.
170. Hsia CC, Johnson RL Jr, Dane DM, Wu EY, Estrera AS, Wagner HE, Wagner PD. The canine spleen in oxygen transport: gas exchange and hemodynamic responses to splenectomy. J Appl Physiol (1985). 2007;103:1496–505.
171. Hsia CC, Johnson RL Jr, Mc Donough P, Dane DM, Hurst MD, Fehmel JL, Wagner HE, Wagner PD. Residence at 3,800-m altitude for 5 mo in growing dogs enhances lung diffusing capacity for oxygen that persists at least 2.5 years. J Appl Physiol (1985). 2007;102:1448–55.
172. Hsia CC, Johnson RL Jr, Wu EY, Estrera AS, Wagner H, Wagner PD. Reducing lung strain after pneumonectomy impairs oxygen diffusing capacity but not ventilation-perfusion matching. J Appl Physiol (1985). 2003;95:1370–8.
173. Hsia CC, Wagner PD, Dane DM, Wagner HE, Johnson RL Jr. Predicting diffusive alveolar oxygen transfer from carbon monoxide-diffusing capacity in exercising foxhounds. J Appl Physiol (1985). 2008;105:1441–7.
174. Huang YC, Caminiti SP, Fawcett TA, Moon RE, Fracica PJ, Miller FJ, Young SL, Piantadosi CA. Natural surfactant and hyperoxic lung injury in primates. I. Physiology and biochemistry. J Appl Physiol (1985). 1994;76:991–1001.
175. Huang YC, Fracica PJ, Simonson SG, Crapo JD, Young SL, Welty-Wolf KE, Moon RE, Piantadosi CA. VA/Q abnormalities during gram negative sepsis. Respir Physiol. 1996;105:109–21.
176. Huang YC, Sane AC, Simonson SG, Fawcett TA, Moon RE, Fracica PJ, Menache MG, Piantadosi CA, Young SL. Artificial surfactant attenuates hyperoxic lung injury in primates. I. Physiology and biochemistry. J Appl Physiol (1985). 1995;78:1816–22.
177. Hubler M, Souders JE, Shade ED, Polissar NL, Schimmel C, Hlastala MP. Effects of vaporized perfluorocarbon on pulmonary blood flow and ventilation/perfusion distribution in a model of acute respiratory distress syndrome. Anesthesiology. 2001;95:1414–21.
178. Hultgren HN. High altitude medicine. Stanford, CA: Hultgren Publications; 1997.
179. Huttemeier PC, Ringsted C, Eliasen K, Mogensen T. Ventilation-perfusion inequality during endotoxin-induced pulmonary vasoconstriction in conscious sheep: mechanisms of hypoxia. Clin Physiol. 1988;8:351–8.
180. Jernudd-Wilhelmsson Y, Hornblad Y, Hedenstierna G. Ventilation-perfusion relationships in interstitial lung disease. Eur J Respir Dis. 1986;68:39–49.

181. Johnson EC, Erickson BK, Podolsky A, Birks EK, Keipert PE, Faithfull NS, Wagner PD. Effects of a perfluorocarbon emulsion for enhanced O_2 solubility on hemodynamics and O_2 transport in dogs. J Appl Physiol (1985). 1995;79:1777–86.
182. Johnson RL Jr, Hsia CC, Cassidy SS, Carlin JI, Wagner PD, Ramanathan M. A post-pneumonectomy comparison of cardiac output and gas exchange in humans and dogs during heavy exercise. In: Sutton JR, Coates G, Remmers JE, editors. Hypoxia: the adaptations. Toronto: B.C. Decker; 1990. p. 148–54.
183. Jonk AM, van den Berg IP, Olfert IM, Wray DW, Arai T, Hopkins SR, Wagner PD. Effect of acetazolamide on pulmonary and muscle gas exchange during normoxic and hypoxic exercise. J Physiol. 2007;579:909–21.
184. Kaiser KG, Davies NJ, Rodriguez-Roisin R, Bencowitz HZ, Wagner PD. Efficacy of high-frequency ventilation in presence of extensive ventilation-perfusion mismatch. J Appl Physiol (1985). 1985;58:996–1004.
185. Kallas HJ, Domino KB, Glenny RW, Anderson EA, Hlastala MP. Pulmonary blood flow redistribution with low levels of positive end-expiratory pressure. Anesthesiology. 1998;88:1291–9.
186. Kapitan KS, Buchbinder M, Wagner PD, Moser KM. Mechanisms of hypoxemia in chronic thromboembolic pulmonary hypertension. Am Rev Respir Dis. 1989;139:1149–54.
187. Kapitan KS, Clausen JL, Moser KM. Gas exchange in chronic thromboembolism after pulmonary thromboendarterectomy. Chest. 1990;98:14–9.
188. Kawai A, Umeda A, Mori M, Takasugi T, Asano K, Tanaka T, Yasuoka M, Yamaguchi K. Role of serotonin in impaired gas exchange during pulmonary embolism. Nihon Kyobu Shikkan Gakkai Zasshi. 1995;33:947–55.
189. Kirkton SD, Wagner H, Landicho MM, Struthers JJ, Busan N, Wagner PD. Effects of IMAGENT on pulmonary hemodynamics and gas exchange in dogs. Ultrasound Med Biol. 2006;32:949–59.
190. Kleinsasser A, Lindner KH, Hoermann C, Schaefer A, Keller C, Loeckinger A. Isoflurane and sevoflurane anesthesia in pigs with a preexistent gas exchange defect. Anesthesiology. 2001;95:1422–6.
191. Kleinsasser A, Olfert IM, Loeckinger A, Prisk GK, Hopkins SR, Wagner PD. Tidal volume dependency of gas exchange in bronchoconstricted pig lungs. J Appl Physiol (1985). 2007;103:148–55.
192. Klingstedt C, Hedenstierna G, Baehrendtz S, Lundqvist H, Strandberg A, Tokics L, Brismar B. Ventilation-perfusion relationships and atelectasis formation in the supine and lateral positions during conventional mechanical and differential ventilation. Acta Anaesthesiol Scand. 1990;34:421–9.
193. Lagerstrand L, Bylin G, Hedenstierna G, Wagner PD. Relationships among gas exchange, spirometry and symptoms in asthma. Eur J Med. 1992;1:145–52.
194. Lagerstrand L, Dahlback M, Hedenstierna G. Gas exchange during simulated airway secretion in the anaesthetized rabbit. Eur Respir J. 1992;5:1215–22.
195. Lagerstrand L, Hedenstierna G. Gas-exchange impairment: its correlation to lung mechanics in acute airway obstruction (studies on a rabbit asthma model). Clin Physiol. 1990;10:363–80.
196. Lagerstrand L, Hjelte L, Jorulf H. Pulmonary gas exchange in cystic fibrosis: basal status and the effect of i.v. antibiotics and inhaled amiloride. Eur Respir J. 1999;14:686–92.
197. Lagerstrand L, Larsson K, Ihre E, Zetterstrom O, Hedenstierna G. Pulmonary gas exchange response following allergen challenge in patients with allergic asthma. Eur Respir J. 1992;5:1176–83.
198. Lagerstrand L, Skedinger M, Ihre E, Hallden G, Zetterstrom O. Spirometry and ventilation-perfusion inequality in patients with mild allergic asthma before and during the pollen season. Clin Physiol. 1995;15:355–64.
199. Lamm WJ, Obermiller T, Hlastala MP, Albert RK. Perfusion through vessels open in zone 1 contributes to gas exchange in rabbit lungs in situ. J Appl Physiol (1985). 1995;79:1895–9.

200. Lee LN, Ueno O, Wagner PD, West JB. Pulmonary gas exchange after multiple airway occlusion by beads in the dog. Am Rev Respir Dis. 1989;140:1216–21.
201. Lemaire F, Matamis D, Lampron N, Teisseire B, Harf A. Intrapulmonary shunt is not increased by 100% oxygen ventilation in acute respiratory failure. Bull Eur Physiopathol Respir. 1985;21:251–6.
202. Lim CM, Domino KB, Glenny RW, Hlastala MP. Effect of increasing perfluorocarbon dose on VA/Q distribution during partial liquid ventilation in acute lung injury. Anesthesiology. 2001;94:637–42.
203. Loeckinger A, Keller C, Lindner KH, Kleinsasser A. Pulmonary gas exchange in coronary artery surgery patients during sevoflurane and isoflurane anesthesia. Anesth Analg. 2002;94:1107–12.
204. Loeckinger A, Kleinsasser A, Hoermann C, Gassner M, Keller C, Lindner KH. Inert gas exchange during pneumoperitoneum at incremental values of positive end-expiratory pressure. Anesth Analg. 2000;90:466–71.
205. Loeckinger A, Kleinsasser A, Keller C, Schaefer A, Kolbitsch C, Lindner KH, Benzer A. Administration of oxygen before tracheal extubation worsens gas exchange after general anesthesia in a pig model. Anesth Analg. 2002;95:1772–6.
206. Loeckinger A, Kleinsasser A, Wenzel V, Mair V, Keller C, Kolbitsch C, Recheis W, Schuster A, Lindner KH. Pulmonary gas exchange after cardiopulmonary resuscitation with either vasopressin or epinephrine. Crit Care Med. 2002;30:2059–62.
207. Loeckinger A, Schutte H, Walmrath D, Seeger W, Grimminger F. Protection against gas exchange abnormalities by pre-aerosolized PGE1, iloprost and nitroprusside in lung ischemia-reperfusion. Transplantation. 2001;71:185–93.
208. Lotterer E. Hepatopulmonary syndrome. Schweiz Rundsch Med Prax. 1994;83:1047–50.
209. Lotterer E, Fleig WE. Hepatopulmonary syndrome. Praxis (Bern 1994). 1997;86:104–8.
210. Lundh R, Hedenstierna G. Ventilation-perfusion relationships during anaesthesia and abdominal surgery. Acta Anaesthesiol Scand. 1983;27:167–73.
211. Lundh R, Hedenstierna G. Ventilation-perfusion relationships during halothane anaesthesia and mechanical ventilation. Effects of varying inspired oxygen concentration. Acta Anaesthesiol Scand. 1984;28:191–8.
212. Lundh R, Hedenstierna G, Johansson H. Ventilation-perfusion relationships during epidural analgesia. Acta Anaesthesiol Scand. 1983;27:410–6.
213. Lynch JP, Mhyre JG, Dantzker DR. Influence of cardiac output on intrapulmonary shunt. J Appl Physiol Respir Environ Exerc Physiol. 1979;46:315–21.
214. Magnusson L, Zemgulis V, Tenling A, Wernlund J, Tyden H, Thelin S, Hedenstierna G. Use of a vital capacity maneuver to prevent atelectasis after cardiopulmonary bypass: an experimental study. Anesthesiology. 1998;88:134–42.
215. Magnusson L, Zemgulis V, Wicky S, Tyden H, Hedenstierna G. Effect of CPAP during cardiopulmonary bypass on postoperative lung function. An experimental study. Acta Anaesthesiol Scand. 1998;42:1133–8.
216. Magnusson L, Zemgulis V, Wicky S, Tyden H, Thelin S, Hedenstierna G. Atelectasis is a major cause of hypoxemia and shunt after cardiopulmonary bypass: an experimental study. Anesthesiology. 1997;87:1153–63.
217. Manier G, Castaing Y. Contribution of multiple inert gas elimination technique to pulmonary medicine—4. Gas exchange abnormalities in pulmonary vascular and cardiac disease. Thorax. 1994;49:1169–74.
218. Manier G, Castaing Y. Influence of cardiac output on oxygen exchange in acute pulmonary embolism. Am Rev Respir Dis. 1992;145:130–6.
219. Manier G, Pillet O, Castaing Y. Influence of cardiac output level on oxygen exchange in chronic obstructive pulmonary disease patients. Clin Physiol Funct Imaging. 2006;26:275–82.

220. Marntell S, Nyman G, Funkquist P, Hedenstierna G. Effects of acepromazine on pulmonary gas exchange and circulation during sedation and dissociative anaesthesia in horses. Vet Anaesth Analg. 2005;32:83–93.
221. Marntell S, Nyman G, Hedenstierna G. High inspired oxygen concentrations increase intrapulmonary shunt in anaesthetized horses. Vet Anaesth Analg. 2005;32:338–47.
222. Martin TR, Castile RG, Fredberg JJ, Wohl ME, Mead J. Airway size is related to sex but not lung size in normal adults. J Appl Physiol (1985). 1987;63:2042–7.
223. Matamis D, Lemaire F, Harf A, Teisseire B, Brun-Buisson C. Redistribution of pulmonary blood flow induced by positive end-expiratory pressure and dopamine infusion in acute respiratory failure. Am Rev Respir Dis. 1984;129:39–44.
224. Maurenbrecher H, Lamy M, Deby-Dupont G, Frascarolo P, Hedenstierna G. An animal model of response and nonresponse to inhaled nitric oxide in endotoxin-induced lung injury. Chest. 2001;120:573–81.
225. Mayoral V, Sabate A. Hepatopulmonary syndrome: meta-analysis. Rev Esp Anestesiol Reanim. 1995;42:159–62.
226. McEvoy RD, Davies NJ, Mannino FL, Prutow RJ, Schumacker PT, Wagner PD, West JB. Pulmonary gas exchange during high-frequency ventilation. J Appl Physiol Respir Environ Exerc Physiol. 1982;52:1278–87.
227. McEvoy RD, Drew MJ, Mahar LJ. The effects of exercise on ventilation-perfusion relationship in patients with mitral stenosis. Bull Eur Physiopathol Respir. 1986;22:239–46.
228. Mead J. Dysanapsis in normal lungs assessed by the relationship between maximal flow, static recoil, and vital capacity. Am Rev Respir Dis. 1980;121:339–42.
229. Melo MF, Harris RS, Layfield JD, Venegas JG. Topographic basis of bimodal ventilation-perfusion distributions during bronchoconstriction in sheep. Am J Respir Crit Care Med. 2005;171:714–21.
230. Melot C. Contribution of multiple inert gas elimination technique to pulmonary medicine. 5. Ventilation-perfusion relationships in acute respiratory failure. Thorax. 1994;49:1251–8.
231. Melot C, Lejeune P, Leeman M, Moraine JJ, Naeije R. Prostaglandin E1 in the adult respiratory distress syndrome. Benefit for pulmonary hypertension and cost for pulmonary gas exchange. Am Rev Respir Dis. 1989;139:106–10.
232. Melot C, Naeije R, Dechamps P, Hallemans R, Lejeune P. Pulmonary and extrapulmonary contributors to hypoxemia in liver cirrhosis. Am Rev Respir Dis. 1989;139:632–40.
233. Melot C, Naeije R, Hallemans R, Lejeune P, Mols P. Hypoxic pulmonary vasoconstriction and pulmonary gas exchange in normal man. Respir Physiol. 1987;68:11–27.
234. Melot C, Naeije R, Hallemans R, Mols P, Lejeune P. Beneficial effects of almitrine bismesylate on pulmonary gas exchange in COPD. Eur J Respir Dis Suppl. 1983;126:249–54.
235. Melot C, Naeije R, Mols P, Hallemans R, Lejeune P, Jaspar N. Pulmonary vascular tone improves pulmonary gas exchange in the adult respiratory distress syndrome. Am Rev Respir Dis. 1987;136:1232–6.
236. Melot C, Naeije R, Mols P, Vandenbossche JL, Denolin H. Effects of nifedipine on ventilation/perfusion matching in primary pulmonary hypertension. Chest. 1983;83:203–7.
237. Moinard J, Manier G, Pillet O, Castaing Y. Effect of inhaled nitric oxide on hemodynamics and VA/Q inequalities in patients with chronic obstructive pulmonary disease. Am J Respir Crit Care Med. 1994;149:1482–7.
238. Mols G, Hermle G, Fries G, Benzing A, Lichtwarck-Aschoff M, Geiger K, Guttmann J. Different strategies to keep the lung open: a study in isolated perfused rabbit lungs. Crit Care Med. 2002;30:1598–604.
239. Mols G, Hermle G, Schubert J, Miekisch W, Benzing A, Lichtwarck-Aschoff M, Geiger K, Walmrath D, Guttmann J. Volume-dependent compliance and ventilation-perfusion mismatch in surfactant-depleted isolated rabbit lungs. Crit Care Med. 2001;29:144–51.
240. Montserrat JM, Barbera JA, Viegas C, Roca J, Rodriguez-Roisin R. Gas exchange response to intravenous aminophylline in patients with a severe exacerbation of asthma. Eur Respir J. 1995;8:28–33.

241. Munoz PA, Gomez FP, Manrique HA, Roca J, Barbera JA, Young IH, Anderson SD, Rodriguez-Roisin R. Pulmonary gas exchange response to exercise- and mannitol-induced bronchoconstriction in mild asthma. J Appl Physiol (1985). 2008;105:1477–85.
242. Naeije R, Melot C, Niset G, Delcroix M, Wagner PD. Mechanisms of improved arterial oxygenation after peripheral chemoreceptor stimulation during hypoxic exercise. J Appl Physiol (1985). 1993;74:1666–71.
243. Neumann P, Berglund JE, Andersson LG, Maripu E, Magnusson A, Hedenstierna G. Effects of inverse ratio ventilation and positive end-expiratory pressure in oleic acid-induced lung injury. Am J Respir Crit Care Med. 2000;161:1537–45.
244. Neumann P, Hedenstierna G. Ventilation-perfusion distributions in different porcine lung injury models. Acta Anaesthesiol Scand. 2001;45:78–86.
245. Neumann P, Hedenstierna G. Ventilatory support by continuous positive airway pressure breathing improves gas exchange as compared with partial ventilatory support with airway pressure release ventilation. Anesth Analg. 2001;92:950–8.
246. Neumann P, Wrigge H, Zinserling J, Hinz J, Maripuu E, Andersson LG, Putensen C, Hedenstierna G. Spontaneous breathing affects the spatial ventilation and perfusion distribution during mechanical ventilatory support. Crit Care Med. 2005;33:1090–5.
247. Nyman G, Bjork M, Funkquist P. Gas exchange during exercise in standardbred trotters with mild bronchiolitis. Equine Vet J. 1999;31:96–101.
248. Nyman G, Hedenstierna G. Ventilation-perfusion relationships in the anaesthetised horse. Equine Vet J. 1989;21:274–81.
249. Nyman G, Lindberg R, Weckner D, Bjork M, Kvart C, Persson SG, Gustafsson H, Hedenstierna G. Pulmonary gas exchange correlated to clinical signs and lung pathology in horses with chronic bronchiolitis. Equine Vet J. 1991;23:253–60.
250. Nyman G, Marntell S, Edner A, Funkquist P, Morgan K, Hedenstierna G. Effect of sedation with detomidine and butorphanol on pulmonary gas exchange in the horse. Acta Vet Scand. 2009;51:22.
251. Nyman G, Röken B, Hedin E-M, Hedenstierna G. Case studies in physiology: ventilation and perfusion in a giraffe-does size matter? J Appl Physiol (1985). 2016;121:1374–8.
252. Ogasawara H. Distribution of ventilation-perfusion ratios in paraquat-induced interstitial pneumonia in dogs. Hokkaido Igaku Zasshi. 1996;71:259–70.
253. Ogura H, Cioffi WG, Offner PJ, Jordan BS, Johnson AA, Pruitt BA Jr. Effect of inhaled nitric oxide on pulmonary function after sepsis in a swine model. Surgery. 1994;116:313–21.
254. Ogura H, Cioffi WG, Okerberg CV, Johnson AA, Guzman RF, Mason AD Jr, Pruitt BA Jr. The effects of pentoxifylline on pulmonary function following smoke inhalation. J Surg Res. 1994;56:242–50.
255. Ogura H, Offner PJ, Saitoh D, Jordan BS, Johnson AA, Pruitt BA Jr, Cioffi WG Jr. The pulmonary effect of nitric oxide synthase inhibition following endotoxemia in a swine model. Arch Surg. 1994;129:1233–9.
256. Ohlsson J, Middaugh M, Hlastala MP. Reduction of lung perfusion increases VA/Q heterogeneity. J Appl Physiol (1985). 1989;66:2423–30.
257. Oku H. Continuous distribution of ventilation-perfusion ratios following lung denervation in dogs by means of inert gas elimination method. Nihon Kyobu Geka Gakkai Zasshi. 1992;40:1693–9.
258. Olfert IM, Balouch J, Kleinsasser A, Knapp A, Wagner H, Wagner PD, Hopkins SR. Does gender affect human pulmonary gas exchange during exercise? J Physiol. 2004;557:529–41.
259. Pappert D, Gilliard N, Heldt G, Merritt TA, Wagner PD, Spragg RG. Effect of N-nitroso-N-methylurethane on gas exchange, lung compliance, and surfactant function of rabbits. Intensive Care Med. 1996;22:345–52.
260. Pappert D, Rossaint R, Slama K, Gruning T, Falke KJ. Influence of positioning on ventilation-perfusion relationships in severe adult respiratory distress syndrome. Chest. 1994;106:1511–6.

261. Perry SF. Reptilian lungs: functional anatomy and evolution. Berlin: Springer Science & Business Media; 2013.
262. Pfeiffer B, Hachenberg T, Feyerherd F, Wendt M. Ventilation-perfusion distribution with volume-reduced, pressure-limited ventilation with permissive hypercapnia. Anasthesiol Intensivmed Notfallmed Schmerzther. 1998;33:367–72.
263. Pillet O, Manier G, Castaing Y. Anticholinergic versus beta 2-agonist on gas exchange in COPD: a comparative study in 15 patients. Monaldi Arch Chest Dis. 1998;53:3–8.
264. Pison U, Lopez FA, Heidelmeyer CF, Rossaint R, Falke KJ. Inhaled nitric oxide reverses hypoxic pulmonary vasoconstriction without impairing gas exchange. J Appl Physiol (1985). 1993;74:1287–92.
265. Podolsky A, Eldridge MW, Richardson RS, Knight DR, Johnson EC, Hopkins SR, Johnson DH, Michimata H, Grassi B, Feiner J, Kurdak SS, Bickler PE, Severinghaus JW, Wagner PD. Exercise-induced VA/Q inequality in subjects with prior high-altitude pulmonary edema. J Appl Physiol (1985). 1996;81:922–32.
266. Popp AJ, Shah DM, Berman RA, Paloski WH, Newell JC, Cooper JA, Rahm RL, Gottlieb ME, Bourke RS, Malik AB. Delayed pulmonary dysfunction in head-injured patients. J Neurosurg. 1982;57:784–90.
267. Powell FL, Gray AT. Ventilation-perfusion relationships in alligators. Respir Physiol. 1989;78:83–94.
268. Powell FL, Hopkins SR. Comparative physiology of lung complexity: implications for gas exchange. News Physiol Sci. 2004;19:55–60.
269. Powell FL, Scheid P. Physiology of gas exchange in the avian respiratory system. In: King AS, McLelland J, editors. Form and function in birds. London: Academic Press; 1989. p. 393–437.
270. Powell FL, Wagner PD. Measurement of continuous distributions of ventilation-perfusion in non-alveolar lungs. Respir Physiol. 1982;48:219–32.
271. Powell FL, Wagner PD. Ventilation-perfusion inequally in avian lungs. Respir Physiol. 1982;48:233–41.
272. Powers SK, Lawler J, Dempsey JA, Dodd S, Landry G. Effects of incomplete pulmonary gas exchange on VO_2 max. J Appl Physiol (1985). 1989;66:2491–5.
273. Prediletto R, Formichi B, Allescia G, Paoletti P, Begliomini E, Fornai E, Pistolesi M, Giuntini C. The assessment of respiratory function in a patient with dyspnoea and severe hypoxaemia. Monaldi Arch Chest Dis. 1993;48:213–20.
274. Prediletto R, Formichi B, Bertolaccini P. Changes in pulmonary gas exchange in chronic obstructive bronchopneumopathies and their modifications induced by therapy. Medicina (Firenze). 1989;9:56–9.
275. Prisk GK, Guy HJ, West JB, Reed JW. Validation of measurements of ventilation-to-perfusion ratio inequality in the lung from expired gas. J Appl Physiol (1985). 2003;94:1186–92.
276. Prisk GK, Olfert IM, Arai TJ, Wagner PD, Hopkins SR. Rapid intravenous infusion of 20 ml/kg saline does not impair resting pulmonary gas exchange in the healthy human lung. J Appl Physiol (1985). 2010;108:53–9.
277. Putensen C, Rasanen J, Downs JB. Effect of endogenous and inhaled nitric oxide on the ventilation-perfusion relationships in oleic-acid lung injury. Am J Respir Crit Care Med. 1994;150:330–6.
278. Putensen C, Rasanen J, Lopez FA. Ventilation-perfusion distributions during mechanical ventilation with superimposed spontaneous breathing in canine lung injury. Am J Respir Crit Care Med. 1994;150:101–8.
279. Putensen C, Rasanen J, Lopez FA, Downs JB. Continuous positive airway pressure modulates effect of inhaled nitric oxide on the ventilation-perfusion distributions in canine lung injury. Chest. 1994;106:1563–9.

280. Putensen C, Rasanen J, Lopez FA, Downs JB. Effect of interfacing between spontaneous breathing and mechanical cycles on the ventilation-perfusion distribution in canine lung injury. Anesthesiology. 1994;81:921–30.
281. Putensen C, Rasanen J, Putensen-Himmer G, Downs JB. Effect of low isoflurane concentrations on the ventilation-perfusion distribution in injured canine lungs. Anesthesiology. 2002;97:652–9.
282. Radermacher P, Falke KJ. The aADN2 in critically ill patients: is it necessary to measure PaN2? Scand J Clin Lab Invest Suppl. 1987;188:53–4.
283. Radermacher P, Santak B, Wust HJ, Tarnow J, Falke KJ. Prostacyclin for the treatment of pulmonary hypertension in the adult respiratory distress syndrome: effects on pulmonary capillary pressure and ventilation-perfusion distributions. Anesthesiology. 1990;72:238–44.
284. Raimondi GA. Gas exchange in acute respiratory distress syndrome. Medicina (B Aires). 2003;63:157–64.
285. Ralph DD, Ott SM, Sherrard DJ, Hlastala MP. Inert gas analysis of ventilation-perfusion matching during hemodialysis. J Clin Invest. 1984;73:1385–91.
286. Ralph DD, Robertson HT, Weaver LJ, Hlastala MP, Carrico CJ, Hudson LD. Distribution of ventilation and perfusion during positive end-expiratory pressure in the adult respiratory distress syndrome. Am Rev Respir Dis. 1985;131:54–60.
287. Reber A, Bein T, Hogman M, Khan ZP, Nilsson S, Hedenstierna G. Lung aeration and pulmonary gas exchange during lumbar epidural anaesthesia and in the lithotomy position in elderly patients. Anaesthesia. 1998;53:854–61.
288. Redding GJ, Standaert TA, Truog WE. Alterations in gas exchange associated with lobar atelectasis in young piglets. Pediatr Res. 1985;19:552–6.
289. Rennotte MT, Reynaert M, Clerbaux T, Willems E, Roeseleer J, Veriter C, Rodenstein D, Frans A. Effects of two inotropic drugs, dopamine and dobutamine, on pulmonary gas exchange in artificially ventilated patients. Intensive Care Med. 1989;15:160–5.
290. Reyes A, Roca J, Rodriguez-Roisin R, Torres A, Ussetti P, Wagner PD. Effect of almitrine on ventilation-perfusion distribution in adult respiratory distress syndrome. Am Rev Respir Dis. 1988;137:1062–7.
291. Rice AJ, Thornton AT, Gore CJ, Scroop GC, Greville HW, Wagner H, Wagner PD, Hopkins SR. Pulmonary gas exchange during exercise in highly trained cyclists with arterial hypoxemia. J Appl Physiol (1985). 1999;87:1802–12.
292. Rivas E, Arismendi E, Agusti A, Gistau C, Wagner PD, Rodriguez-Roisin R. Postural effects on pulmonary gas exchange abnormalities in severe obesity before and after bariatric surgery. Minerva Anestesiol. 2016;82:403–10.
293. Rivas E, Arismendi E, Agusti A, Sanchez M, Delgado S, Gistau C, Wagner PD, Rodriguez-Roisin R. Ventilation/Perfusion distribution abnormalities in morbidly obese subjects before and after bariatric surgery. Chest. 2015;147:1127–34.
294. Robert V, Chabot F, Vial B, Guiot P, Poussel JF, Polu JM. Hepatopulmonary syndrome: physiopathology of impaired gas exchange. Rev Mal Respir. 1999;16:769–79.
295. Robertson HT, Hlastala MP. Microsphere maps of regional blood flow and regional ventilation. J Appl Physiol (1985). 2007;102:1265–72.
296. Robinson NB, Chi EY, Robertson HT. Ventilation-perfusion relationships after hemorrhage and resuscitation: an inert gas analysis. J Appl Physiol Respir Environ Exerc Physiol. 1983;54:1131–40.
297. Robinson TD, Freiberg DB, Regnis JA, Young IH. The role of hypoventilation and ventilation-perfusion redistribution in oxygen-induced hypercapnia during acute exacerbations of chronic obstructive pulmonary disease. Am J Respir Crit Care Med. 2000;161:1524–9.
298. Roca J, Montserrat JM, Rodriguez-Roisin R, Guitart R, Torres A, Agusti AG, Wagner PD. Gas exchange response to naloxone in chronic obstructive pulmonary disease with hypercapnic respiratory failure. Bull Eur Physiopathol Respir. 1987;23:249–54.

299. Roca J, Ramis L, Rodriguez-Roisin R, Ballester E, Montserrat JM, Wagner PD. Serial relationships between ventilation-perfusion inequality and spirometry in acute severe asthma requiring hospitalization. Am Rev Respir Dis. 1988;137:1055–61.
300. Roca J, Wagner PD. Contribution of multiple inert gas elimination technique to pulmonary medicine. 1. Principles and information content of the multiple inert gas elimination technique. Thorax. 1994;49:815–24.
301. Rodriguez-Roisin R. Gas exchange abnormalities in asthma. Lung. 1990;168 (Suppl):599–605.
302. Rodriguez-Roisin R. Pulmonary gas exchange in acute respiratory failure. Eur J Anaesthesiol. 1994;11:5–13.
303. Rodriguez-Roisin R, Ballester E, Roca J, Torres A, Wagner PD. Mechanisms of hypoxemia in patients with status asthmaticus requiring mechanical ventilation. Am Rev Respir Dis. 1989;139:732–9.
304. Rodriguez-Roisin R, Bencowitz HZ, Ziegler MG, Wagner PD. Gas exchange responses to bronchodilators following methacholine challenge in dogs. Am Rev Respir Dis. 1984;130:617–26.
305. Rodriguez-Roisin R, Drakulovic M, Rodriguez DA, Roca J, Barbera JA, Wagner PD. -Ventilation-perfusion imbalance and chronic obstructive pulmonary disease staging severity. J Appl Physiol (1985). 2009;106:1902–8.
306. Rodriguez-Roisin R, Felez MA, Chung KF, Barbera JA, Wagner PD, Cobos A, Barnes PJ, Roca J. Platelet-activating factor causes ventilation-perfusion mismatch in humans. J Clin Invest. 1994;93:188–94.
307. Rodriguez-Roisin R, Ferrer A, Navajas D, Agusti AG, Wagner PD, Roca J. Ventilation-perfusion mismatch after methacholine challenge in patients with mild bronchial asthma. Am Rev Respir Dis. 1991;144:88–94.
308. Rodriguez-Roisin R, Roca J. Contributions of multiple inert gas elimination technique to pulmonary medicine. 3. Bronchial asthma. Thorax. 1994;49:1027–33.
309. Rodriguez-Roisin R, Roca J, Guitart R, Agusti A, Torres A, Wagner PD. Determination of the distribution of ventilation-perfusion ratios: technic of elimination of multiple inert gases. Rev Esp Fisiol. 1986;42:465–82.
310. Romaldini H, Rodriguez-Roisin R, Lopez FA, Ziegler TW, Bencowitz HZ, Wagner PD. The mechanisms of arterial hypoxemia during hemodialysis. Am Rev Respir Dis. 1984;129:780–4.
311. Rossaint R, Hahn SM, Pappert D, Falke KJ, Radermacher P. Influence of mixed venous PO_2 and inspired O_2 fraction on intrapulmonary shunt in patients with severe ARDS. J Appl Physiol (1985). 1995;78:1531–6.
312. Rothen HU, Sporre B, Engberg G, Wegenius G, Hedenstierna G. Airway closure, atelectasis and gas exchange during general anaesthesia. Br J Anaesth. 1998;81:681–6.
313. Rothen HU, Sporre B, Engberg G, Wegenius G, Hedenstierna G. Reexpansion of atelectasis during general anaesthesia may have a prolonged effect. Acta Anaesthesiol Scand. 1995;39:118–25.
314. Rothen HU, Sporre B, Engberg G, Wegenius G, Hogman M, Hedenstierna G. Influence of gas composition on recurrence of atelectasis after a reexpansion maneuver during general anesthesia. Anesthesiology. 1995;82:832–42.
315. Rothen HU, Sporre B, Engberg G, Wegenius G, Reber A, Hedenstierna G. Atelectasis and pulmonary shunting during induction of general anaesthesia–can they be avoided? Acta Anaesthesiol Scand. 1996;40:524–9.
316. Rothen HU, Sporre B, Engberg G, Wegenius G, Reber A, Hedenstierna G. Prevention of atelectasis during general anaesthesia. Lancet. 1995;345:1387–91.
317. Rubinfeld AR, Wagner PD, West JB. Gas exchange during acute experimental canine asthma. Am Rev Respir Dis. 1978;118:525–36.

318. Saba TM, Niehaus GD, Scovill WA, Blumenstock FA, Newell JC, Holman J Jr, Powers SR Jr. Lung vascular permeability after reversal of fibronectin deficiency in septic sheep. Correlation with patient studies. Ann Surg. 1983;198:654–62.
319. Sandek K, Andersson T, Bratel T, Lagerstrand L. Ventilation-perfusion inequality in nocturnal hypoxaemia due to chronic obstructive lung disease (COLD). Clin Physiol. 1995;15:499–513.
320. Sandek K, Bratel T, Lagerstrand L. Effects on diffusing capacity and ventilation–perfusion relationships of budesonide inhalations for 2 months in chronic obstructive pulmonary disease (COPD). Respir Med. 2001;95:676–84.
321. Schaffartzik W, Poole DC, Derion T, Tsukimoto K, Hogan MC, Arcos JP, Bebout DE, Wagner PD. VA/Q distribution during heavy exercise and recovery in humans: implications for pulmonary edema. J Appl Physiol. 1992;72:1657–67.
322. Scheid P, Piiper J. Diffusion. In: Crystal RG, West JB, Barnes PJ, Weibel ER, editors. The lung: scientific foundations. 2nd ed. Philadelphia, PA: Lippincott-Raven; 1997. p. 1681–90.
323. Schermuly R, Schmehl T, Gunther A, Grimminger F, Seeger W, Walmrath D. Ultrasonic nebulization for efficient delivery of surfactant in a model of acute lung injury. Impact on gas exchange. Am J Respir Crit Care Med. 1997;156:445–53.
324. Schermuly RT, Gunther A, Ermert M, Ermert L, Ghofrani HA, Weissmann N, Grimminger F, Seeger W, Walmrath D. Conebulization of surfactant and urokinase restores gas exchange in perfused lungs with alveolar fibrin formation. Am J Physiol Lung Cell Mol Physiol. 2001;280:L792–800.
325. Schermuly RT, Leuchte H, Ghofrani HA, Weissmann N, Rose F, Kohstall M, Olschewski H, Schudt C, Grimminger F, Seeger W, Walmrath D. Zardaverine and aerosolised iloprost in a model of acute respiratory failure. Eur Respir J. 2003;22:342–7.
326. Schermuly RT, Roehl A, Weissmann N, Ghofrani HA, Schudt C, Tenor H, Grimminger F, Seeger W, Walmrath D. Subthreshold doses of specific phosphodiesterase type 3 and 4 inhibitors enhance the pulmonary vasodilatory response to nebulized prostacyclin with improvement in gas exchange. J Pharmacol Exp Ther. 2000;292:512–20.
327. Schmekel B, Hedenstrom H, Kampe M, Lagerstrand L, Stalenheim G, Wollmer P, Hedenstierna G. The bronchial response, but not the pulmonary response to inhaled methacholine is dependent on the aerosol deposition pattern. Chest. 1994;106:1781–7.
328. Schmitt PM, Powell FL, Hopkins SR. Ventilation-perfusion inequality during normoxic and hypoxic exercise in the emu. J Appl Physiol (1985). 2002;93:1980–6.
329. Schumacker PT, Samsel RW, Sznajder JI, Wood LD, Solway J. Gas density dependence of regional VA/V and VA/Q inequality during constant-flow ventilation. J Appl Physiol (1985). 1989;66:1722–9.
330. Schumacker PT, Solway J, Wood LD, Sznajder JI. Lobar contribution to VA/Q inequality during constant-flow ventilation. J Appl Physiol (1985). 1988;65:2132–7.
331. Schumacker PT, Sznajder JI, Nahum A, Wood LD. Ventilation-perfusion inequality during constant-flow ventilation. J Appl Physiol (1985). 1987;62:1255–63.
332. Schwartz J, Katz SA, Fegley RW, Tockman MS. Sex and race differences in the development of lung function. Am Rev Respir Dis. 1988;138:1415–21.
333. Schweiger JW, Downs JB, Smith RA. Chest wall disruption with and without acute lung injury: effects of continuous positive airway pressure therapy on ventilation and perfusion relationships. Crit Care Med. 2003;31:2364–70.
334. Seaman J, Erickson BK, Kubo K, Hiraga A, Kai M, Yamaya Y, Wagner PD. Exercise induced ventilation/perfusion inequality in the horse. Equine Vet J. 1995;27:104–9.
335. Sheel AW, Richards JC, Foster GE, Guenette JA. Sex differences in respiratory exercise physiology. Sports Med. 2004;34:567–79.
336. Shimazu T, Yukioka T, Ikeuchi H, Mason AD Jr, Wagner PD, Pruitt BA Jr. Ventilation-perfusion alterations after smoke inhalation injury in an ovine model. J Appl Physiol (1985). 1996;81:2250–9.

337. Simonson SG, Huang YC, Fracica PJ, Welty-Wolf KE, Moon RE, Young SL, Piantadosi CA. Changes in the lung after prolonged positive pressure ventilation in normal baboons. J Crit Care. 1997;12:72–82.
338. Simonson SG, Welty-Wolf KE, Huang YC, Taylor DE, Kantrow SP, Carraway MS, Crapo JD, Piantadosi CA. Aerosolized manganese SOD decreases hyperoxic pulmonary injury in primates. I. Physiology and biochemistry. J Appl Physiol (1985). 1997;83:550–8.
339. Sinclair SE, Kregenow DA, Starr I, Schimmel C, Lamm WJ, Hlastala MP, Swenson ER. Therapeutic hypercapnia and ventilation-perfusion matching in acute lung injury: low minute ventilation vs inspired CO_2. Chest. 2006;130:85–92.
340. Sorbini CA, Grassi V, Solinas E, Muiesan G. Arterial oxygen tension in relation to age in healthy subjects. Respiration. 1968;25:3–13.
341. Sorensen GK, Redding GJ, Truog WE. Mechanisms of pulmonary gas exchange abnormalities during experimental group B streptococcal infusion. Pediatr Res. 1985;19:922–6.
342. Stewart JH, Rose RJ, Young IH, Costas L. The distribution of ventilation-perfusion ratios in the lungs of a dysmature foal. Equine Vet J. 1990;22:442–6.
343. Stewart JH, Young IH, Rose RJ, Costas L, Barko AM. The distribution of ventilation-perfusion ratios in the lungs of newborn foals. J Dev Physiol. 1987;9:309–24.
344. Sticher J, Scholz S, Boning O, Schermuly RT, Schumacher C, Walmrath D, Hempelmann G. Small-dose nitric oxide improves oxygenation during one-lung ventilation: an experimental study. Anesth Analg. 2002;95:1557–62.
345. Strang CM, Ebmeyer U, Maripuu E, Hachenberg T, Hedenstierna G. Improved ventilation-perfusion matching by abdominal insufflation (pneumoperitoneum) with CO_2 but not with air. Minerva Anestesiol. 2013;79:617–25.
346. Strang CM, Freden F, Maripuu E, Ebmeyer U, Hachenberg T, Hedenstierna G. Improved ventilation-perfusion matching with increasing abdominal pressure during CO(2)-pneumoperitoneum in pigs. Acta Anaesthesiol Scand. 2011;55:887–96.
347. Strang CM, Freden F, Maripuu E, Hachenberg T, Hedenstierna G. Ventilation-perfusion distributions and gas exchange during carbon dioxide-pneumoperitoneum in a porcine model. Br J Anaesth. 2010;105:691–7.
348. Swenson ER, Graham MM, Hlastala MP. Acetazolamide slows VA/Q matching after changes in regional blood flow. J Appl Physiol (1985). 1995;78:1312–8.
349. Swenson ER, Robertson HT, Hlastala MP. Effects of carbonic anhydrase inhibition on ventilation-perfusion matching in the dog lung. J Clin Invest. 1993;92:702–9.
350. Swenson ER, Robertson HT, Hlastala MP. Effects of inspired carbon dioxide on ventilation-perfusion matching in normoxia, hypoxia, and hyperoxia. Am J Respir Crit Care Med. 1994;149:1563–9.
351. Sylvester JT, Cymerman A, Gurtner G, Hottenstein O, Cote M, Wolfe D. Components of alveolar-arterial O_2 gradient during rest and exercise at sea level and high altitude. J Appl Physiol Respir Environ Exerc Physiol. 1981;50:1129–39.
352. Tasaki O, Goodwin CW, Saitoh D, Mozingo DW, Ishihara S, Brinkley WW, Cioffi WG Jr, Pruitt BA Jr. Effects of burns on inhalation injury. J Trauma. 1997;43:603–7.
353. Tasaki O, Mozingo DW, Dubick MA, Goodwin CW, Yantis LD, Pruitt BA Jr. Effects of heparin and lisofylline on pulmonary function after smoke inhalation injury in an ovine model. Crit Care Med. 2002;30:637–43.
354. Tasaki O, Mozingo DW, Ishihara S, Brinkley WW, Johnson AA, Smith RH, Srivastava O, Mason AD Jr, Pruitt BA Jr, Cioffi WG Jr. Effect of Sulfo Lewis C on smoke inhalation injury in an ovine model. Crit Care Med. 1998;26:1238–43.
355. Tedjasaputra V, Sa RC, Arai TJ, Holverda S, Theilmann RJ, Chen WT, Wagner PD, Davis CK, Kim Prisk G, Hopkins SR. The heterogeneity of regional specific ventilation is unchanged following heavy exercise in athletes. J Appl Physiol (1985). 2013;115:126–35.
356. Tenling A, Hachenberg T, Tyden H, Wegenius G, Hedenstierna G. Atelectasis and gas exchange after cardiac surgery. Anesthesiology. 1998;89:371–8.

357. Tenling A, Joachimsson PO, Tyden H, Wegenius G, Hedenstierna G. Thoracic epidural anesthesia as an adjunct to general anesthesia for cardiac surgery: effects on ventilation-perfusion relationships. J Cardiothorac Vasc Anesth. 1999;13:258–64.
358. Thrush DN, Downs JB, Smith RA. Is epinephrine contraindicated during cardiopulmonary resuscitation? Circulation. 1997;96:2709–14.
359. Thurlbeck WM. Postnatal human lung growth. Thorax. 1982;37:564–71.
360. Tokics L, Hedenstierna G, Brismar B, Strandberg A, Lundquist H. Thoracoabdominal restriction in supine men: CT and lung function measurements. J Appl Physiol (1985). 1988;64:599–604.
361. Tokics L, Hedenstierna G, Strandberg A, Brismar B, Lundquist H. Lung collapse and gas exchange during general anesthesia: effects of spontaneous breathing, muscle paralysis, and positive end-expiratory pressure. Anesthesiology. 1987;66:157–67.
362. Tokics L, Hedenstierna G, Svensson L, Brismar B, Cederlund T, Lundquist H, Strandberg A. V/Q distribution and correlation to atelectasis in anesthetized paralyzed humans. J Appl Physiol (1985). 1996;81:1822–33.
363. Tokics L, Strandberg A, Brismar B, Lundquist H, Hedenstierna G. Computerized tomography of the chest and gas exchange measurements during ketamine anaesthesia. Acta Anaesthesiol Scand. 1987;31:684–92.
364. Tornabene VW, Fortune JB, Wagner PD, Halasz NA. Gas exchange after pulmonary fat embolism in dogs. J Thorac Cardiovasc Surg. 1979;78:589–99.
365. Torre-Bueno JR, Wagner PD, Saltzman HA, Gale GE, Moon RE. Diffusion limitation in normal humans during exercise at sea level and simulated altitude. J Appl Physiol (1985). 1985;58:989–95.
366. Torres A, Reyes A, Roca J, Wagner PD, Rodriguez-Roisin R. Ventilation-perfusion mismatching in chronic obstructive pulmonary disease during ventilator weaning. Am Rev Respir Dis. 1989;140:1246–50.
367. Trachsel S, Hambraeus-Jonzon K, Bergquist M, Martijn C, Chen L, Hedenstierna G. No redistribution of lung blood flow by inhaled nitric oxide in endotoxemic piglets pretreated with an endothelin receptor antagonist. J Appl Physiol. 2015;118(6):768–75.
368. Treml B, Neu N, Kleinsasser A, Gritsch C, Finsterwalder T, Geiger R, Schuster M, Janzek E, Loibner H, Penninger J, Loeckinger A. Recombinant angiotensin-converting enzyme 2 improves pulmonary blood flow and oxygenation in lipopolysaccharide-induced lung injury in piglets. Crit Care Med. 2010;38:596–601.
369. Truog WE, Gibson RL Jr, Henderson WR, Redding GJ. Tumor necrosis factor-induced neonatal pulmonary hypertension: effects of dazmegrel pretreatment. Pediatr Res. 1990;27:466–71.
370. Truog WE, Gibson RL, Juul SE, Henderson WR, Redding GJ. Neonatal group B streptococcal sepsis: effects of late treatment with dazmegrel. Pediatr Res. 1988;23:352–6.
371. Truog WE, Hlastala MP, Standaert TA, McKenna HP, Hodson WA. Oxygen-induced alteration of ventilation-perfusion relationships in rats. J Appl Physiol Respir Environ Exerc Physiol. 1979;47:1112–7.
372. Truog WE, Lyrene RK, Standaert TA, Murphy J, Woodrum DE. Effects of PEEP and tolazoline infusion on respiratory and inert gas exchange in experimental meconium aspiration. J Pediatr. 1982;100:284–90.
373. Truog WE, Standaert TA. Effect of high-frequency ventilation on gas exchange and pulmonary vascular resistance in lambs. J Appl Physiol (1985). 1985;59:1104–9.
374. Tsukimoto K, Arcos JP, Schaffartzik W, Wagner PD, West JB. Effect of common dead space on VA/Q distribution in the dog. J Appl Physiol (1985). 1990;68:2488–93.
375. Tsukimoto K, Arcos JP, Schaffartzik W, Wagner PD, West JB. Effects of inspired CO_2, hyperventilation, and time on VA/Q inequality in the dog. J Appl Physiol (1985). 1992;72:1057–63.

376. Tusman G, Suarez-Sipmann F, Bohm SH, Borges JB, Hedenstierna G. Capnography reflects ventilation/perfusion distribution in a model of acute lung injury. Acta Anaesthesiol Scand. 2011;55:597–606.
377. Ueno O, Lee LN, Wagner PD. Effect of N-acetylcysteine on gas exchange after methacholine challenge and isoprenaline inhalation in the dog. Eur Respir J. 1989;2:238–46.
378. Ukita H, Ogasawara H, Masaki Y, Doi I, Ohtsuka Y, Kusaka H, Tanimura K, Munakata M, Kawakami Y, Homma Y. Effect of almitrine in acute canine lung injury induced by paraquat. Nihon Kyobu Shikkan Gakkai Zasshi. 1991;29:1547–52.
379. Valentine DD, Hammond MD, Downs JB, Sears NJ, Sims WR. Distribution of ventilation and perfusion with different modes of mechanical ventilation. Am Rev Respir Dis. 1991;143:1262–6.
380. Van Obbergh L, Carlier M, de Clety SC, Sokal E, Rennotte MT, Veyckemans F, Dekock M, Frans A, Otte JB. Liver transplantation and pulmonary gas exchanges in hypoxemic children. Am Rev Respir Dis. 1993;148:1408–10.
381. Vimlati L, Kawati R, Hedenstierna G, Larsson A, Lichtwarck-Aschoff M. Spontaneous breathing improves shunt fraction and oxygenation in comparison with controlled ventilation at a similar amount of lung collapse. Anesth Analg. 2011;113:1089–95.
382. Voswinckel R, Reichenberger F, Gall H, Schmehl T, Gessler T, Schermuly RT, Grimminger F, Rubin LJ, Seeger W, Ghofrani HA, Olschewski H. Metered dose inhaler delivery of treprostinil for the treatment of pulmonary hypertension. Pulm Pharmacol Ther. 2009;22:50–6.
383. Wagner P, Erickson B, Kubo K, Hiraga A, Kai M, Yamaya Y, Richardson R, Seaman J. Maximum O_2 transport and utilization before and after splenectomy. Equine Vet J. 1994;18:82–9.
384. Wagner PD. Assessment of gas exchange in lung disease: balancing accuracy against feasibility. Crit Care. 2007;11:182.
385. Wagner PD. Diffusion and chemical reaction in pulmonary gas exchange. Physiol Rev. 1977;57:257–312.
386. Wagner PD. Gas exchange. In: Hornbein TF, Schoene R, editors. High altitude an exploration of human adaptation. New York, NY: Marcel Dekker; 2001. p. 199–234.
387. Wagner PD. Gas exchange in chronic pulmonary disease. Clin Physiol. 1985;5(Suppl 3):9–17.
388. Wagner PD. A general approach to the evaluation of ventilation-perfusion ratios in normal and abnormal lungs. The Physiologist. 1977;20:18–25.
389. Wagner PD. HFV and pulmonary physiology. Acta Anaesthesiol Scand Suppl. 1989;90:172–5.
390. Wagner PD. Impairment of gas exchange in liver cirrhosis. Eur Respir J. 1995;8:1993–5.
391. Wagner PD. The multiple inert gas elimination technique (MIGET). Intensive Care Med. 2008;34:994–1001.
392. Wagner PD. Recent advances in pulmonary gas exchange. Int Anesthesiol Clin. 1977;15:81–111.
393. Wagner PD. Ventilation-perfusion inequality and gas exchange during exercise in lung disease. In: Dempsey JA, Reed CE, editors. Muscular exercise and the lung. Madison, WI: University of Wisconsin Press; 1977. p. 345–56.
394. Wagner PD. Ventilation-perfusion matching during exercise. Chest. 1992;101:192S–8S.
395. Wagner PD. Ventilation-perfusion relationships. Annu Rev Physiol. 1980;42:235–47.
396. Wagner PD. Ventilation/perfusion relationships. Clin Physiol. 1981;1:437–51.
397. Wagner PD, Dantzker DR, Dueck R, Clausen JL, West JB. Ventilation-perfusion inequality in chronic obstructive pulmonary disease. J Clin Invest. 1977;59:203–16.
398. Wagner PD, Dantzker DR, Dueck R, de Polo JL, Wasserman K, West JB. Distribution of ventilation-perfusion ratios in patients with interstitial lung disease. Chest. 1976;69:256–7.
399. Wagner PD, Dantzker DR, Iacovoni VE, Tomlin WC, West JB. Ventilation-perfusion inequality in asymptomatic asthma. Am Rev Respir Dis. 1978;118:511–24.

400. Wagner PD, Dueck R. Mechanisms of abnormal gas exchange during anesthesia. Int J Clin Monit Comput. 1984;1:59–71.
401. Wagner PD, Gale GE, Moon RE, Torre-Bueno JR, Stolp BW, Saltzman HA. Pulmonary gas exchange in humans exercising at sea level and simulated altitude. J Appl Physiol (1985). 1986;61:260–70.
402. Wagner PD, Gillespie JR, Landgren GL, Fedde MR, Jones BW, DeBowes RM, Pieschl RL, Erickson HH. Mechanism of exercise-induced hypoxemia in horses. J Appl Physiol (1985). 1989;66:1227–33.
403. Wagner PD, Hedenstierna G, Bylin G. Ventilation-perfusion inequality in chronic asthma. Am Rev Respir Dis. 1987;136:605–12.
404. Wagner PD, Hedenstierna G, Bylin G, Lagerstrand L. Reproducibility of the multiple inert gas elimination technique. J Appl Physiol (1985). 1987;62:1740–6.
405. Wagner PD, Hedenstierna G, Rodriguez-Roisin R. Gas exchange, expiratory flow obstruction and the clinical spectrum of asthma. Eur Respir J. 1996;9:1278–82.
406. Wagner PD, Hsia CC, Goel R, Fay JM, Wagner HE, Johnson RL Jr. Effects of crocetin on pulmonary gas exchange in foxhounds during hypoxic exercise. J Appl Physiol (1985). 2000;89:235–41.
407. Wagner PD, Gillespie JR, Landgren GL, Fedde MR, Jones BW, de Bowes RM, Peischel RL, Erickson HH. Mechanism of exercise-induced hypoxemia in horses. J Appl Physiol. 1989;66:1227–33.
408. Wagner PD, Laravuso RB, Goldzimmer E, Naumann PF, West JB. Distribution of ventilation-perfusion ratios in dogs with normal and abnormal lungs. J Appl Physiol. 1975;38:1099–109.
409. Wagner PD, Laravuso RB, Uhl RR, West JB. Continuous distributions of ventilation-perfusion ratios in normal subjects breathing air and 100 per cent O_2. J Clin Invest. 1974;54:54–68.
410. Wagner PD, Laravuso RB, Uhl RR, West JB. Distributions of ventilation-perfusion ratios in acute respiratory failure. Chest. 1974;65(Suppl):32S–5S.
411. Wagner PD, Rodriguez-Roisin R. Clinical advances in pulmonary gas exchange. Am Rev Respir Dis. 1991;143:883–8.
412. Wagner PD, Sutton JR, Reeves JT, Cymerman A, Groves BM, Malconian MK. Operation Everest II: pulmonary gas exchange during a simulated ascent of Mt. Everest. J Appl Physiol (1985). 1987;63:2348–59.
413. Walmrath D, Pilch J, Scharmann M, Grimminger F, Seeger W. Severe VA/Q mismatch in perfused lungs evoked by sequential challenge with endotoxin and *E. coli* hemolysin. J Appl Physiol (1985). 1994;76:1020–30.
414. Walmrath D, Scharmann M, Konig R, Pilch J, Grimminger F, Seeger W. Staphylococcal alpha-toxin induced ventilation-perfusion mismatch in isolated blood-free perfused rabbit lungs. J Appl Physiol (1985). 1993;74:1972–80.
415. Walmrath D, Schermuly R, Pilch J, Grimminger F, Seeger W. Effects of inhaled versus intravenous vasodilators in experimental pulmonary hypertension. Eur Respir J. 1997;10:1084–92.
416. Walmrath D, Schneider T, Pilch J, Schermuly R, Grimminger F, Seeger W. Effects of aerosolized prostacyclin in severe pneumonia. Impact of fibrosis. Am J Respir Crit Care Med. 1995;151:724–30.
417. Walther SM, Domino KB, Hlastala MP. Effects of posture on blood flow diversion by hypoxic pulmonary vasoconstriction in dogs. Br J Anaesth. 1998;81:425–9.
418. Wang X, Dockery DW, Wypij D, Fay ME, Ferris BG Jr. Pulmonary function between 6 and 18 years of age. Pediatr Pulmonol. 1993;15:75–88.
419. Watremez C, Liistro G, de Kock M, Roeseler J, Clerbaux T, Detry B, Reynaert M, Gianello P, Jolliet P. Effects of helium-oxygen on respiratory mechanics, gas exchange, and ventilation-perfusion relationships in a porcine model of stable methacholine-induced bronchospasm. Intensive Care Med. 2003;29:1560–6.

420. Watremez C, Roeseler J, De Kock M, Clerbaux T, Detry B, Veriter C, Reynaert M, Gianello P, Jolliet P, Liistro G. An improved porcine model of stable methacholine-induced bronchospasm. Intensive Care Med. 2003;29:119–25.
421. Wu W, Halebian PH, Hariri RJ, Cabrales SX, Shires GT, Barie PS. Differential effects of cyclo-oxygenase and thromboxane synthetase inhibition on ventilation-perfusion relationships in acid aspiration-induced acute lung injury. J Trauma. 1992;33:561–7.
422. Yamaguchi K, Mori M, Takashugi T, Oyamada Y, Kawai A, Asano K, Aoki T, Fujita H, Suzuki Y. High oxygen breathing improves inhomogeneities of ventilation-perfusion distributions in acutely injured lungs. Nihon Kyobu Shikkan Gakkai Zasshi. 1994;32:739–46.
423. Yamasawa F, Kawashiro T, Okada Y, Kawai A, Mori M, Yokoyama T. Uneven ventilation-perfusion ratio distribution during acute exacerbation of chronic pulmonary diseases. Kokyu To Junkan. 1992;40:369–73.
424. Yamaya Y, Niizeki K, Kim J, Entin PL, Wagner H, Wagner PD. Effects of Optison on pulmonary gas exchange and hemodynamics. Ultrasound Med Biol. 2002;28:1005–13.
425. Young I, Mazzone RW, Wagner PD. Identification of functional lung unit in the dog by graded vascular embolization. J Appl Physiol Respir Environ Exerc Physiol. 1980;49:132–41.
426. Young IH, Wagner PD. Effect of intrapulmonary hematocrit maldistribution on O_2, CO_2, and inert gas exchange. J Appl Physiol Respir Environ Exerc Physiol. 1979;46:240–8.
427. Ziebart A, Hartmann EK, Thomas R, Liu T, Duenges B, Schad A, Bodenstein M, Thal SC, David M. Low tidal volume pressure support versus controlled ventilation in early experimental sepsis in pigs. Respir Res. 2014;15:101.

Appendix A: Lopez and Wagner Reprint on Gas Chromatography

S.R. Hopkins, P.D. Wagner, *The Multiple Inert Gas Elimination Technique (MIGET)*, Methods in Physiology, https://doi.org/10.1007/978-1-4939-7441-2

Techniques in the Life Sciences, P4/I
Respiratory Physiology, P403 (1984) 1–24
Elsevier Scientific Publishers Ireland Ltd.

P403/1

Gas chromatography techniques in respiratory physiology

PETER D. WAGNER and FRANK A. LOPEZ

Department of Medicine, M-023, University of California, San Diego, La Jolla, CA 92093 (*U.S.A.*)

1. Introduction

In lung disease the structural changes that develop usually result in interference to pulmonary gas exchange. It has long been recognized that the basis for this interference is most often mismatching of ventilation and blood flow. By this is meant that the ratio of the amount of ventilation to the amount of blood flow in

0165-1064/84/$20.00

P403/2

some small lung region is different from the overall ratio for the lung as a whole. In fact, in most diseases, there is a considerable range of ratios of ventilation to blood flow ($\dot{V}_A/\dot{Q}$ ratio). How such $\dot{V}_A/\dot{Q}$ heterogeneity causes inefficient gas exchange has been discussed many times (Farhi and Rahn, 1954–1955; West, 1969; Wagner, 1980). It is important here to note that $\dot{V}_A/\dot{Q}$ mismatch will affect all gases, not just O_2, but quantitatively in ways that depend on the blood solubility of the gas in question. Both shunt (unventilated lung, $\dot{V}_A/\dot{Q} = 0$) and deadspace (unperfused lung, $\dot{V}_A/\dot{Q} = \infty$) are often considered separately from $\dot{V}_A/\dot{Q}$ mismatching, but they can be thought of as the extremes of the possible range of $\dot{V}_A/\dot{Q}$ values.

The problem facing physiologists and physicians alike has long been to measure the degree of $\dot{V}_A/\dot{Q}$ mismatch, or more precisely, the distribution of $\dot{V}_A/\dot{Q}$ ratios over the entire lung. The conceptually direct approach of measuring ventilation and blood flow in all possible small lung regions is infeasible. Even modern day methods involving scintillation counting do not have the spatial resolution to adequately measure the $\dot{V}_A/\dot{Q}$ distribution from the point of view of its quantitative effects on gas exchange.

Consequently, the degree of $\dot{V}_A/\dot{Q}$ heterogeneity is usually estimated not directly, but by mathematical inference from the quantitative disturbance to overall pulmonary gas exchange of one or more gases. To take this indirect approach, however, demands that the quantitative relationships between $\dot{V}_A/\dot{Q}$ mismatching and gas exchange be well understood, and that these relationships are, in fact, adhered to in the in vivo situation. It is beyond the scope of this chapter to discuss these relationships, but any reader who would seek to make use of the practical information in this chapter will need to refer to treatises on these relationships (Farhi, 1967; West and Wagner, 1977; Wagner, 1977).

Inert gases, or more precisely, gases that obey Henry's law (linear relationship between blood content and partial pressure), are well-suited to the quantitative assessment of $\dot{V}_A/\dot{Q}$ mismatching (Farhi and Yokoyama, 1967; Yokoyama and Farhi, 1967; Evans and Wagner, 1977). O_2 and CO_2 have interdependent and non-linear dissociation curves that considerably interfere with their quantitative use in estimating $\dot{V}_A/\dot{Q}$ mismatch. Inert gases can be measured in trace concentrations so as not to affect the lung. Gases with a wide range of solubilities can be used, and this will allow a corresponding $\dot{V}_A/\dot{Q}$ range to be addressed. The usual mode of using O_2 for these purposes involves serial measurements of O_2 exchange at different inspired O_2 levels. The serial nature of these measurements opens the door to problems such as changing total ventilation or cardiac output, rendering the serial data internally inconsistent. Changing inspired O_2 levels potentially will lead to altered vasoconstriction (as alveolar P_{O_2} changes) and also to absorption atelectasis (Briscoe et al., 1960; Dantzker et al., 1975) that change the distribution of $\dot{V}_A/\dot{Q}$ as it is being measured. Inert gases on the other hand, can be administered simultaneously so as to avoid these problems.

This chapter will therefore describe the practical aspects of multiple inert gas measurement by gas chromatography. At the outset it is well worth a brief discussion of the choice of gas chromatography (GC) over the only other currently available approach, namely, mass spectrometry (MS).

Currently, GC is more practical than MS, although with future technological development, MS may well prove superior. The GC is far less expensive than any MS with sufficient sensitivity to make the trace concentration (parts per million) measurements required. In addition, gas chromatographs are much simpler and accordingly more rugged; they are much less liable to malfunction, and what repairs are needed are usually quick and inexpensive to carry out. However, to measure blood gas levels, the gas of interest must first be extracted into the gas phase to be measured by GC (while it is in theory possible to inject blood directly into the GC, the organic components will quickly pollute the column and cause performance to be degraded). On the other hand, with suitable interfaces (Gale and Wagner, 1982), gas can be directly measured in blood by MS. The second advantage of the MS is speed of measurement. Essentially simultaneous measurement of many gases can be made with a quadrupole mass spectrometer in a matter of a few seconds, but each sample takes several minutes for elution by GC, a property that reduces the number of samples that can be processed in any given experiment.

2. Structure and function of the gas chromatograph

2.1. Structure of a basic gas chromatograph

Figure 1 is a simple block diagram of the basic components of a gas chromatograph. Some inert carrier gas (such as helium or N_2) is constantly flowing through the entire system which consists fundamentally of an inlet valve, a column and a detector. There are many different kinds of columns, but they all serve the same function: to separate in time the appearance at the detector of each of the components of a gas mixture introduced at the inlet valve. The detector then senses each component non-specifically as it appears by one of several different

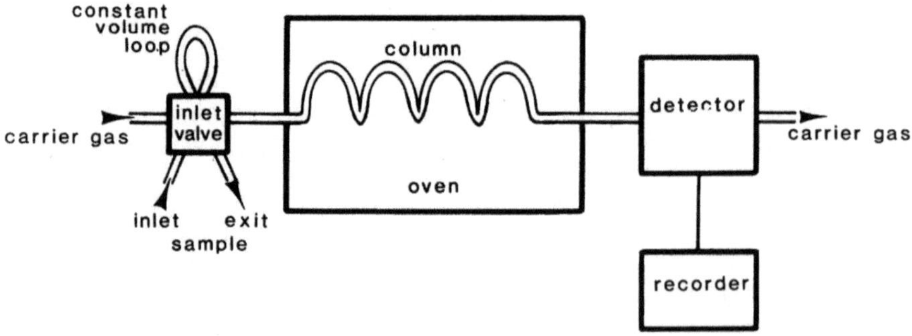

Fig. 1. Block diagram of a typical gas chromatograph. Carrier gas flows through the system by way of the column contained within a heated oven, and then passes on through the detector and out of the system. When a sample is to be measured, it is introduced by filling a constant volume loop and turning the inlet valve to flush the loop into the stream of carrier gas and hence into the column, and then detector. Detector outputs can be treated manually or automatically as discussed in the text.

principles such as thermal conductivity, flame ionization or electron capture (see below). This figure illustrates the simplicity of a basic GC system. To some extent, design of a system is empirical, so that different arrangements can be devised that will all achieve the same desired goals. Thus, while certain equipment configurations and functional specifications will be presented below, the reader should use these as a starting point, and may well expect to end up with a different arrangement due to differences in application and various companies' products, both in hardware and, specifically, column packing material.

The components of the GC as illustrated in Fig. 1 will now be discussed in turn.

2.2. Carrier gas

A carrier gas is needed to provide bulk transport of the sample from the point of introduction (inlet valve) to the detector. Clearly, such a carrier gas should itself be inert, that is, not produce a significant signal as it passes through the detector. The most common carrier gases are N_2 and helium. For detection of inert gases we use N_2 as the carrier for the detection of SF_6 by electron capture and helium as the carrier for the detection of the remaining gases (all hydrocarbons) by flame ionization (see below).

2.3. Sample introduction

Any GC must of course have a means of introducing the gas sample into the flowing stream of carrier gas. There are two principal ways of doing this: manual injection from a syringe into the stream of carrier gas through a plastic membrane (septum). The syringe contains the sample, and a specific volume must be injected. Septa must be replaced after a number of samples due to the multiple perforations produced by prior injections. A major disadvantage is that of reproducibility of introduced sample volumes: since the signal will reflect the total mass of gas injected, variation in sample volume will give rise to variation in signal, even if the concentration of gas in samples is not varying. For these reasons, the constant volume inlet sample loop is preferred. Figure 2 illustrates the ingenuity of such a loop with but two possible positions (far more elegant versions are commercially available for more sophisticated needs). In the "fill" position the carrier gas flows directly through the valve and on into the column as in Fig. 1. In the fill position the sample is flushed into the constant volume loop at the "inject" port in a volume at least as great as the loop itself. In practice, perhaps twice the loop volume is used to ensure adequate purging of gas present prior to injection. Any excess sample passes to the room at the "exit" port. Then, to inject the sample into the carrier gas stream, a handle is turned, putting the system into the "inject" position. This physically connects the constant volume loop in series with the carrier gas which then flushes the sample into the column. The clear-cut advantage of this approach is the constancy of volume of the injected sample. Thus, if two samples give rise to different signals, it must be

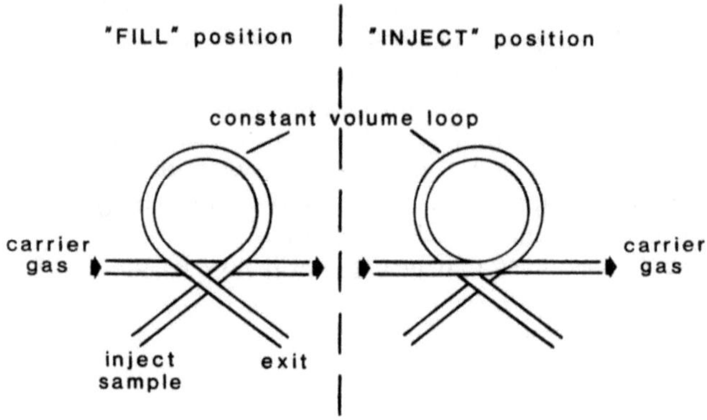

Fig. 2. Principles of the constant volume inlet loop. The loop is simply a hollow stainless steel tube and in the fill position is open to the atmosphere. Thus a sample can be introduced through one end of the tube and the entire tube filled by flushing the loop. In the fill position, carrier gas bypasses the loop and flows straight into the column. In the inject position, at the turning of a valve, the constant volume loop is connected in series with the carrier gas stream so that the contents of the volume loop are flushed into the carrier gas stream.

because of concentration differences and not because of possibly different injected volumes.

Such a constant volume loop injector is considered essential for careful quantitative work as, for example, is required in the multiple inert gas elimination approach.

2.4. *The column*

A column is simply a hollow tube packed with some material that interacts with the gases passing through it. Different gases (injected as a mixture) are retarded to different degrees according to their partitioning into the material, so that they are separated by the time they reach the end of the column and the detector. Separation is, in fact, the sole purpose of the column. Parameters of a column can be selected so as to optimize analysis (optimization being defined essentially as adequate separation in the minimum total elution time). The parameters are divided into three categories: (1) those pertaining to the column casing (length and diameter); (2) those pertaining to the packing material (choice of material, size of particles in the material, density of packing the column); (3) operational parameters (column temperature and carrier gas flow rate). Choice of parameters is a matter of trial and error, and more than one combination may prove acceptable. We happen to use 1.8 m long columns of 3.2 mm diameter stainless steel. They are packed with PORAPAK-T in a particle size of 80–100 mesh. Column temperature is held at ~160°C for hydrocarbon detection by flame ionization and at ~60°C for SF_6 detection by electron capture (see below). We use carrier gas flow rates of about 35 ml/min. These choices should be viewed as an

P403/6

approximate starting point and adjustments may be made to minimize elution time while preserving separation. Separation is improved by increasing column length, lowering column temperature, and lowering column carrier gas flow rate. However, elution time is prolonged by these very same changes, which provide the three principle variables (other than column packing material) for optimizing the analysis. Other potential column materials include molecular sieves and other Porapak series (S, Q).

A gas chromatograph is an extremely robust analytical instrument, but one way to create operational problems is to use column temperatures that are too high. Each column material has an upper limit of temperature before it begins to decompose. For Porapak T, this is 190–200°C. Long-term operation of the column must therefore be below this ceiling, preferably some 20–30°C lower. Consequently, a safe operating limit would be 170–175°C for this material. Under such conditions, column life is essentially indefinite in multiple inert gas analysis.

Another critical operational point is that whenever the column is heated, carrier gas must be kept flowing at all times (even when the instrument is not being used). Failure to do so will also cause column decomposition, which will seriously degrade its performance and also result in detector contamination, reducing signal/noise ratios and stability.

2.5. Detectors

A detector will respond to a particular type of gas in a non-specific way, and thus depends on the prior separation by the column to allow each individual gas in a mixture to be analyzed. The common detectors used in gas chromatography are: (1) thermal conductivity (TCD); (2) flame ionization (FID); (3) electron capture (ECD). Other detectors exist (McNair and Bonelli, 1969) but are much less commonly used. For multiple inert gas detection we need only the FID and ECD, but each of these three detectors will now be briefly described.

2.5.1. Thermal conductivity detector

This operates on the principle that a heated filament will lose heat at a rate that depends on the composition of the surrounding gas. Thus, any gas with a different thermal conductivity from that of the carrier gas is detectable; thermal conductivity is, to a large extent, related inversely to molecular weight. As the filament loses heat, its electrical resistance changes and this is sensed and displayed as an output. Compared to FID and ECD, TCD is about 100–1000-fold less sensitive, and so not as useful for trace gas analysis. It is a good means of detecting large quantities of gases such as O_2 and CO_2 in expired air.

2.5.2. Flame ionization detector

The FID is an excellent detector for hydrocarbon gases, and is characterized by stability, linearity over a wide dynamic range and high sensitivity. It is the cornerstone of trace inert gas analysis in our laboratory, and can detect levels in the range of 1–10 ppm with entirely adequate signal-to-noise ratios. The operating principle is the creation of ions and electrons during the combustion of hydrocarbons in a hydrogen-air flame. These ions and electrons enter an electric field between an anode and a cathode where a polarizing voltage accelerates them to form a measurable current. Consequently, the carrier gas should be non-combustible (N_2, He).

Once the entire GC system has been set-up correctly, the FID is virtually maintenance-free. Its linearity is especially remarkable, with a dynamic linear range of 10^7. Detector temperature can be raised to burn off deposited material, but its operating temperature should be around 200°C for most situations. This is because for our system at least (Beckman GC72-5), linearity is lost at temperatures above about 250°C. Rarely a reduction in signal/noise ratio or baseline instability will develop. In our hands, this only happens when a mistake has been made (e.g. turning off column carrier gas for a while or inserting an unconditioned column (see later) which may bleed into the detector and cause contamination). If baking the detector at say, 300°C overnight fails to cure the problem, most FID's can be easily removed for direct cleaning.

2.5.3. Electron capture detector

This is an exceedingly sensitive detector capable of measuring parts per billion of gases that easily accept electrons. Thus, halogenated compounds are well-suited to analysis by this detector. In the multiple inert gas technique, SF_6 is such a gas, and the sensitivity of the ECD is well-matched to the very low arterial SF_6 levels seen in normal subjects. The principle upon which the ECD operates is as follows. A radioactive source (either tritium or nickel-63) ionizes some of the N_2 carrier gas molecules as they pass through the detector. Electrons are formed and accelerated to an electrode under a voltage to form what is known as the standing current. Then, if a sample containing a gas capable of capturing electrons passes through the detector, this standing current is reduced in proportion to the number of electrons captured. This reduction in current is then displayed as the signal. As can be appreciated, once the standing current is completely reduced, no further signal can be obtained from a sample of higher concentration. Hence, the ECD is inherently alinear, with the signal flattening off above a certain concentration level (Fig. 3). To analyze samples whose concentrations put them on the flat part of the response curve of Fig. 3, prior dilution is necessary to bring the diluted sample down onto the steep part of the response curve. This narrow dynamic range (about 10^2, compared to 10^7 for the FID) leads to intrinsincally greater errors in measurement when prior dilution is necessary, since any such maneuver leads to unavoidable errors. ECD detectors made by different companies vary to some extent in their linearity and attempts have been made to correct for this problem electronically. The safest approach is to determine linearity in the physiological range for the actual gases to be used before committing to any one device. Another problem with the ECD is that it is affected by a fair number of compounds, some of which create definite difficulties in analysis. For example, isoproterenol given to patients with obstructive lung disease causes, after some time, a large slow reduction in standing current. As standing current changes, so

P403/8

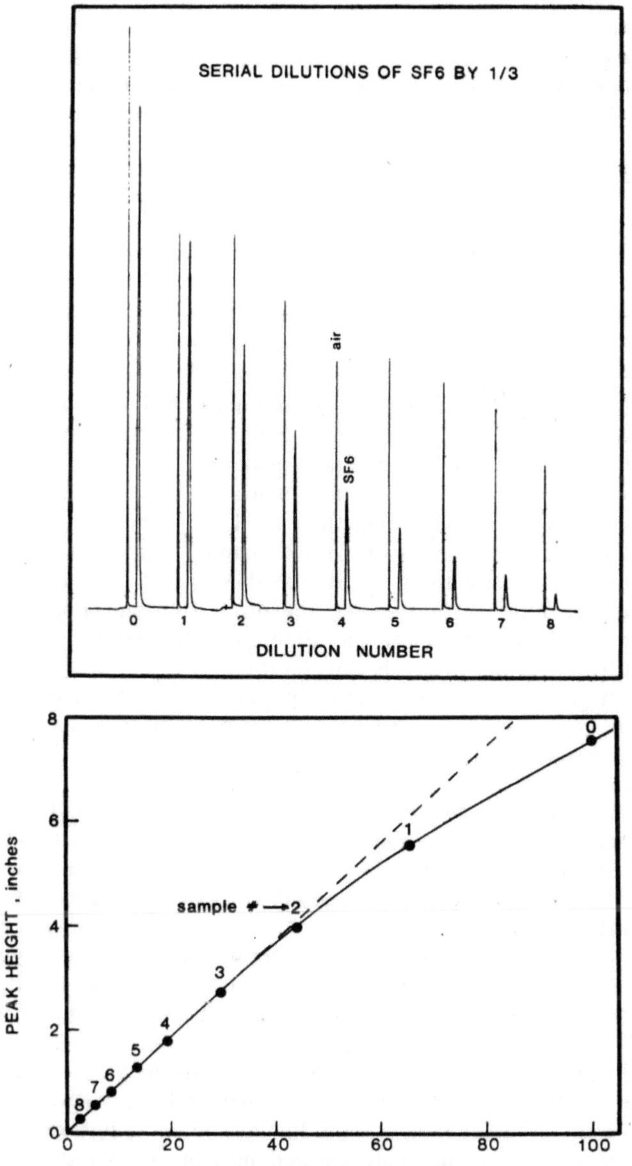

Fig. 3. Serial dilutions of sulfahexafluoride by one-third each time, recorded using the electron captured detector. The upper panel shows the actual chromatograms of eight serial dilutions, and the lower panel is the relationship between peak height and concentration as a percent of the initial value at dilution number 0. Note the alinearity of the relationship with peak height tending to flatten off with more concentrated samples. Optimum performance involves prior dilution of samples to place them on the linear portion of the curve.

does the basic sensitivity of the ECD and all analysis must stop until the contaminant is completely eluted. This annoying problem can be resolved more rapidly if the column temperature is raised from its usual level (50–60°C) to about 140°C for 15 min, with a subsequent cool down time of 30 min. Doing this rapidly elutes the contaminant and returns the standing current to its previous level. The user should be watchful for other halogenated drugs (especially in the ICU setting where polypharmacy is the rule) that might cause similar problems.

2.6. Recording of data

The standard outputs of most gas chromatographs feed into an 0–1 mV full scale analog recorder which produces a permanent record on paper. It is important that any such recorder have a range of paper speeds. Gases eluted early have sharp narrow peaks and require a fast paper speed for visual separation (say, 10–20 cm/min). However, gases eluted later such as enflurane or halothane, are best recorded at 0.6 cm/min to preserve recognizable peak form.

Most companies now offer a line of devices that will automatically read and calculate peak height or peak areas and provide a printed set of values for each sample. These are clearly great time savers, since manual peak reading is laborious. However, our experience with these automated devices has led us to be somewhat wary of their accuracy. In any event, the printed output should be validated against manually read values at the time of initial setting up of equipment, and thereafter at regular intervals (say, 6 monthly).

For the multiple inert gas approach, the FID is operated at a high sensitivity, such that "contaminant" peaks due to air and water vapor produce some baseline interference. This may get in the way of general purpose automated devices performing as desired.

We have, in fact, hard wired our chromatograph (both ECD and FID) to our MINC-11 laboratory computer and written Fortran software specific to the chromatograms obtained in the multiple inert gas approach. This system allows direct transfer of data to the same computer that will subsequently process the inert gas data, so providing rapid inert gas analysis for interactive studies where the next experimental step may depend on the current pattern of inert gas exchange. A critical feature of our system is the real time display of the chromatogram and the peak measurement variables on a graphics terminal to ensure for each sample that the peaks are being correctly processed.

The advantage of the industry-supplied peak scanners is that they are "off the shelf" items, but their disadvantage is that due to their general purpose nature, care must be taken to ensure accuracy in operation. In addition, most such devices require that the printed output be manually entered into one's computer for subsequent analysis of inert gas levels in terms of ventilation/perfusion mismatching.

The advantages of our system are that it costs essentially nothing (given the prior existence of the MINC-11 computer with analog/digital converter and graphics terminal), and is tailored specifically to the particular chromatograms seen in the inert gas approach. Moreover, there is constant visual feedback to ensure accuracy, and no re-entering of peak data into the computer is necessary. The disadvantages of our system are in the past and pertain to the time needed to develop the software for peak processing and interfacing the data with the inert gas analysis programs.

Thus, at least three options for data accumulation exist, and each investigator should choose what best fits his needs.

3. Setting up and checking out a new gas chromatograph system for multiple inert gas studies

This section is intended for the reader interested in developing the capability for making trace inert gas measurement in both blood and gas samples. However, the sequence of steps to be described is, in fact, general and could be applied in the development of any gas chromatograph system. What follows is a practical step-by step approach to establishing the accuracy and reproducibility of the measurement techniques.

The sequence begins with the assumption that a gas chromatograph fitted with a constant volume inlet loop and appropriate column has been connected to the necessary gases (air, H_2 and a carrier gas for the FID, and N_2 as a carrier gas for the ECD) and that a 0–1 mV recorder is connected. Furthermore, it is assumed that the entire system appears to be functioning in that the flame in the FID stays lit, the electronics are healthy and qualitatively reasonable peaks are obtained with sample introduction.

3.1. Step 1: Column conditioning

Before attempting any analysis, the chosen column(s) must be "conditioned" to remove residual species such as solvents that are present in the column material as a result of manufacture. This involves overnight heating of the column to some 20°C above the intended operating temperature (but not so hot as to exceed the recommended maximum temperature). For example, we use a Porapak-T column at 155°C and would condition it at 175°–180°C. Conditioning is performed by connecting the column to the inflowing carrier gas (in the oven compartment), but leaving the effluent end of the column detached from its future connection to the detector. If the column is not left detached, the species eluted by conditioning may well contaminate the detector and degrade its performance. With the column connected to carrier gas (the latter flowing at 20–30 ml/min), the oven is closed and temperature set as stated. The next day the column can be connected to the detector and the system turned on.

3.2. Step 2: Leak detection

The next step is to check for leaks of gas at the many connections between the carrier gas tank upstream and the detector downstream. With carrier gas flowing at some 30 ml/min and the oven not heated, each connection is checked with a low surface tension leak detecting fluid such as "SNOOP", and all leaks corrected. When no more leaks are apparent, a sample of carrier gas should be injected via the constant volume inlet loop and the subsequent chromatogram examined at high sensitivity. There should be no deflections above or below the baseline. Any significant deflection probably indicates leakage and the checking procedure should be repeated.

3.3. Step 3A: Choosing operational settings (FID)

While fine tuning of settings will probably take place later, the next step is to select the combination of carrier gas flow and oven (column) temperature for the gas mixture that will be regularly analyzed. To do this, prepare syringes containing each component of the intended mixture alone. Inject them one at a time and carefully measure the appearance times (that is, onset of peak rise, peak maximum and end of peak times) for each gas. This allows positive identification of each species later. Finally, as a control, inject a sample of room air to characterize the "null" chromatogram. With the FID, an early peak due to air will be seen and possibly also a later small broad peak that seems to be due to water vapor. When the gases have been individually characterized, prepare a syringe containing a mixture of them all in room air and at about physiologic concentrations (10–100 ppm). This should then be injected and the resulting chromatogram inspected for: (a) component gas separation; (b) total elution time. Adjustments are then made to either or both the carrier gas flow rate and the oven temperature. As shown in Figs. 4 and 5, elution time, sensitivity and separability are all sensitive to both variables. A halving of elution time can be achieved by increasing the temperature about 20°C or by doubling the carrier gas flow rate. Note that too high a carrier flow rate will blow out the hydrogen flame!

Every time a flow or temperature adjustment is made, a good 30 min should be allowed for the conditions to stabilize or confusing results will be obtained.

For the gas mix used in the multiple inert gas technique, a Porapak-T column 80/100 mesh, 1.8 m in length and 3.2 mm diameter works well for us at a temperature of 155°C and carrier flow rate of 35 ml/min (Figs. 4 and 5). Because every column and system is different, these parameters will be useful only as initial guidelines and the reader should not be dismayed if different conditions allow adequate separation with minimal elution time. A total elution time of $\leq$4 min is a good goal to reach.

3.4. Step 3B: Choosing operational settings (ECD)

We choose to use a 1.8 m, 3.2 mm diameter 80/100 mesh Porapak-T column (just as for the FID), but keep it in a separate oven at 60°C. This low temperature keeps gases like enflurane from appearing and interfering with rapid sample processing. Under this condition, SF_6 is eluted within 1–2 min. Column flows and temperatures can be adjusted to the operator's desires as with the FID.

P403/12

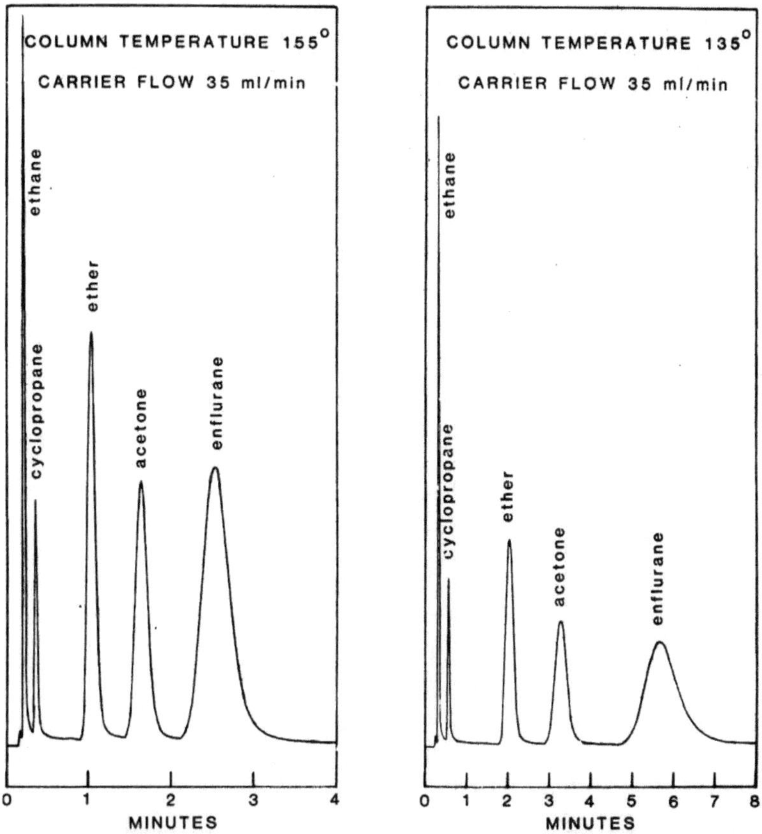

Fig. 4. Effect of column temperature on the gas chromatogram. The five hydrocarbons used in the inert gas technique are shown in a typical tracing at 155°C column temperature on the left and 135°C on the right. Carrier gas flow rates are the same in each case. Note the different time scales and also that the height of the peaks is considerably less at the lower temperature (both chromatograms come from the same gas sample and have been recorded at the same gain). Thus the effect of reducing temperature 20°C is to increase elution time by approximately a factor of 2 for the slowest gas, enflurane, and also reduce peak height by a variable amount from about 10% (ethane) to more than 50% (enflurane). Note that separation of gases is essentially as good at 155°C as at 135°C, and thus it would be far better to run the system at 155°C than 135°C from the point of view of signal-to-noise ratio and elution time.

3.5. Step 4: Stability

After a new system has been set-up, the columns conditioned and the parameters approximately set, it may take 1–3 days to achieve the baseline stability necessary for analysis of parts per million levels of inert gases. The chromatograph should be on, with the flame lit, and at a high sensitivity during

P403/13

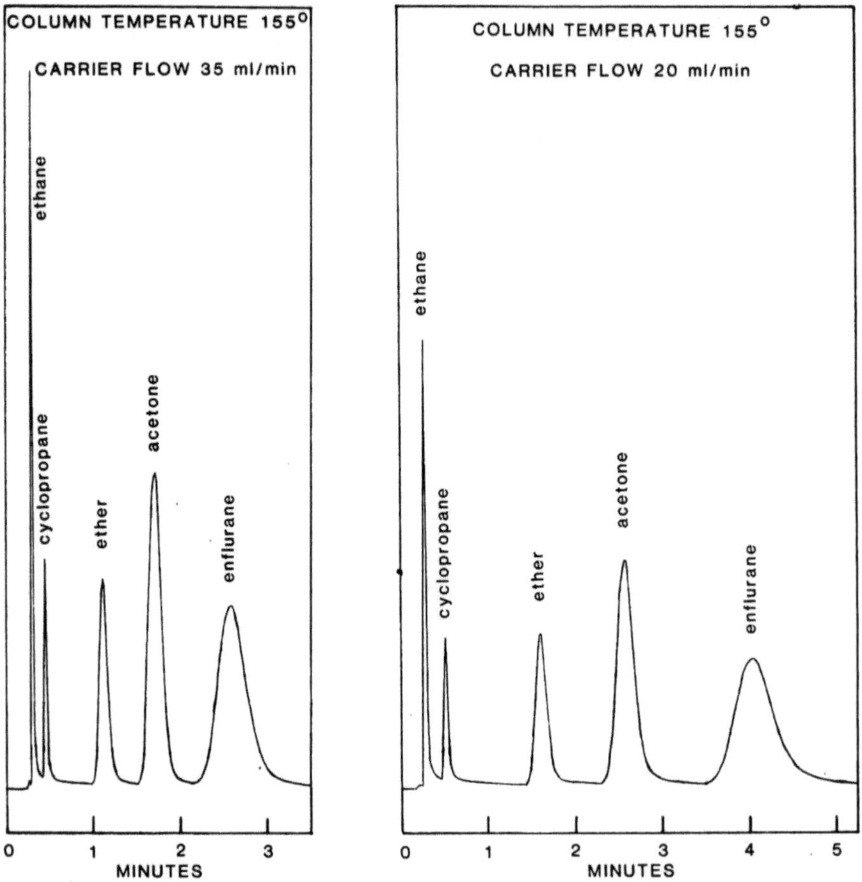

Fig. 5. Similar comparisons as in Fig. 3, but showing the difference between a carrier gas flow rate of 35 ml/min and one of 20 ml/min at the same column temperature of 155°C. The effects of reduction in carrier flow are qualitatively similar to those of reduction in temperature and include an increased elution time and decreased peak height for the same sample (the two chromatograms are from the same sample recorded at the same gains).

this time, and the baseline should be observed at say, hourly intervals during the daytime. After 1–3 days the baseline should be stable over hours with little noise (i.e. limited to pen thickness) at the gains to be used in subsequent studies. It is difficult to be quantitative here, but we never need to reset baselines due to drift over the course of a day's work, and such a level of stability should be sought. It is surprising how long a new system will take to stabilize. There is no point in proceeding to the next steps unless stability is adequate.

P403/14

3.6. *Step 5: Sample volume size and sample memory: ECD and FID*

Assuming stability, a large syringe containing the typical gas mix should be analyzed using a series of sample volumes to fill the constant volume inlet loop: for a typical system, exactly 1 ml is a good starting point, and the chromatogram should be recorded. Next, a large (perhaps 20 ml) sample of carrier gas should be injected to ensure adequate purging of the sample loop. Then, a 2.0 ml sample should be injected and the chromatogram recorded. This sequence of increasing sample volumes (by 1 ml each time) and carrier gas flushes should be repeated until it is clear that the peak height shows no further increase with sample size. By plotting peak height against sample volume, a judgment can easily be made regarding the minimal sample size necessary to completely fill the constant volume loop. All subsequent measurements should be made with sample volumes 1.5 to 2.0 times the minimal size so determined.

Next, any sample memory effect should be investigated. Record the chromatogram from a fairly concentrated mixture of all intended gases and then inject twice the minimum sample volume of carrier gas (ensuring that the source of carrier gas is not contaminated with any of the inert gases) using a gain setting about 50–100-fold more sensitive than that needed to record the concentrated sample. Small peaks will probably be visualized for each gas in the mixture, demonstrating a memory effect. We feel that if such peaks are $\leq 1\%$ of the previous sample values (expressed at the same gains), then the memory effect is insignificant for the multiple inert gas approach (where we never follow a very concentrated sample by a very dilute one) because sequential samples tend not to vary greatly in concentration. Should significant memory effect be seen, you cannot proceed further until the problem has been corrected. The two probable causes are a defective or unsatisfactorily designed sample valve, or having the sample valve heated above room temperature. This diagnosis assumes that the carrier gas sample used in this test is not contaminated by the inert gases!

3.7. *Step 6: Repeatability (FID and ECD)*

Assuming the preceding steps have been satisfactorily undertaken, the next logical step is to determine reproducibility of the chromatograph itself. This is done by preparing a syringe large enough to hold 10 samples, and creating 10 sequential chromatograms from this single syringe as rapidly as possible. Stability of the system is important here, as is the need for a sample syringe that leaks negligibly over the time taken to record the 10 chromatograms (i.e. $\leq 1\%$). Under such conditions, the coefficient of variation of peak height should be less than 1% (and may well be 0.3–0.5%); there should be no systematic change in peak height over the sequence of injections either. Failure to achieve such reproducibility may have to do with stability, sample loss from the holding syringe, or persistent leaks in the gas chromatograph. Appropriate action to take is to allow more time to reach stability, obtain a tighter syringe and correct the leaks, respectively.

3.8. *Step 7: Linearity (FID)*

Most FID's are incredibly linear, through zero, over a dynamic range of 10^6 or 10^7! However, this should be documented for each gas in the intended mixture. This is done by serial dilutions (by 1/3 to 1/2) from a concentrated sample down as far as one wishes to go. The entire 10^7 range need not, of course, be investigated. The physiological range will likely be far narrower (perhaps 10^2 or so) and linearity need only be checked over the working range. A most important practical point in performing serial dilutions is that in any given syringe, serial dilution by more than a total factor of 50 should not be done. Thus, for serial dilutions by factors of 2, only 5–6 such dilutions should be done ($2^5 = 32$, $2^6 = 64$). If a wider range is required, the 50-fold diluted sample should be transferred to a clean syringe and the serial dilutions recommenced. If such a precaution is not taken, concentrations of gases may cease to fall linearly with dilution due to small pockets of residual concentrated gas equilibrating slowly with the diluted sample.

A longtime point of discussion is whether to use peak height or peak area when performing linearity checks, or in fact, when measuring inert gas levels. Most sources prefer peak areas as being theoretically proportional to sample mass, but unless an automated peak area measuring device is available, measurement of peak area is very difficult compared to that of peak height. We therefore prefer peak height; however only after having determined that peak heights behave linearly during serial dilution.

Provided the above precautions are taken, clean syringes with very low leak rates ($\leq$1%/h) are used, and the entire gas chromatograph system has been successfully shepherded through steps 1–6 above, linearity of peak height is virtually assured. It is worth noting that if detector temperatures are kept too high (FID), the inherent linearity may be lost. One should determine the temperature at which this observation first appears and keep the detector at a lower temperature. In our system, linearity is lost above about 280–300°C so that we keep the FID at 200–220°C (except when baking it overnight to clean it as part of regular maintenance; see below).

3.9. *Step 8: Measurements of gases present in liquids*

If steps 1–7 have been successful, one has a stable gas chromatograph whose operational characteristics have been well-defined for gas phase measurements. However, many studies will require measurement of inert gases in blood. While it is possible to inject small amounts of blood into a chromatograph via the heated septa referred to earlier, rapid column contamination by the organic components may occur and we have never attempted such an approach. The alternative is to extract the gases of interest from the blood samples into a gas phase, and use the gas phase as the sample, as above.

Probably the most convenient method is equilibration at one atmosphere (as opposed to extraction under negative pressure). The technique is described more fully elsewhere (Wagner et al., 1974), but it runs briefly as follows: the blood sample is collected anaerobically into a clean, ungreased, matched barrel/plunger glass syringe whose nominal capacity is about twice that of the volume of blood collected. This syringe should have been labelled and then weighed dry, following which it should have been heparinized (leaving only the syringe deadspace containing heparin). Following heparinization, the stopcock should be blown free of excess heparin and the syringe re-weighed to determine the volume of heparin. Once the blood sample has been collected, the stopcock is washed free of excess blood and blown dry, and the syringe weighed for a third time to measure the weight of added blood. Any small bubbles must be retained in the syringe, not ejected as for routine blood:gas measurement. Next, either helium or N_2 is added to the syringe (in a volume about equal to that of the blood sample). The syringe is now placed horizontally in a shaking water bath so that all gases of interest can equilibrate between gas and blood phases. The water bath should be maintained at the body temperature of the animal or subject from whom the sample was obtained. The rate of shaking should be sufficient to promote considerable agitation without causing foaming. The time required for this equilibration procedure will be of the order of 30 min to ensure that all gases will have equilibrated. The necessary time will vary with the conditions, and should be established in each case experimentally in the manner reported previously (Wagner et al., 1974).

Once equilibrated, the gas phase should be separated from the blood phase by transfer into a clean, dry matched barrel/plunger ungreased glass syringe. This transfer should be done rapidly so that the syringe from the water bath has no time to cool down (and thus cause a requilibration of the sample). First, the total gas plus blood volume is read off the scale on the syringe. Next, the syringe is removed from the water bath, the stopcock vigorously blown dry and the clean recipient syringe connected at right angles. The blood-containing syringe should be held vertically, stopcock uppermost, and the recipient syringe will therefore be horizontal. Care should be taken to avoid water on the recipient syringe plunger. The gas phase is then rapidly transferred into this syringe. Apart from speed, the other critical factor in successful transfer of gas is the avoidance of transferring any blood or water into the recipient syringe. This entire procedure, with practice, can be achieved within about 10 s. The recipient syringe is then stored to await chromatography.

By definition, equilibration between gas and blood phases will not remove all of the gas of interest from the blood sample. Thus, while gas and blood tensions (say, P_1) will be the same after equilibration, this tension will be lower than the value we wish to know, that of the blood sample prior to equilibration (say, P_0). To determine P_0 from the measured value of P_1, we apply mass conservation principles:

$$V_B \lambda P_0 = V_B \lambda P_1 + V_H \lambda P_1 + V_G P_1 \tag{1}$$

where V_B is volume of blood in the syringe (weight divided by density, the latter measured directly subsequently from all pooled samples); V_H is volume of heparin in the syringe (=weight, assuming a density of 1.0); V_G is the gas volume at the temperature of the water bath. These volumes must all be in the same units (e.g. ml). In Eqn. 1, λ is the blood:gas partition coefficient of the gas being measured at the water bath temperature. The LHS of Eqn. 1 represents the mass of gas present in the blood sample before equilibration; the RHS represents the same mass now distributed in blood, heparin and gas phases. Because the volume of heparin is usually $\leq 5\%$ that of blood, assuming a λ the same in blood and heparin is sufficiently accurate.

Dividing both sides of Eqn. 1 by $V_B\lambda$ we get:

$$P_0 = P_1\left[1 + \frac{V_H}{V_B} + \frac{V_G}{V_B\lambda}\right] \qquad (2)$$

which allows calculation of the original blood partial pressure in terms of the measured value P_1. Depending on the application, absolute calibrations can be made from reference gas tanks or values can be kept in relative terms if only ratios of tensions are needed. As it happens, in the inert gas method for measuring $\dot{V}_A/\dot{Q}$ mismatch, we require only ratios of tensions (arterial/venous and mixed expired/venous). Hence, no absolute calibrations are necessary.

The entire procedure for measuring gases in blood as outlined must be refined to produce data with a coefficient of variation of 2–3% at physiological levels when determined over 10 samples from a common pool. The exception is SF_6 which will usually be found to be less reproducible (5–6% coefficient of variation) due to the initial dilution of gas samples required because of the ECD alinearity (Fig. 3) and the need to have samples appear on the relatively linear portion of that response curve.

If the above eight steps are satisfactorily accomplished according to the stated criteria, the system is ready to use in investigation of $\dot{V}_A/\dot{Q}$ mismatch.

4. Experimental details in the inert gas elimination technique

4.1. Preparation of infusate

We have come to use the six gases, SF_6, ethane, cyclopropane, enflurane, diethyl ether and acetone. While the greater the number of gases used, the more the information content of the method, this choice of gases appears to be a reasonable compromise between analytical complexity and information content. For non-survival experiments in animals, sterility is not necessary; however, the following description pertains to the constitution of a sterile infusate for use in humans.

A tank of gas is obtained containing approximately 20% SF_6, 20% cyclopropane and 60% ethane. Note that this cannot be pressurized to more than about 200–300 lb/inch^2 to avoid liquefying cyclopropane. A plastic 1-l bag of either normal saline or 5% dextrose in water is connected to this gas tank by

P403/18

plastic I.V. tubing. Somewhere in the line, a 0.22-μm Millipore (TM) filter should be inserted, and all material between that filter and the bag of saline or dextrose should be of sterile disposable materials. Gas from the tank is pumped into the bag until the latter is somewhat distended. The tank is turned off, the bag is vigorously shaken for perhaps 1 min, and the gas in the bag is then vented to the room. A second aliquot of gas from the tank is allowed to fill the bag, and shaking is repeated. Two or three such cycles, with vigorous shaking between, is sufficient to equilibrate the saline or dextrose with the tank gas mixture.

All gas bubbles are allowed to coalesce, and the gas phase is ejected from the saline bag maintaining sterility. All of the liquid is retained in the bag at this point. Next, liquid analytical reagent ether is transferred into the bag (we use 0.2 ml diethyl ether per 100 ml saline), again using sterile disposable materials. Immediately thereafter analytical reagent grade acetone (we use 1.0 ml acetone per 100 ml saline) is similarly added to the bag solution. Finally, 2 ml of liquid enflurane is transferred into the bag in the same way, and the bag is then gently inverted several times to allow the last three gases to distribute in the liquid phase. Since the gas tensions to be measured are those in pulmonary arterial blood, systemic arterial blood and mixed expired gas, it is not important to know the absolute levels of the inert gases in the infusate. It is also not important if some of each gas diffuses through the plastic of the container and escapes.

4.2. Infusion

Once prepared, the infusate is infused into a peripheral superficial vein by means of a roller pump directly from the 1-l plastic bag in which it was made. We use a second 0.22-μm Millipore filter in the line as a precaution against contamination. The infusion should be introduced as smoothly as possible in order to achieve steady state conditions. The rate of infusion depends upon the level of ventilation of the animal or subject. A good rule of thumb is to keep the infusion rate in ml/min at approximately half the ventilation rate in l/min. Thus, for a normal subject ventilating at 6 l/min, an infusion rate of 3 ml/min would be appropriate. During exercise, if ventilation were increased to 60 l/min, some 20–30 ml/min infusion rate would be required in order to achieve the same signal-to-noise ratio as during rest. While a rate of 20–30 ml/min may seem high, most exercise studies only take 10 min, and moreover with sweating during the exercise period, this is a negligible fluid load for subjects who can exercise to this capacity. The critical element in infusion, apart from sterility, is smoothness of flow rate, which should be tested ahead of time. Since the multiple inert gas elimination technique is a steady state method, it is important to allow the infusion to run for several minutes before taking any samples. We routinely wait 30–40 min to allow the lungs to come into a steady state of inert gas exchange.

4.3. Sampling procedures

Arterial and pulmonary arterial samples are drawn simultaneously (ignoring the 3–5 s period required to transport blood from the lungs to the arterial sampling site). The entire sample is collected by slow steady manual withdrawal over approximately 1 min. Such a technique allows time averaging over several respiratory cycles. The volume of the blood sample to be taken should be the smallest volume allowing adequately reproducible results with the particular gas chromatograph system available. In our hands, this is 5–7 ml/sample. Expired gas must also be collected. This should be mixed expired, since any other type of expired sample is not statistically representative of the lung as a whole in the presence of $\dot{V}_A/\dot{Q}$ mismatch. Because some of the inert gases in the technique are extremely soluble in water, plastic and rubber, we choose not to make traditional collections of expired gas in a large bag as is done in measuring oxygen consumption and CO_2 elimination. Thus we connect the animal or patient via a non-return valve to a mixing box constructed principally of metal and kept heated several degrees above body temperature. The mixing box is of such proportions so as to provide a constant concentration of both O_2 and CO_2 at the effluent end, damping out swings between deadspace and alveolar gas. We have for several years used a mixing box which consists of a long coiled copper tube 3.8 cm in diameter and some 69 cm in length. This achieves the stated goals of mixing, but produces a lag time that must be allowed for in the sampling procedure. In other words, expired gas at the mouth will take a finite time to reach the end of this long tube at which point it would be sampled. This lag time is calculated as the ratio of the volume of the mixing box to the minute ventilation. Thus, mixed expired gas samples are collected from the subject at a point in time later than those of arterial and pulmonary arterial blood, according to the above logic.

4.4. Measurements of inert gas concentrations

Measurements of inert gas levels in blood and gas are carried out exactly as described above in the section on setting up and checking out the gas chromatography system. Although the gas chromatography check-out is generally done prior to instituting animal or human experiments, it is important to re-establish that the check-out procedures were done using gases of concentrations appropriate to those obtained physiologically. Thus, linearity and reproducibility measurements in particular may have to be carried out again once some experience with animal experimentation has been obtained in order to ensure that the chromatography system is functioning adequately under the conditions of the particular experiment.

4.5. Solubility measurements

There is significant variation in blood gas partition coefficients (solubilities) between animals, between normal subjects and between species. These can amount to differences of 20% or greater, and suggest that direct measurement of partition coefficients for each animal or subject on a given day is a worthwhile enterprise. This has certainly become a part of our routine. The measurement of partition coefficients is accomplished in much the same manner as the measurement of gases in blood as mentioned above. However, rather than a single equilibration of blood with nitrogen or helium in the shaking water bath, two sequential equilibrations are performed. Thus, we begin with a blood sample that contains reasonable amounts of each of the inert gases. This blood sample is equilibrated with helium or nitrogen and the gas phase transferred into a clean syringe and measured by gas chromatography at this point. All gas bubbles must be eliminated to assure reproducible and accurate values, and the weight of the blood (needed for volume measurement) must also be determined. This same blood sample after the initial equilibration has a second aliquot of helium or nitrogen added to it and a second equilibration is performed in the water bath exactly as above. At the end of the second equilibration, the volumes of blood and gas in the syringe must be determined (as described above) and the gas phase drawn off and analyzed by chromatography.

In this manner, for each gas, two measurements are made. In essence, Eqn. 2 is invoked to calculate the blood:gas partition coefficient (λ), knowing the gas and blood volumes and the ratio of the peaks in the two successive equilibrations. Simple rearrangement of Eqn. 2 will give the formula for partition coefficient in terms of those variables. It has been our practice to measure these in duplicate and to take the average.

For the gases SF_6, ethane, cyclopropane and enflurane, the two series of equilibrations are performed using approximately equal volumes of blood and gas in the syringe. However, for the two soluble gases, ether and acetone, equal blood and gas volumes would be inappropriate. This is because on the second equilibration there would be so much of the gas left in the blood after the first equilibration that the ratio of the two peaks would be very close to unity and thus cause large errors in the calculated partition coefficient. Accordingly, in the second of the two equilibrations for ether and acetone, we use a gas volume of 20–40 ml and a blood volume of 0.25–0.5 ml. This large gas to blood volume ratio then produces significant reduction in peak height on the second equilibration and a more accurate estimate of solubility.

As with any aspect of this approach, the repeatability of solubility measurements in a given subject or animal should be determined at the time the chromatograph is initially set up.

4.6. Accounting for experimental errors in gas chromatography measurements in the estimation of $\dot{V}_A/\dot{Q}$ maldistribution

The data obtained during inert gas elimination studies (arterial/venous and mixed expired/venous partial pressure ratios for each of six gases) is fed into a computer program that estimates the distribution of ventilation/perfusion ratios necessary to account for that quantitative pattern of gas tension ratios (Wagner, 1977; Evans and Wagner, 1977). As described elsewhere, this involves the solution of six simultaneous linear equations, one for each gas. It is critically important that the numerical contribution, or weight, of each gas in this system is appropriate to the error of measurement for the tension ratios for each gas. We define retention, R, as the arterial/venous ratio and excretion, E, as the mixed expired/venous ratio. We seek an estimate of the variance of these ratios by which to weight each of the six equations.

Two scenarios are possible. One involves the situation where arterial (P_A) and mixed expired (P_E) gas levels are measured along with cardiac output ($\dot{Q}_T$), minute ventilation ($\dot{V}_E$) and partition coefficient (λ) for each gas. Mixed venous ($P\bar{v}$) inert gas levels are then computed by mass balance:

$$\dot{V}_E P_E = \lambda \dot{Q}_T (P\bar{v} - Pa) \tag{3}$$

Equation 3, for elimination of infused gases, expresses steady state mass balance from which $P\bar{v}$ is calculated, knowing all remaining variables. Thus,

$$P\bar{v} = Pa + \frac{\dot{V}_E P_E}{\dot{Q}_T \lambda} \quad \text{and hence} \quad R = \frac{1}{1 + \left[\frac{\dot{V}_E P_E}{\dot{Q}_T \lambda Pa}\right]} \tag{4}$$

If we define $\delta\dot{V}_E$, δP_E, $\delta\dot{Q}_T$, $\delta\lambda$ and δPa as the errors in measurement of $\dot{V}_E$, P_E, $\dot{Q}_T$, λ and Pa, we have:

$$R = \frac{1}{1 + \left[\dfrac{\dot{V}_E P_E \left(1 + \frac{\delta\dot{V}_E}{\dot{V}_E}\right)\left(1 + \frac{\delta P_E}{P_E}\right)}{\dot{Q}_T \lambda Pa \left(1 + \frac{\delta\dot{Q}_T}{\dot{Q}_T}\right)\left(1 + \frac{\delta\lambda}{\lambda}\right)\left(1 + \frac{\delta Pa}{Pa}\right)}\right]}$$

where $\dot{V}_E$, P_E, $\dot{Q}_T$, λ and Pa are the true values. Binomial expansion ignoring 2nd order terms gives:

$$R = \frac{1}{1 + \frac{\dot{V}_E P_E}{\dot{Q}_T \lambda Pa}\left(1 + \frac{\delta\dot{V}_E}{\dot{V}_E} + \frac{\delta P_E}{P_E} - \frac{\delta\dot{Q}_T}{\dot{Q}_T} - \frac{\delta\lambda}{\lambda} - \frac{\delta Pa}{Pa}\right)}$$

$$= \frac{1}{\left(1 + \frac{\dot{V}_E P_E}{\dot{Q}_T \lambda Pa}\right) + \frac{\dot{V}_E P_E}{\dot{Q}_T \lambda Pa}\left(\frac{\delta\dot{V}_E}{\dot{V}_E} + \frac{\delta P_E}{P_E} - \frac{\delta\dot{Q}_T}{\dot{Q}_T} - \frac{\delta\lambda}{\lambda} - \frac{\delta Pa}{Pa}\right)}$$

If we let

$$\alpha = \frac{\dot{V}_E P_E}{\dot{Q}_T \lambda Pa}$$

P403/22

we have

$$R = \left[\frac{1}{(1+\alpha)}\right] \times \left[\frac{1}{1+\frac{\alpha}{1+\alpha}\left(\frac{\delta\dot{V}_E}{\dot{V}_E}+\frac{\delta P_E}{P_E}-\frac{\delta\dot{Q}_T}{\dot{Q}_T}-\frac{\delta\lambda}{\lambda}-\frac{\delta Pa}{Pa}\right)}\right]$$

$$= \frac{1-\left(\frac{\alpha}{1+\alpha}\right)\left(\frac{\delta\dot{V}_E}{\dot{V}_E}+\frac{\delta P_E}{P_E}-\frac{\delta\dot{Q}_T}{\dot{Q}_T}-\frac{\delta\lambda}{\lambda}-\frac{\delta Pa}{Pa}\right)}{1+\alpha}$$

The variance V in R is thus:

$$V = \frac{\alpha^2}{(1+\alpha)^4}\left(\left(\frac{\delta\dot{V}_E}{\dot{V}_E}\right)^2+\left(\frac{\delta P_E}{P_E}\right)^2+\left(\frac{\delta\dot{Q}_T}{\dot{Q}_T}\right)^2+\left(\frac{\delta\lambda}{\lambda}\right)^2+\left(\frac{\delta Pa}{Pa}\right)^2\right) \tag{5}$$

For each gas, V is calculated from Eqn. 5 and each of the six equations is then divided by the square root of the corresponding variance before the distribution is estimated by algorithms described elsewhere (Wagner, 1977).

The second scenario occurs when λ, Pa, P_E and $P\bar{v}$ are all measured directly. Similar principles are used, but the analysis for determining variance is more complicated. This is because given λ, Pa, P_E and $P\bar{v}$, two partly independent estimates of retention (R) are available, and we thus choose an appropriately weighted average of the two. We proceed as follows.

First, the ratio $\dot{V}_E/\dot{Q}_T$ is obtained as a weighted average of values for the six gases from Eqn. 3:

$$\dot{V}_E/\dot{Q}_T = \frac{\sum W_i\lambda_i(P\bar{v}_i - Pa_i)/P_{E_i}}{\sum W_i} \tag{6}$$

(where W_i are estimated in an analogous manner to that above in the first scenario). With several gases contributing to the mean $\dot{V}_E/\dot{Q}_T$, we assume the variance of this ratio is much smaller than the variance in gas levels for any one gas. The two estimates of R referred to above are thus:

$$R = \frac{Pa}{P\bar{v}} \quad \text{and} \quad R = 1 - \frac{\dot{V}_E P_E}{\lambda \dot{Q}_T P\bar{v}}$$

Suppose the weighting is t, $0 \leq t \leq 1$ for each gas:

$$R^* = t\frac{Pa}{P\bar{v}} + (1-t)\left(1-\frac{\dot{V}_E P_E}{\lambda\dot{Q}_T P\bar{v}}\right) \tag{7}$$

where R^* is the weighted average of the two estimates. We must choose t to minimize the variance in R^*.

Proceeding as before, R^* can be expressed as:

$$R^* = t\frac{Pa}{P\bar{v}}\left(\frac{1+\dfrac{\delta Pa}{Pa}}{1+\dfrac{\delta P\bar{v}}{P\bar{v}}}\right) + (1-t)\left(1-\frac{\dot{V}_E P_E}{\dot{Q}_T\lambda P\bar{v}}\left(\frac{1+\dfrac{\delta P_E}{P_E}}{\left(1+\dfrac{\delta\lambda}{\lambda}\right)\left(1+\dfrac{\delta P\bar{v}}{P\bar{v}}\right)}\right)\right)$$

$$= tR\left(1+\frac{\delta Pa}{Pa}-\frac{\delta P\bar{v}}{P\bar{v}}\right)+(1-t)\left(1-\frac{\dot{V}_E E}{\dot{Q}_T\lambda}\left(1+\frac{\delta P_E}{P_E}-\frac{\delta\lambda}{\lambda}-\frac{\delta P\bar{v}}{P\bar{v}}\right)\right)$$

$$= tR + (1-t)\left(1-\frac{\dot{V}_E P_E}{\dot{Q}_T\lambda}\right) + tR\frac{\delta Pa}{Pa} - tR\frac{\delta P\bar{v}}{P\bar{v}}$$

$$-(1-t)\frac{\dot{V}_E E}{\dot{Q}_T\lambda}\frac{\delta P_E}{P_E} + (1-t)\frac{\dot{V}_E E}{\dot{Q}_T\lambda}\frac{\delta\lambda}{\lambda} + (1-t)\frac{\dot{V}_E E}{\dot{Q}_T\lambda}\frac{\delta P\bar{v}}{P\bar{v}}$$

the variance V in R^* is thus

$$V = t^2R^2\left(\frac{\delta Pa}{Pa}\right)^2 + \left[-tR+(1-t)\frac{\dot{V}_E E}{\dot{Q}_T\lambda}\right]^2\left(\frac{\delta P\bar{v}}{P\bar{v}}\right)^2 + (1-t)^2\left(\frac{\dot{V}_E E}{\dot{Q}_T\lambda}\right)^2\left(\frac{\delta\lambda}{\lambda}\right)^2$$

$$+(1-t)^2\left(\frac{\dot{V}_E E}{\dot{Q}_T\lambda}\right)^2\left(\frac{\delta P_E}{P_E}\right)^2 \qquad (8)$$

When Eqn. 8 is differentiated with respect to t to give conditions for minimum variance, the associated value of t is given by:

$$t = (Y+Z)/(X+Y+2Z)$$

where

$$X = R^2\left[\left(\frac{\delta Pa}{Pa}\right)^2 + \left(\frac{\delta P\bar{v}}{P\bar{v}}\right)\right]^2$$

$$Y = \left(\frac{\dot{V}_E E}{\dot{Q}_T\lambda}\right)^2\left(\left(\frac{\delta P_E}{P_E}\right)^2 + \left(\frac{\delta\lambda}{\lambda}\right)^2 + \left(\frac{\delta P\bar{v}}{P\bar{v}}\right)^2\right) \quad \text{and} \quad Z = \frac{\dot{V}_E E R}{\dot{Q}_T\lambda}\left(\frac{\delta P\bar{v}}{P\bar{v}}\right)^2$$

This t value is substituted into Eqn. 8 to give the minimum variance from which the weight is calculated as its inverse square root, as above. The minimum variance retention is itself calculated from Eqn. 7 using the above t value, and these data are then fed into the algorithm for estimating the distribution.

5. Summary

This chapter has attempted to give the reader a reasonably broad, but at the same time reasonably detailed overview of equipment requirements, operational requirements, and technical aspects of the analytical methods needed to measure

P403/24

a mixture of inert gases in both the gas phase and the blood phase using gas chromatography. Important practical points have been emphasized in order to make the reader aware of potential problems and issues that should be concentrated upon to minimize the error in the measurements. Particular attention has been paid to a sequential approach to the initial set-up and checking out of a gas chromatograph system since it is the authors' experience that many workers interested in making multiple gas measurements have difficulty in the beginning in achieving satisfactory chromatography performance. It is hoped that this strict series of steps will obviate some of the problems that may develop. Finally, all of the technical aspects described are those functioning in the author's laboratory, but it is entirely possible that different arrangements could lead to an equally acceptable result. This comment applies particularly to issues such as gas chromatograph functional specifications (column temperatures and flow rates and arrangement of gas chromatograph ovens and analyzers). As long as the final objectives of reproducible and accurate measurements can be made at biological concentrations, it seems of little importance as to how one has arrived at this successful result.

References

Briscoe, W.A., Cree, E.M., Filler, J., Houssay, H.E.J. and Cournand, A. (1960) Lung volume, alveolar ventilation and perfusion interrelationships in chronic pulmonary emphysema. J. Appl. Physiol. 15, 785–795.

Dantzker, D.R., Wagner, P.D. and West, J.B. (1975) Instability of lung units with low $\dot{V}_A/\dot{Q}$ ratios during O_2 breathing. J. Appl. Physiol. 38, 886–895.

Evans, J.W. and Wagner, P.D. (1977) Limits on $\dot{V}_A/\dot{Q}$ distributions from analysis of experimental inert gas elimination. J. Appl. Physiol. 42, 889–898.

Farhi, L.E. (1967) Elimination of inert gas by the lung. Respir. Physiol. 3, 1–11.

Farhi, L.E. and Rahn, H. (1954–1955) A theoretical analysis of the alveolar-arterial O_2 difference with special reference to the distribution effect. Special Communication. J. Appl. Physiol. 7, 699–703.

Farhi, L.E. and Yokoyama, T. (1967) Effects of ventilation-perfusion inequality on elimination of inert gases. Respir. Physiol. 3, 12–20.

Gale, G.E. and Wagner, P.D. (1982) Continuous on-line measurement of inert gases in arterial and mixed venous blood. The Physiologist 25, 267.

McNair, H.M. and Bonelli, E.J. (1969) In: Basic Gas Chromatography, 5th edn., Varian Aerograph, Walnut Creek.

Wagner, P.D. (1977) Measurement of the distribution of ventilation-perfusion ratios. In: G.G. Davies and C. Barnes (Eds.), Research Topics in Physiology, Vol. 1, Regulation of Respiration and Gas Exchange, chap. 8, Academic Press, New York, pp. 217–260.

Wagner, P.D. (1980) Ventilation-perfusion relationships. Annu. Rev. Physiol. 42, 235–283.

Wagner, P.D., Naumann, P.F. And Laravuso, R.B. (1974) Simultaneous measurement of eight foreign gases in blood by gas chromatography. J. Appl. Physiol. 36, 600–605.

West, J.B. (1969) Ventilation-perfusion inequality and overall gas exchange in computer models of the lung. Respir. Physiol. 7, 88–110.

West, J.B. And Wagner, P.D. (1977) Pulmonary gas exchange. In: J.B. West (Ed.), Bioengineering Aspects of the Lung, Vol. 3, chap. 5, Marcel Dekker, New York/Basel, pp. 361–457.

Yokoyama, T. and Farhi, L.E. (1967) Study of ventilation-perfusion ratio distribution in the anesthetized dog by multiple inert gas washout. Respir. Physiol. 3, 166–176.

See also: Techniques in the Life Sciences, B2/II. *Metabolic Research* – Part II:B215, Application of high-performance liquid chromatography (HPLC) to biochemical analysis.

Appendix B: Short Program PDF

```
      PROGRAM SHORT
C
C      This is the FORTRAN data entry program for the steady state
       multiple inert
C      gas elimination technique (MIGET).  It creates a raw data
       file & an input file
C      for VQBOHR, the program which estimates ventilation and
       blood
C      flow distributions from the MIGET data.
C      Entry of data is format-free using DOS.
C      The reason we have two sequential, separate programs for
       MIGET is to allow the user
C      to review the data for errors (ie, review after running
       this, "SHORT" program)
C      before trying to process the data into VA/Q distributions.
C
C      Data files are named from the keyboard.
C
C      Updated by PDW  on DECEMBER 14, 1990 annotated 6/22/17
C
      CHARACTER IQ,IADDON,IOK,IFLAG,NAMRAW
      DIMENSION S(10),PC(10),PA(10),PE(10),PV(10),R(10),E(10)
      DIMENSION WT(10),ERRA(10),ERRE(10),ERRV(10),ERRPC(10),RMV(10)
      DIMENSION EMV(10),VARQT(10),QTCALC(10),PAC(10),PVC(10),PEC(10)
      DIMENSION NAMRAW(60),GA(10),GE(10),GV(10),PCC(10)
      DIMENSION HUMSLO(6),DOGSLO(6),HORSLO(6),PCFACT(6),PCBODY(6)
      DATA GA/10*1.0/,GE/10*1.0/,GV/10*1.0/
      DATA HUMSLO/2950.,1374.,2025.,3016.,4066.,805./
```


S.R. Hopkins, P.D. Wagner, *The Multiple Inert Gas Elimination Technique (MIGET)*, Methods in Physiology, https://doi.org/10.1007/978-1-4939-7441-2

```
      DATA DOGSLO/2263.,2396.,2909.,4435.,3281.,1391./
      DATA HORSLO/1251.,2466.,3262.,2696.,3877.,2412./
C
C       The program starts by offering options that appear as ques-
        tion on the screen on file names,
C     and whether results are to be written to disk, to screen, to
      printer
C
        WRITE(*,1)
   1    FORMAT(' WRITE OUTPUT TO DISK (0) OR TO PRINTER (1) ?')
        READ(*,*)IOUT
        IF(IOUT.EQ.0) WRITE(*,2)
   2    FORMAT(' ENTER  DISK:NAME.EXT  OF OUTPUT DATA FILE'/)
        IF(IOUT.EQ.0) OPEN(3,FILE=' ',STATUS='NEW')
        IF(IOUT.EQ.1) OPEN(3,FILE='PRN')
        WRITE(*,3)
   3    FORMAT(' ARE DATA FOR MAN(1), DOG(2), OR HORSE(3) ?')
        READ(*,*)ISPEC
        WRITE(*,4)
   4    FORMAT(' WANT TO CORRECT FOR BODY/BATH TEMP DIFF ? 1=YES,
        0=NO')
        READ(*,*)IBATH
        WRITE(*,1234)
 1234   FORMAT(' WANT TO CORRECT FOR EXPIRED ACETONE LOSS ? (YES=1/
        NO=0)')
        READ(*,*)IACET
C
C       below are initial "dummy" values of main variables; they are
        of course replaced by
C       real values from the keyboard when entered
C
        VO2=300.0
        VCO2=240.0
        TOL=99000.0
        TEMP=37.0
        PICO2=0.0
        Y1=30.0
        Y2=45.0
        X3=40.0
        X4=55.0
        SO2=0.003
        APCO2=30.0
        BPCO2=60.0
        LCASE=1
C
```

```
C     99.99% of the time we use retention-weighting of the data, so
      enter 0
C
   10  WRITE(*,20)
   20 FORMAT(' SHORT WILL WEIGHT EITHER RETENTIONS OR EXCRETIONS',/,
     +' ENTER 0 FOR R (USUAL WAY) OR 1 FOR E, BUT NOT <CR>')
       READ(*,*) IRORE
C************
C     ASK WHETHER ARTERIAL OR PERIPHERAL VENOUS BLOOD IS TO BE USED
C************
       WRITE(*,25)
   25  FORMAT(' USING ARTERIAL (1) OR PERIPHERAL VENOUS (2) BLOOD
       ?'/)
       READ(*,*)IPAPV
C
C      The Hewlett-Packard GC has an integrator or software to
       account for
C      gain/attenuation changes; the old Beckman required manually
       changing'
C      attentuations to keep peaks on scale and large enough to
       measure, which
C      required entering the attenuation used for every peak!!!
C
       WRITE (*,30)
   30  FORMAT(' ARE DATA FROM HP(0) OR FROM BECKMAN (1) GC ?',/,
     1'  <CR> NOT ALLOWED')
       READ(*,*)IGC
       IF(IRORE.NE.0 .AND. IRORE.NE.1) GO TO 10
   40  FORMAT(' ENTER LABEL FOR DATA SET, UP TO 60 CHRS')
   50  FORMAT(60A1)
C
C      now comes the option of starting a new file or adding on to
       an existing file
C
       WRITE (*,60)
   60  FORMAT(' ARE RAW DATA ALREADY ON FILE ? (Y/N)')
       READ (*,70) IQ
   70  FORMAT(A1)
       IF(IQ.EQ.'N') WRITE(*,40)
       IF(IQ.EQ.'N') READ(*,50)(NAMRAW(I),I=1,60)
       IF(IQ.NE.'Y') GO TO 130
       LCASE=3
       WRITE(*,80)
   80  FORMAT(' ENTER   DISK:FILE.EXT   FOR OLD RAW DATA FILE'/)
       OPEN(8,FILE=' ')
```

```
      READ(8,50)(NAMRAW(I),I=1,60)
      WRITE (*,90) (NAMRAW(I),I=1,60)
  90  FORMAT(' DO YOU WANT TO ADD ON TO FILE: ',/,60A1'?? (Y/N)')
      READ (*,70) IADDON
      READ(8,190) NRUNS,PBSEA,ELECT,TBATH,RER,SO2
C*************
C     IF(IADDON.NE.'Y') GO TO 150
      IF(IADDON.NE.'Y') GO TO 130
C*************
      LCASE=2
 100  WRITE (*,110) NRUNS
 110  FORMAT(I4' DATA SETS ARE ON FILE (IF INCORRECT EDIT # IN
      FILE|)'
     +/' ENTER # OF SETS TO ADD')
      READ (*,190) NADDON
      NTOT=NRUNS+NADDON
      WRITE (*,120) NRUNS,NADDON,NTOT
 120  FORMAT(I3' SETS IN OLD FILE +'I3' NEW SETS ='I3' TOTAL, OK?
     + (Y/N)')
      READ (*,70) IOK
      IF(IOK.EQ.'N') GO TO 100
 130  CONTINUE
      IF(LCASE.EQ.3) NTOT=NRUNS
C
C     below, you need to name the RAW data file that saves the data
      you enter.
C     traditionally, we use "name.RAW"  (where "name" is as you
      want it for ID and
C     where the file extension "RAW" indicates it is the raw data
      file
C
      WRITE(*,140)
 140  FORMAT(' ENTER    DISK:FILE.EXT    FOR NEW RAW DATA FILE'/)
      OPEN(7,FILE=' ',STATUS='NEW')
      WRITE(7,50)(NAMRAW(I),I=1,60)
 150  CONTINUE
C
C     below, you need to name the PROCESSED data file that is the
      input file for VQDIST
C     traditionally, we use "name.VQ"  (where "name" is as you
      want it for ID and
C     where the file extension "VQ" indicates it is the data file to
      be used as input for
C     program two - VQBOHR
C
```

```
      WRITE(*,160)
 160  FORMAT(' ENTER    DISK:FILE.EXT    FOR VQDIST INPUT FILE'/)
      OPEN(9,FILE=' ',STATUS='NEW')
      WRITE(3,170)(NAMRAW(I),I=1,60)
 170  FORMAT(' LABEL FOR DATA SET  =  ',60A1/)
      WRITE(9,50)(NAMRAW(I),I=1,60)
      IF(IQ.EQ.'Y') GO TO 200
      WRITE (*,180)
 180  FORMAT(' ENTER # OF DATA SETS, PB SEA LEVEL,',/,
     1' BLOOD GAS ELECTRODE TEMP, & INERT GAS BATH TEMP ',/,
     2' FRACTIONAL CV OF GASES OTHER THAN SF6',/,
     3' AND O2 SOLUBILITY, ML/100 ML/TORR'/)
C
C     RER, THE (FRACTIONAL) MEASURED CV FOR GASES OTHER THAN SF6,
C     IS NORMALLY 0.03, AND IS TWICE THIS FOR SF6.
C     HOWEVER, YOU CAN ENTER ANY VALUE THAT YOU WISH
C
C     RER is NOT respiratory exchange ratio!!
C
C     SO2 is the water solubility of O2, needed in case you want a
      different number
C     than standard (which is 0.003 ml/100ml/mm Hg) because(for
      example) you have added
C     perfluorocarbons to blood to increase solubility
C
      READ(*,*)NRUNS,PBSEA,ELECT,TBATH,RER,SO2
      WRITE(7,190) NRUNS,PBSEA,ELECT,TBATH,RER,SO2
 190  FORMAT(I4,F7.1,2F6.1,F8.2,F8.4)
C***********
C 200  IF(LCASE.EQ.2) GO TO 220
 200  IF(LCASE.EQ.2.OR.LCASE.EQ.3) GO TO 220
C***********
      WRITE(9,190)NRUNS,PBSEA,ELECT,TBATH,RER,SO2
      WRITE(3,210)NRUNS,PBSEA,ELECT,TBATH,RER,SO2
 210  FORMAT(' NUMBER OF RUNS =',I5,';    SEA LEVEL PB =',F7.1,/,
     1' ELECTRODE TEMP =',F5.1,';   H2O BATH TEMP =',F7.1,/,
     2' FRACTIONAL MEASUREMENT CV (EXCEPT SF6) =',F8.2,/,
     3' AND O2 SOLUBILITY, ML/100 ML/Torr      = ',F9.4/)
      GO TO 230
 220  WRITE(7,190)NTOT,PBSEA,ELECT,TBATH,RER,SO2
      WRITE(9,190)NTOT,PBSEA,ELECT,TBATH,RER,SO2
      WRITE(3,210)NTOT,PBSEA,ELECT,TBATH,RER,SO2
 230  IF(LCASE.EQ.1) GO TO 270
      READ(8,240) NGASES,NVAQS,ZZ,VQLO,VQHI
      READ(8,250) (PC(I),I=1,NGASES)
```

```
 240   FORMAT(2I4,1X,F5.1,2X,F5.3,2X,F5.1)
 250   FORMAT(1X,6(1PE12.3))
 260   FORMAT(1X,1PE12.3,5(1PE12.3))
C************
       IF(LCASE-2) 310,310,310
C       IF(LCASE-2) 310,310,320
C************
C
C      now enter number of MIGET gases used (NGASES, usually 6)
C      number of VA/Q compartments in the distribution (NVAQS, usu-
       ally 50)
C      smoothing coefficient (Z, uslly 40)
C      lowest VA/Q ratio separable from shunt (VQLO, usually 0.005)
C      highest VA/Q ratio separable from deadspace (VQHI, usually
       100.0)
C
 270   WRITE (*,280)
 280   FORMAT(' ENTER NGASES,NVAQS,Z,VQLO & VQHI')
       READ (*,*) NGASES,NVAQS,ZZ,VQLO,VQHI
       WRITE (*,240) NGASES,NVAQS,ZZ,VQLO,VQHI
C
C      Partition coefficient (dimensionless) = solubility
       (ml/100ml/mm Hg) x (PB-PH2O)/100
C
       WRITE (*,290) NGASES
 290   FORMAT(' ENTER',I3,'   PARTITION COEFFICIENTS, BATH TEMP')
       READ (*,*) (PC(I),I=1,NGASES)
       WRITE (*,250) (PC(I),I=1,NGASES)
       WRITE (*,300)
 300   FORMAT(' OK OR NOT ? (Y/N)')
       READ (*,70) IOK
       IF(IOK.EQ.'N') GO TO 270
 310   WRITE(7,240) NGASES,NVAQS,ZZ,VQLO,VQHI
       WRITE(7,250) (PC(I),I=1,NGASES)
 320   WRITE(3,330)NGASES,NVAQS,ZZ,VQLO,VQHI
 330   FORMAT(' # OF GASES=',I2,';   # OF COMPTS=',I3,';   Z=',F7.2,
      1';  VQLO=',F6.3,';  VQHI=',F6.1)
       WRITE(3,340)(PC(I),I=1,NGASES)
 340   FORMAT(' PARTITION COEFFICIENTS, BATH TEMP =',
      1F7.5,2F7.3,F6.2,F6.1,F8.1)
       WRITE(3,350)
 350   FORMAT(80('-'))
C
C      initializing some numbers to zero to ensure no junk is there
C
```

```
      INDEX=0
      LOGAS=1
      IF(PC(1).LT.0.01) LOGAS=2
      L1=1
      IAGAIN=0
 360  DO 1350 LM=L1,NRUNS
 370  CONTINUE
      DO 380 I=1,NGASES
      E(I)=0.0
      R(I)=0.0
      EMV(I)=0.0
      RMV(I)=0.0
      WT(I)=0.0
      QTCALC(I)=0.0
 380  CONTINUE
      INDEX=INDEX+1
C************
C      IF(IQ.EQ.'Y') GO TO 610
      IF(IQ.EQ.'Y'.AND.IAGAIN.EQ.0) GO TO 610
C************
      IF(IAGAIN.EQ.1) GO TO 420
C
C     here enter the actual MIGET data as prompted- arterial,
      expired, venous
C     if thereare no venous samples, just enter 0 for each venous
      sample
C
C     the code below talks about dilutions. We pre-calculate peaks
      accounting for dilutions,
C     which is usually needed only for SF6, but may be needed for
      other gases if there
C     is too little sample to run thru the GC, such that it must
      be diluted with N2
C
C     Thus, if the peak was diluted 1:4 and was 31.2 units in
      height, we enter 124.8
C
 390  IF(IPAPV.EQ.1) WRITE (*,400) LM
 400  FORMAT(//9X'SET #'I3//3X'ENTER ARTERIAL PEAKS (in mm)')
      IF(IPAPV.EQ.2) WRITE (*,401) LM
  401  FORMAT(//9X'SET #'I3//3X'ENTER PERIPHERAL VENOUS PEAKS
       (in mm)')
      READ (*,*) (PA(I),I=1,NGASES)
      IF(IGC.EQ.1.AND.IPAPV.EQ.1) WRITE (*,410) NGASES
```

```
 410   FORMAT(' ENTER',I3'  ARTERIAL SF6 DILUTION/FID GAINS (deci-
       mal)')
       IF(IGC.EQ.1.AND.IPAPV.EQ.2) WRITE (*,411) NGASES
  411  FORMAT('  ENTER',I3'   PERIPHERAL VENOUS SF6 DILUTION/FID
       GAINS')
       IF(IGC.EQ.1) READ (*,*) (GA(I),I=1,NGASES)
 420   WRITE (*,1200)(PA(I),I=1,NGASES)
       WRITE (*,1210)(GA(I),I=1,NGASES)
       WRITE (*,300)
       READ (*,70) IOK
       IF(IOK.EQ.'N') GO TO 390
       IF(IAGAIN.EQ.1) GO TO 460
 430   WRITE (*,440)
 440   FORMAT(3X,'ENTER EXPIRED PEAKS (decimal in mm)')
       READ (*,*) (PE(I),I=1,NGASES)
       IF(IGC.EQ.1) WRITE (*,450) NGASES
 450   FORMAT(' ENTER',I3'  EXPIRED SF6 DILUTION/FID GAINS (deci-
       mal)')
       IF(IGC.EQ.1) READ(*,*)(GE(I),I=1,NGASES)
 460   WRITE (*,1200)(PE(I),I=1,NGASES)
       WRITE (*,1210)(GE(I),I=1,NGASES)
       WRITE (*,300)
       READ (*,70) IOK
       IF(IOK.EQ.'N') GO TO 430
       IF(IAGAIN.EQ.1) GO TO 500
 470   WRITE (*,480)
 480   FORMAT(3X,'ENTER VENOUS PEAKS (decimal in mm)')
       READ(*,*)(PV(I),I=1,NGASES)
       IF(IGC.EQ.1) WRITE (*,490) NGASES
  490  FORMAT(' ENTER',I3'  VENOUS SF6 DILUTION/FID GAINS (deci-
       mal)')
       IF(IGC.EQ.1) READ(*,*)(GV(I),I=1,NGASES)
 500   WRITE (*,1200) (PV(I),I=1,NGASES)
       WRITE (*,1210) (GV(I),I=1,NGASES)
       WRITE (*,300)
       READ (*,70) IOK
       IF(IOK.EQ.'N') GO TO 470
       IF(IAGAIN.EQ.1) GO TO 530
C
C      now enter ancillary data
C      VE=ventilation at temp measutred, l/min
C      QT=cardiac output as measured, l/min (enter 0 if not measured)
C      PB=barometric pressure, mm Hg
C      tempb=bodt temp, deg C
C      tempr=temp at which VE measured, deg C
```

```
C     VGA=gas headspace volume in arterial sample syringe, ml
C     VBA=blood volume in arterial sample syringe, ml
C     VHA=heparin volume in arterial sample syringe, ml
C     VGV=gas headspace volume in venous sample syringe, ml (enter
      0 if no sample)
C     VBV=blood volume in venous sample syringe, ml (enter 1 if no
      sample)
C     VHV=heparin volume in venous sample syringe, ml (enter 0 if
      no sample)
C
 510  WRITE (*,520)
 520  FORMAT(' ENTER VE,QT,PB,TEMPB,TEMPR,VGA,VBA,VHA,VGV,VBV,VHV
     1 (FREE FORMAT)')
      READ(*,*)VE,QT,PB,TEMPB,TEMPR,VGA,VBA,VHA,VGV,VBV,VHV
 530  WRITE (*,535) VE,QT,PB,TEMPB,TEMPR,VGA,VBA,VHA,VGV,VBV,VHV
 535  FORMAT(F9.2,F6.2,2X,F5.1,2F6.1,6F7.2)
      WRITE (*,300)
      READ (*,70) IOK
      IF(IOK.EQ.'N') GO TO 510
      VEO=VE
      IF(IAGAIN.EQ.1) GO TO 600
 540  IF(PV(1).EQ.0.0) GO TO 560
      WRITE (*,550)
 550  FORMAT('    ENTER    HB,HCRIT,PVO2,PVCO2,PHV,FIO2,FICO2,P50,
      PMAO2,PMACO2
     1,PHA,VO2 & VCO2',/,' FREE FORMAT')
      GO TO 590
 560  CONTINUE
C
C     enter remaining ancillary data
C     HB=hemoglobin, ml/100ml blood
C     HCRIT=hematocrit, %
C     VO2=O2 consumption, ml/min
C     VCO2=CO2 production, ml/min
C     TOL=99000 normally (see a few lines below)
C     FIO2=fractional inspired O2 concentration, eg, 0.21
C     FICO2=fractional inspired CO2 concentration, almost always
      use 0.0
C     p50=hemoglobin standard P50
C     PMAO2=measured arterial PO2, mm Hg at electrode temp
C     PMACO2=measured arterial PCO2, mm Hg at electrode temp
C     PHA=measured arterial pH
C
      WRITE (*,570)
```

```
 570   FORMAT(' ENTER HB,HCRIT,VO2,VCO2,TOL,FIO2,FICO2,P50,PMAO2,
       PMACO2, 1PHA, FREE FORMAT')
      WRITE (*,580)
 580  FORMAT(' TOL=20.0 will iterate mixed venous gases to match
     + VO2 & VCO2'/' TOL=99000. will use Fick calculated mixed
     + venous tensions'//)
 590  CONTINUE
       READ(*,*)HB,HCRIT,PVO2,PVCO2,PHV,FIO2,FICO2,P50,PMAO2,
       PMACO2,PHA,
     1VO2DUM,VCO2DM
 600  WRITE (*,1230)HB,HCRIT,PVO2,PVCO2,PHV,FIO2,FICO2,P50,PMAO2,
      PMACO2,
     1PHA,VO2DUM,VCO2DM
      WRITE (*,300)
      READ (*,70) IOK
      IF(IOK.EQ.'N') GO TO 540
      GO TO 620
 610  CONTINUE
      READ(8,1200) (PA(I),I=1,NGASES)
      READ(8,1210) (GA(I),I=1,NGASES)
      READ(8,1200) (PE(I),I=1,NGASES)
      READ(8,1210) (GE(I),I=1,NGASES)
      READ(8,1200) (PV(I),I=1,NGASES)
      READ(8,1210) (GV(I),I=1,NGASES)
      READ(8,1220) VEO,QT,PB,TEMPB,TEMPR,VGA,VBA,VHA,VGV,VBV,VHV
       READ(8,1230)HB,HCRIT,PVO2,PVCO2,PHV,FIO2,FICO2,P50,PMAO2,
       PMACO2,
     1PHA,VO2DUM,VCO2DM
 620  CONTINUE
C
C     ABILITY TO COPE WITH 1 FEWER GAS ON ANY RUN
C
      NGSS=NGASES
      IGAS=0
      DO 630 I=1,NGASES
      PCC(I)=PC(I)
      IF(PA(I).GT.0.0) GO TO 630
      IF(PE(I).GT.0.0) GO TO 630
      IF(PV(I).GT.0.0) GO TO 630
      IGAS=I
 630  CONTINUE
      IF(IGAS.GT.0) NGSS=NGASES-1
      DP50=P50-26.8
      REXP=9.0578-2290.5/(273.15+TEMPR)
      BEXP=9.0578-2290.5/(273.15+TEMPB)
```

```
      CEXP=9.0578-2290.5/(273.15+TBATH)
      SVPB=10.0**BEXP
      SVPBTH=10.0**CEXP
      IF(TEMPR.LT.TEMPB) SVPR=10.0**REXP
      IF(TEMPR.GE.TEMPB) SVPR=SVPB
      EXPFAC=(PB-SVPB)/(PBSEA-SVPB)
      PIO2=(PB-SVPB)*FIO2
      PICO2=(PB-SVPB)*FICO2
      X5=PB-SVPB-PIO2
      TEMP=TEMPB
      VC=0.01*HCRIT
      TBODY=TEMPB-(ELECT-37.0)
C
C     TEMPCR, called here,  is the subroutine that corrects blood
      PO2/PCO2/pH from electrode
C     to body temperature
C
     IF(PHV.GT.6.5.AND.PHV.LT.8.0) CALL TEMPCR(PHV,PVCO2,PVO2,VC,
     &TBODY,PHVBT,PVCOBT,PVO2BT,BX)
      IF(PHA.GT.6.5.AND.PHA.LT.8.0) CALL TEMPCR(PHA,PMACO2,PMAO2,
VC,
     &TBODY,PHABT,PACOBT,PAO2BT,BX)
      FACT=(PBSEA-SVPB)/100.0
      WRITE (*,640) INDEX
 640  FORMAT(30X,'  RUN NUMBER  ',I3,/)
      IF(TEMPB.GT.TEMPR) CORR=(273.0+TEMPB)*(PB-SVPR)/
     & ((273.0+TEMPR)*(PB-SVPB))
      IF(TBATH.GT.TEMPR) COR1=(273.0+TBATH)*(PB-SVPR)/
     & ((273.0+TEMPR)*(PB-SVPBTH))
      IF(TEMPB.LE.TEMPR) CORR=(273.0+TEMPB)/(273.0+TEMPR)
      IF(TBATH.LE.TEMPR) COR1=(273.0+TBATH)/(273.0+TEMPR)
      VE=VEO*CORR
C**********
      VESTAR=VEO*COR1
C**********
      IF(PA(1).GT.0.0.AND.IPAPV.EQ.1)
     1ERRA(1)=(2.0*RER)*(2.0*RER) + 0.212*0.212/(PA(1)*PA(1))
      IF(PA(1).GT.0.0.AND.IPAPV.EQ.2)
     1ERRA(1)=(4.0*RER)*(4.0*RER) + 0.212*0.212/(PA(1)*PA(1))
      IF(PE(1).GT.0.0) ERRE(1)=(2.0*RER)*(2.0*RER) +
     10.212*0.212/(PE(1)*PE(1))
      IF(PV(1).GT.0.0) ERRV(1)=(2.0*RER)*(2.0*RER) +
     10.212*0.212/(PV(1)*PV(1))
      ERRPC(1) = (4.0*RER)*(4.0*RER)/2.0
      DO 650 I=LOGAS,NGASES
```

```
      IF(IGAS.EQ.I) GO TO 650
      IF(PA(I).GT.0.0.AND.IPAPV.EQ.1)
     1ERRA(I)=RER*RER + 0.212*0.212/(PA(I)*PA(I))
      IF(PA(I).GT.0.0.AND.IPAPV.EQ.2)
     1ERRA(I)=(2.0*RER)*(2.0*RER) + 0.212*0.212/(PA(I)*PA(I))
      IF(PE(I).GT.0.0) ERRE(I)=RER*RER + 0.212*0.212/(PE(I)*PE(I))
      IF(PV(I).GT.0.0) ERRV(I)=RER*RER + 0.212*0.212/(PV(I)*PV(I))
C
C      PA(I) ETC MUST BE IN MM OF CHART PAPER, NOT INCHES|||||
C      AND MUST BE PRIOR TO ANY GAIN OR DILUTION MULTIPLICATION
C
      ERRPC(I) = (2.0*RER)*(2.0*RER)/2.0
 650  CONTINUE
      DO 670 I=1,NGASES
C
C     CORRECTIONS WHEN BODY AND INERT GAS BATH TEMP ARE DIFFERENT
C
      IF(IBATH.EQ.0) PCFACT(I)=1.0
      IF(IBATH.EQ.0) PCBODY(I)=PC(I)
      IF(IBATH.EQ.0) GO TO 659
      TT1=1.0/(TEMPB+273.0)
      TT2=1.0/(TBATH+273.0)
      IF(ISPEC.EQ.1) SLP=HUMSLO(I)
      IF(ISPEC.EQ.2) SLP=DOGSLO(I)
      IF(ISPEC.EQ.3) SLP=HORSLO(I)
      PCFACT(I)=EXP(SLP*(TT1-TT2))
      PCBODY(I)=PC(I)*PCFACT(I)
 659  CONTINUE
      S(I)=PCBODY(I)/FACT
      PEC(I)=GE(I)*PE(I)*EXPFAC
      IF(PA(I).EQ.0.0) GO TO 660
      IF(IPAPV.EQ.1) PAPV=1.0
      IF(IPAPV.EQ.2) PAPV=0.95
      PAC(I) = GA(I)*PA(I)*(1.0 + VHA/VBA + VGA/(VBA*PC(I)))/
     1(PCFACT(I)*PAPV)
 660  IF(PV(I).GT.0.0)
       1PVC(I) = GV(I)*PV(I)*(1.0 + VHV/VBV + VGV/(VBV*PC(I)))/
       PCFACT(I)
C
C       CORRECT FOR EXPIRED ACETONE LOSS BASED ON ETHER EXP/ART
        RATIO:
C      ACETONE "BOHR" VDVT MUST BE NO GREATER THAN THAT OF ETHER,
       THUS:
C
C      (Pa-PE)/Pa (acetone)  =  (Pa-PE)/Pa (ether)
```

```
C
C     OR
C
C      PE (acetone)  =  Pa (acetone) * PE (ether) / Pa (ether)
C
C      HENCE, REPLACE MEASURED PE (acetone) BY THE ABOVE IF THE
       ABOVE
C     NUMBER is larger than measured PE (acetone)
C
      IF(IACET.EQ.0) GO TO 670
      IF(I.LT.NGASES) GO TO 670
      IF(PAC(NGASES-1).EQ.0.0.OR.PAC(NGASES).EQ.0.0) GO TO 670
      IF(PEC(NGASES-1).EQ.0.0.OR.PEC(NGASES).EQ.0.0) GO TO 670
      ETHER=PEC(NGASES-1)/PAC(NGASES-1)
 669  FORMAT(/' CORRECTING FOR EXPIRED ACETONE LOSS BY ',F7.3,/)
       IF(PV(NGASES).GT.0.0)      ACEXP=0.5*ETHER*(PAC(NGASES)+PVC
       (NGASES))
      IF(PV(NGASES).EQ.0.0) ACEXP=    ETHER*PAC(NGASES)
      IF(PEC(NGASES).GE.ACEXP) GO TO 670
      ACEFAC= ACEXP/PEC(NGASES)
      WRITE(*,669)ACEFAC
      WRITE(3,669)ACEFAC
      PEC(NGASES)=ACEXP
 670  CONTINUE
C
C      now start the actual calculations, beginning with printer
       headings
C
      IF(PA(1).EQ.0.0) GO TO 690
      IF(PE(1).EQ.0.0) GO TO 710
      IF(PV(1).EQ.0.0) GO TO 730
      WRITE (*,680)
 680  FORMAT(/' PA, PE, AND PV ALL MEASURED IN THIS SET'/)
      GO TO 750
 690  WRITE (*,700)
 700  FORMAT(/' ONLY PE AND PV HAVE BEEN MEASURED - PA IS DERIVED'/)
      GO TO 750
 710  WRITE (*,720)
 720  FORMAT(/' ONLY PA AND PV HAVE BEEN MEASURED - PE IS DERIVED'/)
      GO TO 750
 730  IF(IPAPV.EQ.1) WRITE(*,740)
 740  FORMAT(/' ONLY PA AND PE HAVE BEEN MEASURED - PV IS DERIVED'/)
           IF(IPAPV.EQ.2) WRITE(*,741)
       741  FORMAT(/' PERIPHERAL VENOUS AND PE ARE MEASURED - PV IS
     DERIVED'/
```

```
      1' ARTERIAL = PERIPHERAL VENOUS/0.95, AND COEFF VAR ARE
      DOUBLED'/)
 750  CONTINUE
      IF(NGASES.EQ.6) WRITE (*,760)
 760  FORMAT(19X,'SF6',6X,'ETHANE',5X,'CYCLO',4X,'ENFLURANE',4X,
     1'ETHER',5X,'ACETONE')
      WRITE (*,770) (S(I),I=1,NGASES)
      WRITE (*,780) (PC(I),I=1,NGASES)
      WRITE (*,781) (PCBODY(I),I=1,NGASES)
 770  FORMAT(' SOLUBILITY ', 8F11.5)
 780  FORMAT(' PC, BATH T ', 8F11.5)
 781  FORMAT(' PC, BODY T ', 8F11.5)
C
C     now calculate the weights for retention of each of the
      6 gases.
C     This uses error propagation concepts to find the combination of
C     arterial blood and expired gas data from each gas to use to
      minimize
C     overall variance in the retention of that gas. Dividing
      actual measured
C     retention by the square root of the minimal variance so
      found creates a
C     weighted retention for each gas which has, for each gas, a
      variance of 1.0
C     This then gives all six gases appropriate weight in the
      determination of the
C     VA/Q distribution calculated in the ensuing program VQBOHR
C
      IF(PA(1).EQ.0.0) GO TO 940
      IF(PE(1).EQ.0.0) GO TO 980
      IF(PV(1).EQ.0.0) GO TO 1010
      VO2=VO2DUM
      VCO2=VCO2DM
      SUMTOP=0.0
      SUMBOT=0.0
      DO 790, I=1,NGASES
      IF(I.EQ.IGAS) GO TO 790
      R(I) = PAC(I)/PVC(I)
      E(I) = PEC(I)/PVC(I)
      QTCALC(I) = VESTAR*E(I)/(PC(I)*(1.0-R(I)))
      V = ERRE(I) + ERRPC(I) + ERRV(I)/((1.0-R(I))**2) +
     1ERRA(I)*R(I)*R(I)/((1.0-R(I))**2)
      VARQT(I) = QTCALC(I)*QTCALC(I)*V
      IF(QTCALC(I).LE.0.0) VARQT(I)=10.0**6
      IF (PC(I).GT.50.0) VARQT(I)=10.0**6
```

```
      SUMTOP=SUMTOP+QTCALC(I)/VARQT(I)
      SUMBOT=SUMBOT+1.0/VARQT(I)
 790  CONTINUE
      QT=SUMTOP/SUMBOT
      DO 810, I=1,NGASES
      IF(I.EQ.IGAS) GO TO 810
      IF(IRORE.NE.0) GO TO 800
      X=R(I)*R(I)*(ERRA(I)+ERRV(I))
       Y=((VE*E(I)/(QT*PCBODY(I)))**2)*(ERRE(I) + ERRPC(I) + ERRV
(I))
      Z=VE*E(I)*R(I)*ERRV(I)/(QT*PCBODY(I))
      TOPT=(Y+Z)/(X + Y + 2.0*Z)
      RMV(I)=TOPT*R(I) + (1.0-TOPT)*(1.0-VE*E(I)/(QT*PCBODY(I)))
      EMV(I)=QT*PCBODY(I)*(1.0-RMV(I))/VE
      VRCE=Y - 2.0*TOPT*(Y+Z) + TOPT*TOPT*(X+Y+2.0*Z)
      WT(I) = 1.0/SQRT(VRCE)
      GO TO 810
 800  AA=QT*PCBODY(I)/VE
      ASQ=AA*AA
      RSQ=R(I)*R(I)
      ESQ=E(I)*E(I)
      BB=ESQ*ERRPC(I)+ASQ*RSQ*ERRA(I)
      CC=ASQ*ERRV(I)
      TNUM=BB+CC*R(I)
      DENOM=BB+ESQ*ERRE(I)+CC
      TOPT=TNUM/DENOM
      VRCE=TOPT*TOPT*DENOM-2.0*TOPT*TNUM+BB+CC*RSQ
      EMV(I)=TOPT*E(I)+(1.0-TOPT)*AA*(1.0-R(I))
      RMV(I)=1.0-EMV(I)/AA
      WT(I)=1.0/SQRT(VRCE)
 810  CONTINUE
      WRITE (*,819)(PCFACT(I),I=1,NGASES)
 819  FORMAT(' BODY/BATH PC',F10.5,7F11.5)
      IF(IRORE.NE.0) GO TO 870
      WRITE (*,820)(R(I),I=1,NGASES)
 820  FORMAT(' MEASURED R ',8F11.5)
      WRITE (*,830)(RMV(I),I=1,NGASES)
 830  FORMAT(' MIN VAR R  ',8F11.5)
      WRITE (*,840)(E(I),I=1,NGASES)
 840  FORMAT(' MEASURED E ',8F11.5)
      WRITE (*,850)(EMV(I),I=1,NGASES)
 850  FORMAT(' ASSOCIATED E ',F9.5,7F11.5)
      WRITE (*,860)(WT(I),I=1,NGASES)
 860  FORMAT(' WEIGHTS (R)',8F11.1)
      GO TO 910
```

```
870   WRITE (*,840) (E(I),I=1,NGASES)
      WRITE (*,880) (EMV(I),I=1,NGASES)
880   FORMAT(' MIN VAR E   ',8F11.5)
      WRITE (*,820) (R(I),I=1,NGASES)
      WRITE (*,890) (RMV(I),I=1,NGASES)
890   FORMAT(' ASSOCIATED R ',F9.5,7F11.5)
      WRITE (*,900) (WT(I),I=1,NGASES)
900   FORMAT(' WEIGHTS (E)',8F11.1)
910   WRITE (*,920) (QTCALC(I),I=1,NGASES)
920   FORMAT(' PREDICTED QT ',F9.2,7F11.2)
      WRITE (*,930) VE,QT
930  FORMAT(/' MINUTE VENTILATION, BTPS ='F8.2,'   MEAN FICK CARDIAC
     1OUTPUT =',F7.2)
      GO TO 1050
940   CONTINUE
      DO 960, I=1,NGASES
      IF(I.EQ.IGAS) GO TO 960
      PAC(I) = PVC(I) - VE*PEC(I)/(QT*PCBODY(I))
      R(I) = PAC(I)/PVC(I)
      RMV(I) = R(I)
      E(I) = PEC(I)/PVC(I)
      EMV(I) = E(I)
      IF(IRORE.NE.0) GO TO 950
      V = ERRE(I) + ERRV(I) + ERRPC(I)
      V = V*((VE*E(I)/(QT*PCBODY(I)))**2)
      WT(I) = SQRT(1.0/V)
      GO TO 960
950   V=E(I)*E(I)*(ERRE(I)+ERRV(I))
      WT(I)=SQRT(1.0/V)
960   CONTINUE
      WRITE (*,819) (PCFACT(I),I=1,NGASES)
      WRITE (*,820) (R(I),I=1,NGASES)
      WRITE (*,840) (E(I),I=1,NGASES)
      IF(IRORE.EQ.0) WRITE (*,860) (WT(I),I=1,NGASES)
      IF(IRORE.NE.0) WRITE (*,900) (WT(I),I=1,NGASES)
      WRITE (*,970) VE,QT
970   FORMAT(/' MEASURED MINUTE VENTILATION =',F9.2,
     1'   AND MEASURED CARDIAC OUTPUT =',F8.2)
      GO TO 1050
980   CONTINUE
      DO 1000, I=1,NGASES
      IF(I.EQ.IGAS) GO TO 1000
      R(I) = PAC(I)/PVC(I)
      RMV(I) = R(I)
      PEC(I) = QT*PCBODY(I)*(PVC(I)-PAC(I))/VE
```

```
      E(I)=PEC(I)/PVC(I)
      EMV(I)=E(I)
      IF(IRORE.NE.0) GO TO 990
      V=R(I)*R(I)*(ERRA(I)+ERRV(I))
      WT(I) = SQRT(1.0/V)
      GO TO 1000
 990  ASQ=(QT*PCBODY(I)/VE)**2
      V=ASQ*(ERRPC(I)+R(I)*R(I)*(ERRA(I)+ERRV(I)+ERRPC(I)))
      WT(I)=SQRT(1.0/V)
1000  CONTINUE
      WRITE (*,819) (PCFACT(I),I=1,NGASES)
      WRITE (*,820) (R(I),I=1,NGASES)
      WRITE (*,840) (E(I),I=1,NGASES)
      IF(IRORE.EQ.0) WRITE (*,860) (WT(I),I=1,NGASES)
      IF(IRORE.NE.0) WRITE (*,900) (WT(I),I=1,NGASES)
      WRITE (*,970) VE,QT
      GO TO 1050
1010  CONTINUE
      TOL=PHV
      VO2=PVO2
      VCO2=PVCO2
      IF(TOL.GT.50.0) GO TO 1020
      TOL=20.0
      PVO2=40.0
      PVCO2=45.0
1020  CONTINUE
      DO 1040 I=1,NGASES
      IF(I.EQ.IGAS) GO TO 1040
      PVC(I) = PAC(I)+ VE*PEC(I)/(QT*PCBODY(I))
      R(I)=PAC(I)/PVC(I)
      RMV(I)=R(I)
      E(I)=PEC(I)/PVC(I)
      EMV(I)=E(I)
      IF(IRORE.NE.0) GO TO 1030
      X=QT*PCBODY(I)*PAC(I)
      Y=VE*PEC(I)
      W = X*Y/((X+Y)**2)
      V=ERRA(I) + ERRE(I) + ERRPC(I)
      V = V*W*W
      WT(I) = SQRT(1.0/V)
      GO TO 1040
1030  ESQ=E(I)*E(I)
      RSQ=R(I)*R(I)
      V=ESQ*ERRPC(I)+ESQ*RSQ*(ERRA(I)+ERRPC(I)+ERRE(I))
      WT(I)=SQRT(1.0/V)
```

```
1040  CONTINUE
      WRITE (*,819) (PCFACT(I),I=1,NGASES)
      WRITE (*,820) (R(I),I=1,NGASES)
      WRITE (*,840) (E(I),I=1,NGASES)
      IF(IRORE.EQ.0) WRITE (*,860) (WT(I),I=1,NGASES)
      IF(IRORE.NE.0) WRITE (*,900) (WT(I),I=1,NGASES)
      WRITE (*,970) VE,QT
1050  CONTINUE
C
C      NOW WRITE DATA TO DISK FOR THE VQBOHR PROGRAM
C
C*************
C      IF(IQ.EQ.'Y') GO TO 1070
C*************
      WRITE (*,1060)
1060  FORMAT(' IS THIS SET OK OR NOT ? (Y/N)')
      READ (*,70) IFLAG
      IF(IFLAG.NE.'N') GO TO 1070
      INDEX=INDEX-1
      IAGAIN=1
      VE=VEO
      GO TO 370
1070  CONTINUE
      IAGAIN=0
      WRITE(3,640)INDEX
      WRITE(3,1075)
1075  FORMAT(22X,'GAS 1',5X,'GAS 2',5X,'GAS 3',5X,'GAS 4',5X,'GAS 5',
     15X,'GAS 6'/)
      WRITE(3,1080)(PA(I),I=1,NGASES)
1080  FORMAT(' ARTERIAL PEAKS  ',10F10.1)
      WRITE(3,1090)(GA(I),I=1,NGASES)
1090  FORMAT(' AND GAIN FACTORS',10F10.1)
      WRITE(3,1100)(PE(I),I=1,NGASES)
1100  FORMAT(' EXPIRED PEAKS   ',10F10.1)
      WRITE(3,1090)(GE(I),I=1,NGASES)
      WRITE(3,1110)(PV(I),I=1,NGASES)
1110  FORMAT(' VENOUS PEAKS    ',10F10.1)
      WRITE(3,1090)(GV(I),I=1,NGASES)
      WRITE(3,1119)(PCFACT(I),I=1,NGASES)
1119  FORMAT(' BODY/BATH PC   ',10F10.4)
      WRITE(3,1120)VE,QT,PB,SVPR,SVPB,TEMPB,TEMPR,VGA,VBA,VHA,
      VGV,VBV,1VHV
1120  FORMAT(
     1' VE  =',F7.2,'; QT   =',F7.2,'; PB   =',F5.1,/,
```

```
      2' SVPR=',F7.1,'; SVPB =',F7.1,'; TMPB =',F5.1,'; TMPR =',
      F6.1,';'/
      3' VGa =',F7.2,'; VBa  =',F7.2,'; VHa  =',F5.2,'; VGv  =',F6.2,
      4';  VBv =',F5.2,'; VHv=',F4.2)
      IF(PV(1).EQ.0.0) GO TO 1140
      WRITE(3,1130)HB,HCRIT,PVO2BT,PVCOBT,PHVBT,P50,PIO2,PICO2,
      PAO2BT,
           1PACOBT,PHABT,VO2,VCO2,VCO2/VO2
1130  FORMAT(
      1' HB   =',F7.1,'; HCT   =',F7.1,';',13X,' PvO2 =',F6.1,';
      PvCO2=',
      2F5.1,'; PHv=',F4.2,/,
      3' P50 =',F7.1,'; PIO2 =',F7.1,'; PICO2=',F5.1,'; PaO2 =',
      F6.1,
      4'; PaCO2=',F5.1,'; PHa=',F4.2,/,
      5' VO2 =',F7.1,'; VCO2 =',F7.1,';   R  =',F5.2,';')
           GO TO 1160
1140  WRITE(3,1150)HB,HCRIT,PHV,P50,PIO2,PICO2,PAO2BT,PACOBT,
      PHABT,
      1PVO2,PVCO2,PVCO2/PVO2
1150  FORMAT(
      1' HB  =',F7.1,'; HCT  =',F6.1,'; TOLERANCE =',F7.1,/,
      2' P50 =',F7.1,'; PIO2 =',F6.1,'; PICO2 =',F6.1,'; PaO2 =',
      F6.1,
      3'; PaCO2=',F6.1,'; PHa =',F5.2,/,
      4' VO2 =',F7.1,'; VCO2 =',F6.1,';   R  =',F6.2)
1160  WRITE(3,1170)EXPFAC
1170  FORMAT(' HYPO/HYPERBARIC CORRECTION FACTOR FOR EXPIRED GAS'
      1' VALUES=',F6.3)
      IPF=0
      IF(PV(1).EQ.0.0 .OR.PA(1).EQ.0.0 .OR.PE(1).EQ.0.0) IPF=1
      IF(IPF.EQ.0) WRITE(3,680)
      IF(PV(1).EQ.0.0.AND.IPAPV.EQ.1) WRITE(3,740)
      IF(PV(1).EQ.0.0.AND.IPAPV.EQ.2) WRITE(3,741)
      IF(PA(1).EQ.0.0) WRITE(3,700)
      IF(PE(1).EQ.0.0) WRITE(3,720)
      WRITE(3,760)
      WRITE(3,770)(S(I),I=1,NGASES)
      WRITE(3,780)(PC(I),I=1,NGASES)
      WRITE(3,781)(PCBODY(I),I=1,NGASES)
      WRITE(3,819)(PCFACT(I),I=1,NGASES)
      IF(IRORE.NE.0) GO TO 1180
      WRITE(3,820)(R(I),I=1,NGASES)
      IF(IPF.EQ.0) WRITE(3,830)(RMV(I),I=1,NGASES)
      WRITE(3,840)(E(I),I=1,NGASES)
```

```
      IF(IPF.EQ.0) WRITE(3,850)(EMV(I),I=1,NGASES)
      WRITE(3,860)(WT(I),I=1,NGASES)
      GO TO 1190
1180  WRITE(3,840)(E(I),I=1,NGASES)
      IF(IPF.EQ.0) WRITE(3,880)(EMV(I),I=1,NGASES)
      WRITE(3,820)(R(I),I=1,NGASES)
      IF(IPF.EQ.0) WRITE(3,890)(RMV(I),I=1,NGASES)
      WRITE(3,900)(WT(I),I=1,NGASES)
1190  IF(IPF.EQ.0) WRITE(3,920)(QTCALC(I),I=1,NGASES)
      IF(IPF.EQ.0) WRITE(3,930)VE,QT
      IF(IPF.EQ.1) WRITE(3,970)VE,QT
C************
C       IF(LCASE.EQ.3) GO TO 1240
C************
      WRITE(7,1200) (PA(I),I=1,NGASES)
      WRITE(7,1210) (GA(I),I=1,NGASES)
      WRITE(7,1200) (PE(I),I=1,NGASES)
      WRITE(7,1210) (GE(I),I=1,NGASES)
      WRITE(7,1200) (PV(I),I=1,NGASES)
      WRITE(7,1210) (GV(I),I=1,NGASES)
      WRITE(7,1220) VEO,QT,PB,TEMPB,TEMPR,VGA,VBA,VHA,VGV,VBV,VHV
     IF(PV(1).GT.0.0) WRITE(7,1230) HB,HCRIT,PVO2,PVCO2,PHV,FIO2,
     &FICO2,P50,PMAO2,PMACO2,PHA,VO2,VCO2
       IF(PV(1).EQ.0.0) WRITE(7,1230) HB,HCRIT,VO2,VCO2,TOL,FIO2,
       FICO2,
     & P50,PMAO2,PMACO2,PHA
1200  FORMAT(6(1X,F6.1))
1210  FORMAT(1X,6(F5.0,2X))
1220  FORMAT(F7.2,F6.2,2X,F5.1,2F6.1,6F7.2)
1230   FORMAT(2(2X,F4.1),2F7.1,F9.2,1X,2F6.4,F6.1,/,2F7.1,
       F6.2,2F8.1)
1240  CONTINUE
      PICO2=(PB-SVPB)*FICO2
      WRITE(9,1250) VO2,VCO2,PIO2,PB,TEMP,HB,HCRIT,PICO2,BX,DP50
1250  FORMAT(F7.1,F8.1,F8.2,F7.1,3F6.1,F7.2,2F6.2)
      IF(PV(1).GT.0.0) GO TO 1280
      WRITE(9,1260) PVO2,PVCO2,Y1,Y2,X3,X4,X5,SO2
      WRITE(9,1270) PHABT,PHV,APCO2,BPCO2,PAO2BT,PACOBT
1260  FORMAT(F7.1,6F7.1,F9.5)
1270  FORMAT(F9.2,F9.2,2F6.1,2F7.1)
      GO TO 1290
1280  WRITE(9,1260) PVO2BT,PVCOBT,Y1,Y2,X3,X4,X5,SO2
      WRITE(9,1270) PHABT,PHVBT,APCO2,BPCO2,PAO2BT,PACOBT
1290  WRITE(9,1300) NGSS,NVAQS,ZZ,VQLO,VQHI,VE,SVPB,QT,TOL
1300  FORMAT(I3,I4,F6.1,F7.3,F7.1,2F8.2,F6.2,F8.0)
```

```
      IF(IGAS.EQ.0) GO TO 1320
      DO 1310 I=IGAS,NGSS
      PCC(I)=PCBODY(I+1)
      RMV(I)=RMV(I+1)
      EMV(I)=EMV(I+1)
      WT(I)=WT(I+1)
1310  CONTINUE
1320  CONTINUE
      IF(IGAS.NE.0) WRITE(9,260) (PCC(I),I=1,NGSS)
      IF(IGAS.EQ.0) WRITE(9,260) (PCBODY(I),I=1,NGSS)
      WRITE(9,1330) (RMV(I),I=1,NGSS)
      WRITE(9,1330) (EMV(I),I=1,NGSS)
      WRITE(9,1340) (WT(I),I=1,NGSS)
1330  FORMAT(F12.6,7F12.6)
1340  FORMAT(F11.2,7(F11.2,1X))
      WRITE(3,350)
1350  CONTINUE
      IF(LCASE.NE.1) CLOSE(8)
      IF(LCASE.NE.2) GO TO 1360
      L1=NRUNS+1
      NRUNS=NTOT
      IQ='N'
      LCASE=1
      GO TO 360
1360  CLOSE(9)
      IF(LCASE.NE.3) CLOSE(7)
      STOP
      END
      SUBROUTINE TEMPCR(PHP,PCO2,PO2,VC,T,PHPT,PCO2T,PO2T,BEB)
C
C     CORRECTS BLOOD PO2 & PCO2 & PH VALUES TO BODY TEMP & CALCS
      BASE EXCESS
C      this comes from the Kelman O2/CO2/pH algorithms from the
       1960's
C
      REAL N,N1,MIN,MAX
      A=15.
      B=2045.
      C=-2000.
      D=2400.
      E=31100.
      F=2.4E+6
      PP=75.0
      N=-PO2*10.**(.48*(PHP-7.4))
      SO2=(N**4+A*N**3+B*N**2+C*N)/(N**4+A*N**3+D*N**2+E*N+F)
```

```
      DSO2BB=-VC*(1-SO2)*(8.5+PCO2*(.01-.05526*10.**(PHP-7.4)))
      RH=.627-.440*(PHP-7.4)
      PK=6.099-.04167*(PHP-7.4)
      HBO2NH=20.1*VC*.2658*PCO2/(.8238*PCO2+10.**(8.-PHP)/RH*
     &(8.611+10.**(8.-PHP)/RH))
      PB=(1.-VC)*PP*(.2413+.104*(PHP-7.4))
      PHC=PHP+ALOG10(RH)
      HB=20.1*VC*(10.625*PHC-.5*PHC**2-48.46)
      HCO3=.0306*PCO2*10.**(PHP-PK)*(1.+VC*(.7676*RH-1.))
      BBB=HCO3+HB+PB+HBO2NH+DSO2BB
      BEB=BBB-24.48-31.395*VC-.2413*(1.-VC)*PP
      N1=-PO2*10.**(.48*(PHP-7.4)-.0013*BEB)
      PCO2T=PCO2*10.**(.019*(T-37.))
      PHPT=PHP-(T-37.)*(.0147+.00654*(PHP-7.4))
      SO237=(N1**4+A*N1**3+B*N1**2+C*N1)/(N1**4+A*N1**3+D*N1**2+
     &E*N1+F)
      SM=-PO2*10.**(.48*(PHPT-7.4)-.0013*BEB-.024*(T-37.))
      SP=1.+ALOG10(T/37.)+.00012*(T-37.)**2
      ST=342.18*VC/(PO2*(.02114+.00516*VC))
      SA=A/SM+SP*(ST*(1.-SO237)-1.)
      SB=(D/SM+A*SP*(ST*(1.-SO237)-1.))/SM
      SC=(E/SM+SP*(B*ST-D*(1.+ST*SO237)))/SM**2
      SD=(F/SM+SP*(C*ST-E*(1.+ST*SO237)))/SM**3
      SE=-F*SP*(1.+ST*SO237)/SM**4
      MIN=0.
      MAX=1.
      CALCX=T-37.
      IF(CALCX)20,10,70
10    X=1.
      GO TO 120
20    DO 50, J=1,24
      XT=(MAX+MIN)/2.
      Y=XT**5+SA*XT**4+SB*XT**3+SC*XT**2+SD*XT+SE
      IF(Y) 40,60,30
30    MAX=XT
      GO TO 50
40    MIN=XT
50    CONTINUE
60    X=XT
      GO TO 120
70    DO 100, K=1,24
      RXT=(MAX+MIN)/2.
      Y=1.+SA*RXT+SB*RXT**2+SC*RXT**3+SD*RXT**4+SE*RXT**5
      IF(Y)80,110,90
80    MAX=RXT
```

```
      GO TO 100
90    MIN=RXT
100   CONTINUE
110   X=1./RXT
120   PO2T=X*PO2
      SL=SM*X
      SO2T=(SL**4+A*SL**3+B*SL**2+C*SL)/(SL**4+A*SL**3+D*
     &SL**2+E*SL+F)
      RETURN
      END
```

Appendix C: VQ Bohr Program PDF

```
C    THIS PROGRAM COMBINES FINDING THE VA/Q DISTRIBUTION FROM
     MIGET DATA
c    WITH BOHR INTEGRATION (USING THE KELMAN COMPUTERIZED DISSOCI-
     ATION CURVES)
c    TO ACCOUNT FOR POSSIBLE O2 DIFFUSION LIMITATION
c
C    THE PROGRAM BASICALLY FOLLOWS HAMMOND AND HEMPLEMAN, RESP.
     PHYSIOL.
c    69:129-147, 1987
C
C    THIS VERSION FINDS DLO2 ITERATIVELY BY QUADRATIC INTERPOLA-
     TION
C    USING THREE STARTING GUESSES AT DLO2 ("DM") TO CALCULATE THE
     THREE
C    COEFFICIENTS (A,B,C) OF:  PO2ERR = A.DM.DM + B.DM + C
C    PO2ERR IS THE DIFFERENCE BETWEEN MEASURED AND PREDICTED ARTE-
     RIAL PO2
C    THESE GUESSES ARE 0.010, 0.015 AND 0.020 OF THE MEASURED VO2.
C    DLO2 IS IN ML/MIN/TORR AND VO2 IS IN ML/MIN. HOWEVER, TO
     ENSURE
C    NUMERICAL STABILITY, THE FIRST VALUE (0.010) IS STEPPED DOWN
     IF
C    ITS PO2ERR IS POSITIVE, UNTIL PO2ERR JUST BECOMES NEGATIVE,
     WITH DM
C    BEING 0.8 OF ITS PREVIOUS VALUE EACH STEP. SIMILARLY, THE
     THIRD
C    VALUE (0.020) IS STEPPED UP TO 1.2 OF ITS PREVIOUS VALUE IF
     ITS
```


S.R. Hopkins, P.D. Wagner, *The Multiple Inert Gas Elimination Technique (MIGET)*, Methods in Physiology, https://doi.org/10.1007/978-1-4939-7441-2

```
C     PO2ERR IS NEGATIVE, UNTIL PO2ERR JUST BECOMES POSITIVE.
C     FOR EACH OF THESE GUESSES, THE ACTUAL PO2ERR IS COMPUTED BY
      RUNNING
C     THE BOHR INTEGRATION, TO GIVE 3 EQNS IN 3 UNKNOWNS (A,B,C):
C
C           PO2ERR(1) = A.DM(1).DM(1)  +  B.DM(1)  +  C
C           PO2ERR(2) = A.DM(2).DM(2)  +  B.DM(2)  +  C
C           PO2ERR(3) = A.DM(3).DM(3)  +  B.DM(3)  +  C
C
C     NEXT, THE VALUE OF DM THAT GIVES PO2ERR=0 IS CALCULATED FROM
C     THIS QUADRATIC, AND ITS ACTUAL PO2ERR COMPUTED BY RUNNING THE
C     BOHR INTEGRATION WITH THIS NEW VALUE OF DM.
C
C     THE FIRST (DM,PO2ERR) PAIR IS NOW DROPPED, THE MOST RECENT
      PAIR
C     RETAINED, FORMING AN UPDATED SET OF THREE (DM,PO2ERR) PAIRS,
      AND
C     A NEW SET OF QUADRATIC COEFFICIENTS IS COMPUTED FROM THEM.
C
C     THIS PROCESS IS REPEATED TO CONVERGENCE TO WITHIN THE DESIRED
C     TOLERANCE, CURRENTLY STATED AS PO2ERR < 0.1 TORR ABSOLUTE.
C
      PROGRAM MAIN
      CHARACTER NAME(60),IO2,ISKO2
      DIMENSION IFLAG(24),X(4),Y(4)
      DIMENSION VSAVE(20,500),QSAVE(20,500)
      COMMON/VQ/V(500),Q(500),QQ(500),VAQ(500)
      COMMON/INERT/NVAQS,VT,QT,NGASES,FACT,Z,VQLO,VQHI
      COMMON/OXY1/HB,HCRIT,TEMP,APH,BPH,APCO2,BPCO2,KOUNT,DP50,
      PIO2,SO2
      COMMON/OXY2/PB,FIO2,PICO2,FICO2,VS,QS,SKEW,GVO2,GVCO2,TOL,
     +PVO2,PVCO2,FMVO2,FMVCO2,PVN2,ALPHA,SHUNT,QR,GVAQ,FVQ,PAO2,
     +PACO2,O2CON,CCO2,PO2,PCO2,FVO2,FVCO2,RZ,RM,ARTO2C,ARTCO2,
      AMO2C
      COMMON/BOHR/DM,VC,BETA,RNT,DDLO2,DDLCO2,PMAO2,PMACO2,PO2B,
      PCO2B,
     +IBOHR,CAIII,IPRN,ARTPM,IWARN,IWARN2
      COMMON/O2ARAY/OO2CON(500),OCCO2(500),FVQQ(500),PN22(500),
     +PCO22(500),PO22(500),RZZ(500),PBO2(500),PBCO2(500)
C
C     define and open data files first. Remember that this is program
      VQBOHR
C     and it is designed to read a MIGET data file already produced
      by the
C     preceding program, "SHORT"
```

```
C
      IPRN=0
      DO 10, I=1,24
 10   IFLAG(I)=0
      WRITE(*,310)
      OPEN (8,FILE=' ')
      WRITE(*,320)
      READ(*,*)IDP
      IF(IDP.LE.0) WRITE(*,330)
      IF(IDP.LE.0) OPEN(3,FILE=' ',STATUS='NEW')
      IF(IDP.GE.1) OPEN(3,FILE='PRN')
      WRITE(*,390)
      READ(*,*)IPLTVQ
      WRITE(*,395)
      READ(*,*)IPLTRE
      IF(IPLTVQ.EQ.1) OPEN(9,FILE='VQPLOT.FYL',STATUS='NEW')
      IF(IPLTRE.EQ.1) OPEN(6,FILE='REPLOT.FYL',STATUS='NEW')
      IF(IPLTVQ.EQ.1) WRITE(*,410)
      IF(IPLTRE.EQ.1) WRITE(*,415)
      READ(8,20)(NAME(I),I=1,60)
 20   FORMAT(60A1)
      READ(8,30) NRUNS,PBSEA
 30   FORMAT(I4,F7.1)
C
C     now there are options to choose from that are self-explanatory
C
 40   WRITE(*,50)
 50   FORMAT(' ARE THESE SETS RETENTION OR EXCRETION WEIGHTED ?'/
     + '  (0 FOR R (USUAL WAY) OR 1 FOR E)')
      READ(*,*)IRORE
      IF(IRORE.NE.0 .AND. IRORE.NE.1) GO TO 40
      WRITE(*,60)
 60   FORMAT('  ENTER SET NUMBERS TO BE SKIPPED, 20I3 FORMAT'/)
      READ(*,65)(IFLAG(I),I=1,NRUNS)
  65  FORMAT(20I3)
      WRITE(*,70)
 70   FORMAT(' DO YOU WANT O2/CO2 CALCS, D/Q INFINITE (Y/N) ?')
      READ(*,90)IO2
      IF(IO2.NE.'Y') GO TO 110
      WRITE(*,80)
 80   FORMAT(' DO YOU WANT BOHR INTEGRATION CALCULATIONS ? (Y/N)')
      READ(*,90)ISKO2
 90   FORMAT(A1)
      IF(ISKO2.NE.'Y') GOTO 110
      WRITE(*,100)
```

```
100   FORMAT(' ENTER # OF INTEGRATION STEPS (20 - 100), FREE FOR-
       MAT',/,
      1' AND FLAG SCREEN-PRINT OPTION (1=YES, 0=NO)')
      READ(*,*)NT,IPRN
      RNT=NT
110   CONTINUE
      DO 300 KK=1,NRUNS
      ISKIP=0
      DO 120 KIJ=1,NRUNS
120   IF(IFLAG(KIJ).EQ.KK) ISKIP=1
      IF(ISKIP.EQ.0) WRITE(3,130)KK,(NAME(I),I=1,60)
      IF(ISKIP.EQ.0) WRITE(*,130)KK,(NAME(I),I=1,60)
130   FORMAT(/////,' SET NUMBER :'I3'  FROM FILE: '60A1)
      IF(ISKIP.EQ.0 .AND. IRORE.EQ.0) WRITE(3,140)
      IF(ISKIP.EQ.0 .AND. IRORE.NE.0) WRITE(3,150)
140   FORMAT(' FITTING A BLOOD FLOW DISTRIBUTION TO RETENTIONS'/)
150   FORMAT(' FITTING A VENTILATION DISTRIBUTION TO EXCRETIONS'/)
      READ(8,160) GVO2,GVCO2,PIO2,PB,TEMP,HB,HCRIT,PICO2,BX,DP50
160   FORMAT(F7.1,F8.1,F8.2,F7.1,3F6.1,F7.2,2F6.2)
      READ(8,170) X(1),Y(1),X(2),Y(2),X(3),Y(3),PVN2,SO2
170   FORMAT(7F7.1,F9.5)
      ALPHA=0.0017
      READ(8,180) PHA,PHV,APCO2,BPCO2,PMAO2,PMACO2
180   FORMAT(2F9.2,2F6.1,2F7.1)
      READ(8,190) NGASES,NVAQS,Z,VQLO,VQHI,VT,PH2O,QT,TOL
190   FORMAT(I3,I4,F6.1,F7.3,F7.1,2F8.2,F6.2,F8.0)
      IF(IO2.EQ.'N') GO TO 200
      Y1=0.003*HB*(100.0-SATURA(PMAO2,PMACO2,PHA))/100.0
      PHX=7.59+Y1-0.2741*ALOG(PMACO2/20.0)
      DELPH=PHA-PHX
      APH=7.59+DELPH-0.2741*ALOG(APCO2/20.0)
      BPH=7.59+DELPH-0.2741*ALOG(BPCO2/20.0)
      IF(PHV.LE.6.0.OR.PHV.GE.8.0) GO TO 200
      Y1=0.003*HB*(100.0-SATURA(X(1),Y(1),PHV))/100.0
      PHX=7.59+Y1-0.2741*ALOG(Y(1)/20.0)
      DELPH=PHV-PHX
      APHV=7.59+DELPH-0.2741*ALOG(APCO2/20.0)
      BPHV=7.59+DELPH-0.2741*ALOG(BPCO2/20.0)
      HA1=EXP(-APH)
      HA2=EXP(-APHV)
      HB1=EXP(-BPH)
      HB2=EXP(-BPHV)
      HA=(HA1+HA2)/2.0
      HBAV=(HB1+HB2)/2.0
      APH=-ALOG(HA)
```

```
      BPH=-ALOG(HBAV)
200   CONTINUE
      FACT = (PBSEA-PH2O)/100.0
      P50=26.8+DP50
      DSPCE=0.0
      KRUN=KK
C
C     CALCVQ is the master subroutine that ends up determining the
      50-compartment
C     VA/Q distribution from the MIGET data. All O2/CO2 calculations
      come later
C     by means of other routines, see below]
C
      CALL CALCVQ(IPLTRE,ISKIP,DSPCE,SUMQ,KRUN,IRORE)
      IF(SUMQ.EQ.0.0) ISKIP=1
      IF(ISKIP.EQ.1) GO TO 300
      IF(IO2.EQ.'N') GO TO 300
C
C     now we start the O2/CO2 calculations, which are designed to
      predict the arterial
C     PO2 and PCO2 expected on the basis of the particular VA/Q
      distribution MIGET has recovered.
C     This prediction is always done first assuming NO O2 diffusion
      limitation.
C
C     However, if measured arterial PO2 is less than this predicted
      arterial PO2,
C     you have the option (see above) of running the program to ask
      what finite value of
C     O2 diffusing capacity must exist to drop the predicted PO2
      down to equal the measured value.
C
      KOUNT=0
      WRITE(3,210)
210   FORMAT(30X,'GAS EXCHANGE')
      FIO2=PIO2/(PB-PH2O)
      FICO2=PICO2/(PB-PH2O)
      WRITE(3,220)GVO2,GVCO2,PIO2,FIO2,PICO2,FICO2,PB,TEMP,HB,
      HCRIT,
     + P50,BX
220   FORMAT(/'    GVO2',4X,'GVCO2',3X,'PIO2',3X,'FIO2',2X,'PICO2',
     +2X,'FICO2'2X'PB'4X'TEMP'3X'HB'2X'HCRIT'2X'P50'2X'BX'/,
     +2F8.1,1X,F6.1,F7.4,F5.1,2X,F6.2,1X,F6.1,3(1X,F5.1),2F5.1,/)
      CALL BLOOD(PMAO2,PMACO2,AMO2C,AMCO2C)
      WRITE(3,230)APH,APCO2,PMAO2,AMO2C,BPH,BPCO2,PMACO2,AMCO2C
```

```
230    FORMAT(4X,'FIRST  BLOOD PH =',F4.2,4X,'PCO2 =',F5.2,6X,
      +'PMAO2  =',F6.2,5X,'CMAO2  =',F6.2/,4X,'SECOND  BLOOD  PH  =',
       F4.2,
      +4X,'PCO2 =',F5.2,6X,'PMACO2=',F6.2,5X,'CMACO2=',F6.2/)
       WRITE(3,240)VT,QT,SO2
240    FORMAT(/,25X,'TOTAL VENTILATION=',F12.2,/,
      + 25X,'TOTAL BLOOD FLOW =',F12.2,/,
      + 25X,'O2 SOLUBILITY    =',F12.4)
       WRITE(3,250)TOL
250    FORMAT(/,25X,'TOLERANCE        =',F12.2/)
       QR=QT
       IF(X(1).NE.GVO2 .OR. Y(1).NE.GVCO2) GO TO 280
       CALL BLOOD(PMAO2,PMACO2,AMO2C,AMCO2C)
       FMVO2=-GVO2/(10.0*QT)+AMO2C
       FMVCO2=GVCO2/(10.0*QT)+AMCO2C
       IF(FMVO2.GT.0.0) GO TO 270
       WRITE(3,260) FMVO2,FMVCO2
       WRITE(*,260)FMVO2,FMVCO2
260    FORMAT(' FICK CALC MIXED VENOUS O2 CONTENT='F5.1,5X,'CO2
      + CONTENT='F5.1,6X,'THEREFORE O2 CALCS IMPOSSIBLE'////////)
       GO TO 300
270    CALL FNDTEN(PVO2,PVCO2,FMVO2,FMVCO2)
       X(1)=PVO2
       Y(1)=PVCO2
280    CONTINUE
C      FNDMVP is the subroutine that itself calkls other routines
C      and ends up with the predicted arterial PO2 and PCO2
C
C      WRITE is as the name suggests an output routine for results
C
       DO 290 IBOHR=1,2
       CALL FNDMVP(KRUN,X,Y)
       CALL WRITE
       IF(IBOHR.EQ.2) GO TO 286
       DO 285 J=1,NVAQS
       VSAVE(KK,J)=V(J)
285    QSAVE(KK,J)=Q(J)
286    IF(ISKO2.EQ.'N') GO TO 300
290    CONTINUE
300    CONTINUE
310    FORMAT(' ENTER FILE NAME.EXT FOR VQ INPUT DATA',/)
320    FORMAT(' WRITE OUTPUT TO DISK (0) OR PRINTER (1) ?')
330    FORMAT(' ENTER FILE NAME.EXT OF YOUR CHOICE FOR RESULTS')
C
C      WRITE DISTRIBUTION TO DISK FOR PLOTTING
```

```
C
      IF(IPLTVQ.EQ.0) GO TO 400
      ZERO=0.0
      VAQ(1)=0.001
      VAQ(NVAQS)=1000.
      DO 370 KK=1,NRUNS
      DO 340 KIJ=1,NRUNS
340   IF(IFLAG(KIJ).EQ.KK) GO TO 370
      IF(IPLTVQ.EQ.1.AND.KK.EQ.1) WRITE(3,355)
      IF(IPLTRE.EQ.1.AND.KK.EQ.1) WRITE(3,350)
350   FORMAT(' R & E PLOT VARIABLES FIELD IS 8 COLS, EACH 10 CHARS')
355   FORMAT('   V/Q PLOT VARIABLES FIELD IS 5 COLS, EACH 10 CHARS')
      WRITE(9,380)VAQ(1),ZERO,ZERO,Q(1),ZERO
      DO 360 J=2,NVAQS-1
360   WRITE(9,380)VAQ(J),VSAVE(KK,J),QSAVE(KK,J),ZERO,ZERO
      WRITE(9,380)VAQ(NVAQS),ZERO,ZERO,ZERO,V(NVAQS)
370   CONTINUE
380   FORMAT(5F10.3)
390   FORMAT(' WANT TO HARDCOPY PLOT THE V/Q CURVES?  0=NO, 1=YES')
395   FORMAT(' WANT TO HARDCOPY PLOT THE R & E CURVES?  0=NO,
      1=YES')
400   CONTINUE
410   FORMAT(' DATA FOR HARDCOPY VA/Q  PLOTS ARE IN:     C:VQPLOT.
      FYL')
415   FORMAT(' DATA FOR HARDCOPY R & E PLOTS ARE IN:     C:REPLOT.
      FYL')
      STOP
      END
      SUBROUTINE CALCVQ(IPLTRE,ISKIP,DSPCE,SUMQ,KRUN,IRORE)
C
C     master subroutine for MIGET inert gas conversion to VA/Q dis-
      tribution
C
      DIMENSION S(10),PC(10),RDATA(10),EDATA(10),AD(10)
      DIMENSION WEIGHT(10),RAWDAT(10),RAD(10)
      DIMENSION SOL(50),CALCDR(50),CALCDE(50),RHOMO(50),EHOMO(50)
      DIMENSION DKK(10,10),DQ(11),DV(11)
      CHARACTER IPLCHR
      COMMON/VQ/V(500),Q(500),QQ(500),VAQ(500)
      COMMON/INERT/NVAQS,VT,QT,NGASES,FACT,Z,VQLO,VQHI
      NGAS1=NGASES+1
      READ(8,10) (PC(I),I=1,NGASES)
10    FORMAT(1X,6(1PE12.3))
      PC(NGAS1)=0.0
```

```
      READ(8,20) (RDATA(I),I=1,NGASES)
      READ(8,20) (EDATA(I),I=1,NGASES)
20    FORMAT(8F12.6)
      READ(8,30) (WEIGHT(I),I=1,NGASES)
30    FORMAT(8(F11.2,1X))
      IF(ISKIP.EQ.1) RETURN
      DO 40, I=1,NGAS1
40    S(I)=PC(I)/FACT
      DO 50 I=1,NGASES
      IF(IRORE.EQ.0) RAWDAT(I)=RDATA(I)
50    IF(IRORE.NE.0) RAWDAT(I)=EDATA(I)
      RAWDAT(NGAS1)=1.0
      RAWDAT(NGAS1+1)=1.0
      WRITE(3,60)NGASES,NVAQS,Z
60    FORMAT(' NUMBER OF GASES ='I3/' NUMBER OF VA/Q COMPARTMENTS ='
     & I4,/,' SMOOTHING COEFFICIENT Z =',F8.2/)
            WRITE(3,70)
70    FORMAT(8X,'GAS'10X'SOL'12X'PC'14X'R'14X'E')
      DO 80, I=1,NGASES
80    WRITE(3,90)I,S(I),PC(I),RDATA(I),EDATA(I)
90    FORMAT(I10,4F15.5)
      NV=NVAQS-1
      RNV=NV
      DVQ=(ALOG(VQHI/VQLO))/RNV
      DO 110, J=1,NVAQS
      TJ=J
110   VAQ(J) = VQLO*EXP(DVQ*(TJ-1.0))
      VAQ(1)=0.0
      VAQ(NVAQS) = 10000.0
      IC=1
C
C     subroutine SMOOTH is the code that determines the
      50-compartment VA/Q
C     distribution from the inert gas retentions
C
120   IF(IRORE.EQ.0) CALL SMOOTH(E,PC,RDATA,RAWDAT,WEIGHT,AD,IC,
     + SUMQ,IRORE)
      IF(IRORE.NE.0) CALL SMOOTH(E,PC,EDATA,RAWDAT,WEIGHT,AD,IC,
     + SUMV,IRORE)
      IF(IRORE.NE.0) GO TO 184
      VTNEW=0.0
      V(1)=0.0
      DO 140, J=1,NV
      QQ(J)=QQ(J)/SUMQ
      V(J) = QT*VAQ(J)*QQ(J)
```

```
140     VTNEW=VTNEW+ V(J)
        V(NVAQS)=VT-VTNEW
        DO 150, J=1,NVAQS
150     V(J)=V(J)/VT
C
C       all the commented WRITE statements that follow are not used
        routinely.
C       They can be activated as diagnostic flags to help see
        wheresomething has gone wrong
C
C        WRITE(3,155)
C155     FORMAT(4X,'I'7X,'PC',8X,'AD/WEIGHT',8X,'RAD')
        DO 170, I=1,NGASES
        RAD(I)=0.0
        DO 160, J=1,NV
160     RAD(I)=RAD(I)+QQ(J)*PC(I)/(VAQ(J)+PC(I))
C        WRITE(3,180)I,PC(I),AD(I)/WEIGHT(I),RAD(I)
170     CONTINUE
C180     FORMAT(I5,F11.4,2F15.8)
C        WRITE(3,182) IC,QT,VT,VTNEW,VTNEW/VT
C182     FORMAT(I5,3X,'QT:'F6.3,3X,'VT:'F7.3,3X,'CALC:'F9.5,3X,
C       & 'CALC/MEAS:'F8.5)
        QQ(NVAQS)=0.0
        GO TO 198
184     QTNEW=0.0
        QQ(NVAQS)=0.0
        V(1)=0.0
        DO 186, J=2,NV
        V(J)=V(J)/SUMV
        QQ(J) = V(J)*VT/VAQ(J)
186     QTNEW=QTNEW+ QQ(J)
        V(NVAQS)=V(NVAQS)/SUMV
        QQ(1)=QT-QTNEW
        DO 188, J-1,NVAQS
188     QQ(J)=QQ(J)/QT
C        WRITE(3,189)
C189         FORMAT(4X'I'7X'PC'8X'AD/WEIGHT'8X'EXC'11X'RETEN'11X'RAD')
        DO 192, I=1,NGASES
        RAD(I)=0.0
        EXC=0.0
        DO 190, J=1,NV
        EXC=EXC+ V(J)*PC(I)/(VAQ(J)+PC(I))
190     RAD(I)=RAD(I)+QQ(J)*PC(I)/(VAQ(J)+PC(I))
        RETEN=1.0- VT*EXC/(PC(I)*QT)
C        WRITE(3,194) I,PC(I),AD(I)/WEIGHT(I),EXC,RETEN,RAD(I)
```

```
192     CONTINUE
C194     FORMAT(I5,F11.4,4F15.8)
C        WRITE(3,182) IC,QT,VT,QTNEW,QTNEW/QT
C
C       following summs up VA and Q into amounts in each VA/Q decade
C
198     SHNT=QQ(1)
        DSPCE=V(NVAQS)
        SMQPO1=0.0
        SMVPO1=0.0
        SMQP1=0.0
        SMVP1=0.0
        SMQ1=0.0
        SMV1=0.0
        SMQ10=0.0
        SMV10=0.0
        SMQ100=0.0
        SMV100=0.0
        DO 240, J=2,NV
        IF(VAQ(J).GT.0.01) GO TO 200
        SMQPO1=SMQPO1 + QQ(J)
        SMVPO1=SMVPO1 + V(J)
        GO TO 240
200     IF(VAQ(J).GT.0.1) GO TO 210
        SMQP1=SMQP1 + QQ(J)
        SMVP1=SMVP1 + V(J)
        GO TO 240
210     IF(VAQ(J).GT.1.0) GO TO 220
        SMQ1=SMQ1 + QQ(J)
        SMV1 =SMV1 + V(J)
        GO TO 240
220     IF(VAQ(J).GT.10.0) GO TO 230
        SMQ10=SMQ10 + QQ(J)
        SMV10=SMV10 + V(J)
        GO TO 240
230     SMQ100=SMQ100 + QQ(J)
        SMV100=SMV100 + V(J)
240     CONTINUE
        WRITE(3,250)
250     FORMAT(//6X'RANGE'17X'BLOOD FLOW'11X'VENTILATION'/)
        WRITE(3,260)SHNT
        WRITE(3,270)SMQPO1,SMVPO1
        WRITE(3,280)SMQP1,SMVP1
        WRITE(3,290)SMQ1,SMV1
        WRITE(3,300)SMQ10,SMV10
```

```
      WRITE(3,310)SMQ100,SMV100
      WRITE(3,320)DSPCE
260   FORMAT(' VA/Q OF ZERO       ',F16.3,8X,'        ZERO')
270   FORMAT(' VA/Q RANGE    0  TO  .01'F11.3,F20.3)
280   FORMAT(' VA/Q RANGE   .01 TO  .1',F12.3,F20.3)
290   FORMAT(' VA/Q RANGE   .1 TO  1.',F13.3,F20.3)
300   FORMAT(' VA/Q RANGE  1.0 TO 10.',F13.3,F20.3)
310   FORMAT(' VA/Q RANGE 10. TO 100.',F13.3,F20.3)
320   FORMAT(' VA/Q OF INFINITY               ZERO     ',F15.3/)
C
C     follwing calculate the first, second and third moments of the
      distribution
C     for both VA and Q curves separately
C
      SUMQVQ=0.0
      SUMVVQ=0.0
      SUMQ=0.0
      SUMV=0.0
      DO 330, I=2,NV
      SUMQVQ=SUMQVQ + QQ(I)*ALOG(VAQ(I))
      SUMVVQ=SUMVVQ + V(I)*ALOG(VAQ(I))
      SUMQ=SUMQ + QQ(I)
330   SUMV=SUMV + V(I)
      IF(SUMQ.LE.0.0.OR.SUMV.LE.0.0) RETURN
      FQ=SUMQVQ/SUMQ
      FV=SUMVVQ/SUMV
      QBAR=EXP(FQ)
      VBAR=EXP(FV)
      SUMVRQ=0.0
      SUMVRV=0.0
      SUMSKQ=0.0
      SUMSKV=0.0
      DO 340, I=2,NV
      SUMVRQ=SUMVRQ + QQ(I)*(ALOG(VAQ(I)) - FQ)**2
      SUMVRV=SUMVRV + V(I)*(ALOG(VAQ(I)) - FV)**2
      SUMSKQ=SUMSKQ + QQ(I)*(ALOG(VAQ(I)) - FQ)**3
      SUMSKV=SUMSKV + V(I)*(ALOG(VAQ(I)) - FV)**3
340   CONTINUE
      QSD=SQRT(SUMVRQ/SUMQ)
      VSD=SQRT(SUMVRV/SUMV)
      QSKEW = SUMSKQ/SUMQ
      VSKEW = SUMSKV/SUMV
      WRITE(3,350)QBAR,QSD,QSKEW,VBAR,VSD,VSKEW
350   FORMAT(/,'        MEAN OF BLOOD FLOW  DISTRIBUTION =',F7.2,/,
     1' 2nd MOMENT OF BLOOD FLOW  DISTRIBUTION =',F7.2,/,
```

```
     2' 3rd MOMEMT OF BLOOD FLOW  DISTRIBUTION =',F7.2,//,
     3'       MEAN OF VENTILATION DISTRIBUTION =',F7.2/,
     4' 2nd MOMENT OF VENTILATION DISTRIBUTION =',F7.2/,
     5' 3rd MOMENT OF VENTILATION DISTRIBUTION =',F7.2,//)
      DO 360, I=1,NVAQS
      V(I)=V(I)*VT
360   QQ(I)=QQ(I)*QT
      VA=VT-V(NVAQS)
C
C      Calculation of max possible deadspace ventilation
C      by John Evans' recursion method
C
      DQ(NGAS1)=QT
      DV(1)=0.0
      DO 370, LL=1,NGASES
      DKK(NGASES,LL)=RAD(LL)*QT
370   CONTINUE
      DO 380, KK=1,NGASES-1
      K=NGASES-KK
      K1=K+1
      DQ(K1)=(DQ(K1+1)-DKK(K1,K1))/DKK(K1,K1)
      DO 380,I=1,K
      DKK(K,I)=(PC(I)*(DKK(K1,I)-DKK(K1,K1)))/(PC(I)*DKK(K1,K1)-
     1 PC(K1)*DKK(K1,I))
      IF(K.NE.1) GO TO 380
      DQ(1)=(DQ(K1)-DKK(K,K))/DKK(K,K)
380   CONTINUE
      DO 390, L=1,NGASES
      L1=L+1
      DV(L1)=DKK(L,L)*(DV(L)+PC(L)*(DQ(L)+1.0)*DQ(L))
390   CONTINUE
      VD=VT-DV(NGAS1)
      DED=VD/VT
C*********
C     FOLLOWING IS DISPERSION DIRECTLY FROM DATA, AS THE SUM OF
      SQUARES
C     OF THE DIFFERENCES BETWEEN HOMOGENEOUS RETENTIONS AND THE BEST
      FIT
C     VALUES OUT OF SMOOTH.
C*********
      WRITE(3,400)
400   FORMAT(' GAS',4X,'PC',7X,'R',7X,'RH',4X,'R - RH',5X,'E',
     &7X,'E*',6X,'EH',3X,'EH - E*',1X,'R - E*')
      DISR=0.0
      DISE=0.0
```

```
      DISRE=0.0
      DO 410, I=1,NGASES
      RH=PC(I)/(PC(I)+VA/QT)
      EH=RH
      ADD=RAD(I)
      EE=PC(I)*QT*(1.0-ADD)/VT
      EXC=EE/(1.0-DED)
      DIFFR=ADD-RH
      DIFFE=EH - EXC
      DIFFRE=ADD - EXC
      DISR=DISR+DIFFR*DIFFR
      DISE=DISE+DIFFE*DIFFE
      DISRE=DISRE+DIFFRE*DIFFRE
      WRITE(3,420)I,PC(I),ADD,RH,DIFFR,EE,EXC,EH,DIFFE,DIFFRE
410   CONTINUE
420   FORMAT(I2,F10.5,8F8.5)
      DISPR=100.0*SQRT(DISR/(FLOAT(NGASES)))
      DISPE=100.0*SQRT(DISE/(FLOAT(NGASES)))
      DISPRE=100.0*SQRT(DISRE/(FLOAT(NGASES)))
      WRITE(3,430) VD,DED
430   FORMAT(/' MAX POSSIBLE DEADSPACE VENTILATION =',F7.1,
     1' L/MIN,  OR AS A FRACTION =',F5.3/)
      WRITE(3,440)DISPR,DISPE,DISPRE
440   FORMAT(' DISPERSION DIRECTLY FROM DIFFERENCES BETWEEN:',//,
     1'  BEST FIT   RETENTIONS & HOMOGENEOUS RETENTIONS IS:',F7.2,/,
     2' HOMOGENEOUS EXCRETIONS &  BEST FIT   EXCRETIONS IS:',F7.2,/,
     3'  BEST FIT   RETENTIONS *  BEST FIT   EXCRETIONS IS:',F7.2///)
C
C     following is a coarse printer plot of the retention & excre-
      tion curves
C     and of the ventilation and blood flow curves
C
      IF(IC.EQ.2)  GO TO 450
      IF(IRORE.EQ.0 .AND. V(NVAQS).GE.0.0) GO TO 450
      IF(IRORE.NE.0 .AND. QQ(1).GE.0.0) GO TO 450
      IC=2
      GO TO 120
450   CONTINUE
      YMAX=0.0
      DO 460, J=2,NVAQS
      IF(YMAX.GE.QQ(J)) GO TO 460
      YMAX=QQ(J)
460   CONTINUE
      DO 470, J=1,NV
      IF(YMAX.GE.V(J)) GO TO 470
```

```
      YMAX=V(J)
470   CONTINUE
      YMAX=1.25*YMAX
      YMAXV=YMAX
      VAQ(1)=0.0002
      VAQ(NVAQS)=990.0
      LINES=36
      IF(YMAX.EQ.0.0) GO TO 480
      IPLCHR='*'
      CALL PLOT(1,VAQ,QQ,YMAX,LINES,NVAQS,1,2,1,0,IPLCHR,KRUN)
      IPLCHR='O'
      CALL PLOT(1,VAQ,V,YMAXV,LINES,NV,2,2,1,0,IPLCHR,KRUN)
480   CONTINUE
      VAQ(1)=0.0
      SOLHI=1000.0
      SOLLO=0.0001
      DS=(ALOG10(SOLHI/SOLLO))/49.0
      DO 490, J=1,50
      TJ=J
490   SOL(J) = SOLLO*(10.0**(DS*(TJ-1.0)))
      DO 510 I=1,50
      CALCDR(I)=0.0
      CALCDE(I)=0.0
      DO 500, J=1,NV
      CALCDR(I)=CALCDR(I) + QQ(J)*SOL(I)/(SOL(I)+VAQ(J))
500   CALCDE(I)=CALCDE(I) + V(J)*SOL(I)/(SOL(I)+VAQ(J))
      CALCDR(I)=CALCDR(I)/QT
      CALCDE(I)=CALCDE(I)/VT
      VA=VT-V(NVAQS)
      RHOMO(I)=SOL(I)/(SOL(I) + VA/QT)
      EHOMO(I) = RHOMO(I)*VA/VT
510   CONTINUE
      LINES=36
      IPLCHR='.'
      CALL PLOT(2,SOL,CALCDR,1.0,LINES,50,1,6,1,0,IPLCHR,KRUN)
      IPLCHR='*'
      CALL PLOT(2,SOL,RHOMO,1.0,LINES,50,2,6,1,0,IPLCHR,KRUN)
      IPLCHR='O'
      IF(IRORE.EQ.0)CALL PLOT(2,PC,RAWDAT,1.0,LINES,NGASES,3,6,1,0,
     + IPLCHR,KRUN)
      IF(IRORE.NE.0)CALL PLOT(2,PC,RDATA,1.0,LINES,NGASES,3,6,1,0,
     + IPLCHR,KRUN)
      IPLCHR='.'
      CALL PLOT(2,SOL,CALCDE,1.0,LINES,50,4,6,1,0,IPLCHR,KRUN)
      IPLCHR='*'
```

```
      CALL PLOT (2,SOL,EHOMO,1.0,LINES,50,5,6,1,0,IPLCHR,KRUN)
      IPLCHR='O'
      IF(IRORE.EQ.0) CALL PLOT(2,PC,EDATA,1.0,LINES,NGASES,6,6,1,0,
     + IPLCHR,KRUN)
      IF(IRORE.NE.0)       CALL       PLOT(2,PC,RAWDAT,1.0,LINES,
      NGASES,6,6,1,0,
     + IPLCHR,KRUN)
      IF(IPLTRE.EQ.0) GO TO 540
      ZERO=0.0
      DO 520 J=1,50
      IF(J.LE.NGASES) WRITE(6,530)SOL(J),RHOMO(J),CALCDR(J),EHOMO(J),
     1CALCDE(J),PC(J),RAWDAT(J),EDATA(J)
      IF(J.GT.NGASES) WRITE(6,530)SOL(J),RHOMO(J),CALCDR(J),EHOMO(J),
     1CALCDE(J),ZERO,ZERO,ZERO
520   CONTINUE
530   FORMAT(8F10.3)
540   CONTINUE
      RETURN
      END
      SUBROUTINE SMOOTH(E,PC,DATA,RAWDAT,WEIGHT,AD,IC,SUMQV,IRORE)
      DIMENSION WEIGHT(10),PC(10),AD(10),DATA(10),RAWDAT(10)
      DIMENSION IFLOW(500),Y(500)
      COMMON/VQ/V(500),Q(500),QQ(500),VAQ(500)
      DIMENSION A(500,10),WT(500),FLOW(500),H(500)
      DIMENSION C(10,10),RBAR(10),RD(10)
      COMMON/INERT/NVAQS,VT,QT,NGASES,FACT,Z,VQLO,VQHI
      DOUBLE PRECISION A,WT,FLOW,H,C,RBAR,RD
C
C     this routine does the actiual equation solving to convert MIGET
      data
C     into the VA/Q distribution
C
C     first GENERATE A, THE ORIGINAL CHEBYSHEV MATRIX:  S/(S+VAQ)
C     where S is poartition coefficient for each gas (1-6) and VAQ is
      VAQ ratio, 1-50
C
      NV=NVAQS-1
      QTVT=QT/VT
      VTQT=VT/QT
      JLO=1
      JHI=NV
      JC=1
      IF(IRORE.EQ.0) GO TO 10
      JLO=2
      JHI=NVAQS
```

```
      JC=NVAQS
 10   SSQPRE=1000000.0
      SSLOOP=1000000.0
      NEQNS=NGASES+IC
      NEQN1=NEQNS+1
      NEQN2=NEQNS+2
      DO 20 I=1,NGASES
      A(NVAQS,I)=0.0
      DO 20  J=1,NV
 20   A(J,I) = PC(I)/(PC(I) + VAQ(J))
      DO 30, J=JLO,JHI
      A(J,NGASES+1) = 1.0
      IF(IC.EQ.2 .AND. IRORE.EQ.0) A(J,NEQNS)=VAQ(J)*QTVT
 30   IF(IC.EQ.2 .AND. IRORE.NE.0) A(J,NEQNS)=VTQT/VAQ(J)
      DO 40 J=1,IC
      DATA(NGASES+J) = 1.0
 40   WEIGHT(NGASES+J) = 20000.0
      ICL=(IC-1)*NGASES + 1
      DO 50, I=ICL,NEQNS
 50   DATA(I)=DATA(I)*WEIGHT(I)
      DO 60, I=1,NEQNS
      DO 60, J=JLO,JHI
 60   A(J,I)=A(J,I)*WEIGHT(I)
      DO 70, J=2,NV
      IFLOW(J)=1
      IF(IRORE.EQ.0) WT(J)=SQRT(Z*(1.0 + QTVT*QTVT*VAQ(J)*VAQ(J)))
      IF(IRORE.NE.0) WT(J)=SQRT(Z*(1.0 + VTQT*VTQT*
     & (1.0/VAQ(J))*(1.0/VAQ(J))))
 70   CONTINUE
      IFLOW(JC)=1
      WT(1)=1.0
      WT(NVAQS)=1.0
      DO 80, I=1,NEQNS
      DO 80, J=JLO,JHI
 80   A(J,I)=A(J,I)/WT(J)
      IF(IC.EQ.2 .AND. IRORE.EQ.0) WRITE(3,90)
 90   FORMAT(' SMOOTH RERUN WITH A NON-NEGATIVE DEADSPACE CONSTRAINT -
     & AN 8th EQN WITH TOTAL V=1')
      IF(IC.EQ.2 .AND. IRORE.NE.0) WRITE(3,100)
100   FORMAT(' SMOOTH RERUN WITH A NON-NEGATIVE SHUNT CONSTRAINT -
     & AN 8th EQN WITH TOTAL Q=1')
      IF(IRORE.EQ.0) WRITE(3,110)
110   FORMAT(/3X,'ITN',2X,'LOOP',9X,'TOTAL SSQ',12X,'FIT TO R',13X,
     1'SUM Q*Q')
      IF(IRORE.NE.0) WRITE(3,120)
```

```
120    FORMAT(/3X,'ITN',2X,'LOOP',9X,'TOTAL SSQ',12X,'FIT TO E',13X,
      1'SUM V*V')
       ITER=0
       IREP=1
C
C      GENERATE THE UPPER HALF OF A*A TRANSPOSE, NOTING THAT IT IS
       SYMMETRIC
C
       LOOP=0
130    LOOP=LOOP+1
       IF(LOOP.GT.1) SSLOOP=SSQ
       DO 150 I=1,NEQNS
       DO 150 J=1,I
       C(J,I)=0.0
       DO 140 K=2,NV
       IF(IFLOW(K).EQ.0) GO TO 140
       C(J,I) = C(J,I) + A(K,I)*A(K,J)
140    CONTINUE
       IF(I.EQ.J) C(J,I) = 1.0 + C(J,I)
150    CONTINUE
C
C      THE SMOOTHING FACTOR IS CALLED Z. WE GENERATE MATRIX  C, WHICH IS
C      IDENTITY PLUS A x AT (ie., A x A transpose) DIVIDED BY Z, PLUS A
       DATA COLUMN & A SHUNT
C      OR DEADSPACE SPECIAL COMPARTMENT COLUMN
C
       DO 160 I=1,NEQNS
       C(I,NEQN1) = DATA(I)
       IF(IFLOW(JC).GT.0) C(I,NEQN2) = A(JC,I)
160    CONTINUE
C
C      NOW SOLVE THE UNCONSTRAINED SYSTEM C x RD = DATA, USING GAUSSIAN
C      ELIMINATION WITH BACK-SUBSTITUTION
C
       DO 180, I=1,NEQNS-1
       I1=I+1
       DO 180 J=I1,NEQNS
       DO 170 K=J,NEQN1
       C(J,K) = C(J,K) -  C(I,J)*C(I,K)/C(I,I)
170    CONTINUE
       IF(IFLOW(JC).GT.0) C(J,NEQN2)=C(J,NEQN2)-
      & C(I,J)*C(I,NEQN2)/C(I,I)
180    CONTINUE
       RD(NEQNS) = C(NEQNS,NEQN1)/C(NEQNS,NEQNS)
       IF(IFLOW(JC).GT.0) RBAR(NEQNS)=C(NEQNS,NEQN2)/C(NEQNS,NEQNS)
```

```
      DO 200 I=1,NEQNS-1
      RD(NEQNS-I) = C(NEQNS-I,NEQN1)
      IF(IFLOW(JC).GT.0) RBAR(NEQNS-I) = C(NEQNS-I,NEQN2)
      NN1 = NEQN1 -I
      DO 190, K=NN1,NEQNS
      RD(NEQNS-I) = RD(NEQNS-I) - C(NEQNS-I,K)*RD(K)
      IF(IFLOW(JC).GT.0) RBAR(NEQNS-I) = RBAR(NEQNS-I) -
     & C(NEQNS-I,K)*RBAR(K)
190   CONTINUE
      RD(NEQNS-I) = RD(NEQNS-I)/C(NEQNS-I,NEQNS-I)
      IF(IFLOW(JC).GT.0) RBAR(NEQNS-I)=
     & RBAR(NEQNS-I)/C(NEQNS-I,NEQNS-I)
200   CONTINUE
      IF(IFLOW(JC).EQ.0) GO TO 230
      A1=0.0
      A2=0.0
      DO 210, I=1,NEQNS
      A1=A1 + A(JC,I)*RD(I)
      A2=A2 + A(JC,I)*RBAR(I)
210   CONTINUE
      A0=A1/A2
      DO 220, I=1,NEQNS
220   RD(I)=RD(I)-A0*RBAR(I)
C
C     NOW COMES CALCULATION OF THE Q (OR V) VALUES
C
230   DO 240 J=JLO,JHI
      FLOW(J)=0.0
      DO 240 I=1,NEQNS
      FLOW(J)=FLOW(J) + RD(I)*A(J,I)
240   CONTINUE
      IF(IFLOW(JC).GT.0) FLOW(JC) = A0
      SUM1=0.0
      DO 250 I=1,NEQNS
      SUM1=SUM1+RD(I)*RD(I)
250   CONTINUE
      SUM2=0.0
      DO 260 J=2,NV
      IF(IFLOW(J).EQ.0) GO TO 260
      SUM2=SUM2+FLOW(J)*FLOW(J)
260   CONTINUE
      SSQ=SUM1+SUM2
      WRITE(3,270)ITER,LOOP,SSQ,SUM1,SUM2
270   FORMAT(2I5,3F20.6)
C
```

```
C       printing used only for troubleshooting
C
C       WRITE(3,272)(IFLOW(JC),IFLOW(J),J=2,NV)
C272    FORMAT(10(I8,4X))
C       WRITE(3,274)FLOW(JC),((FLOW(JJ)/WT(JJ)),JJ=2,NV)
C274    FORMAT(10F12.5)
C
C       HAVING CALCULATED Q OR V VALUES, WE NOW NEED TO ENFORCE THE
C       NON-NEGATIVITY CONSTRAINT:
C
        IF(IREP.EQ.1) GO TO 280
        GO TO 300
280     IREP=2
        DO 290, I=JLO,JHI
        IF(IFLOW(I).EQ.1 .AND. FLOW(I).LE.0.0) IREP=1
        IF(IFLOW(I).EQ.1 .AND. FLOW(I).LE.0.0) IFLOW(I)=0
290     CONTINUE
        IF(IREP.EQ.1) GO TO 130
300     IREP=3
        XMIN=1.0
        DO 320 I=JLO,JHI
        IF(IFLOW(I).GT.0 .AND. FLOW(I).LE.0.0) GO TO 310
        GO TO 320
310     IREP=2
        Y(I)=H(I)/(H(I)-FLOW(I))
        IF(XMIN.GT.Y(I)) XMIN=Y(I)
320     CONTINUE
        IF(IREP.EQ.3) GO TO 350
        DO 340 I=JLO,JHI
        IF(IFLOW(I).EQ.0) GO TO 340
        IF(FLOW(I).GT.0.0) GO TO 330
        Y(I)=H(I)/(H(I)-FLOW(I))
        IF(Y(I).EQ.XMIN) IFLOW(I)=0
330     H(I)= (1.0-XMIN)*H(I)+XMIN*FLOW(I)
340     CONTINUE
        GO TO 130
350     CONTINUE
        ITER=ITER+1
        SUM1=0.0
        DO 360 I=1,NEQNS
        SUM1=SUM1+RD(I)*RD(I)
360     CONTINUE
        SUM2=0.0
        DO 370 J=2,NV
        IF(IFLOW(J).EQ.0) GO TO 370
```

```
      SUM2=SUM2+FLOW(J)*FLOW(J)
370   CONTINUE
      SSQ=SUM1 + SUM2
C******************************
      IF(SSQ.GE.SSQPRE) GO TO 390
C******************************
      SSQPRE=SSQ
      DO 380 I=JLO,JHI
      IF(IFLOW(I).GT.0) H(I)=FLOW(I)
      IF(IFLOW(I).GT.0) IFLOW(I)=2
      IF(IFLOW(I).EQ.0 .AND. FLOW(I).GT.0.0) IREP=1
      IF(IFLOW(I).EQ.0 .AND. FLOW(I).GT.0.0) IFLOW(I)=1
      IF(IFLOW(I).EQ.0) H(I)=0.0
380   CONTINUE
      IF(ITER.EQ.99) GO TO 410
      IF(IREP.EQ.1) GO TO 130
390   CONTINUE
C
C     CALCULATE THE ERROR AND THE APPROXIMATING DATA
C
      DO 400 I=2,NV
400   FLOW(I)=FLOW(I)/WT(I)
410   WRITE(3,420)ITER
420   FORMAT(/' ITERATION NUMBER =',I3)
      SUMQV=0.0
      SUME8=0.0
      IF(IRORE.NE.0) GO TO 470
      DO 430 J=1,NV
      IF(IFLOW(J).EQ.0) FLOW(J)=0.0
      QQ(J)=SNGL(FLOW(J))
      SUMQV=SUMQV + QQ(J)
      SUME8=SUME8+QQ(J)*VAQ(J)*QTVT
430   CONTINUE
      WRITE(3,440) SUMQV
440   FORMAT(' TOTAL BLOOD FLOW ='F10.6)
      IF(IC.EQ.2) WRITE(3,450) SUME8
450   FORMAT(' TOTAL VENTILATION='F10.6)
      WRITE(3,460)
460   FORMAT(/6X,'PC'7X,'RETENTIONS',5X,'BEST FIT'7X,'ERROR',
     15X,'RAW DATA',7X,'ERROR')
      GO TO 520
470   DO 480 J=2,NVAQS
      IF(IFLOW(J).EQ.0) FLOW(J)=0.0
      V(J)=SNGL(FLOW(J))
      SUMQV=SUMQV + V(J)
```

```
      SUME8=SUME8+V(J)*VTQT/VAQ(J)
480   CONTINUE
      WRITE(3,490)SUMQV
490   FORMAT(' TOTAL VENTILATION='F10.6)
      IF(IC.EQ.2) WRITE(3,500) SUME8
500   FORMAT(' TOTAL BLOOD FLOW ='F10.6)
      WRITE(3,510)
510   FORMAT(/6X,'PC'7X,'EXCRETIONS',5X,'BEST FIT'7X,'ERROR',
     15X,'RAW DATA',7X,'ERROR')
520   E=0.0
      DO 530 I=1,NEQNS
      E=E+RD(I)*RD(I)
      AD(I)=DATA(I)-RD(I)
530   CONTINUE
      DO 540 I=1,NEQNS
      S=PC(I)
      IF(I.GT.NGASES) S=0.0
      RAWERR=AD(I)/WEIGHT(I) - RAWDAT(I)
      WRITE(3,550)S,DATA(I),AD(I),RD(I),RAWDAT(I),RAWERR
540   CONTINUE
550   FORMAT(F10.4,F15.3,F13.3,F12.3,F13.5,F12.5)
      WRITE(3,560)E
560   FORMAT(/' REMAINING SUM OF SQUARES =',1PE10.2)
      RETURN
      END
      SUBROUTINE PLOT(IM,XX,Y,YMAX,LINES,LAST,NO,MOST,LOGX,
     1LOGY,ISYMBO,KRUN)
C
C     this is the printer plot for the retention/solubility abd VA/Q
      curves
C
      DIMENSION XX(500),Y(500),ZX(8),IGRAPH(61,40),X(500)
      DIMENSION LRUN(10),INAM1(15),INAM2(15)
      DIMENSION IHEAD1(36),IHEAD2(36)
      CHARACTER INAM1,INAM2,IHEAD1,IHEAD2,IHEAD,IOK,IBLANK,IBORDE,
     1LRUN,ISYMBO,IGRAPH,IPOINT
      DATA IHEAD1/' ',' ','C','O','M','P',' ',' ',' ','V','E','N','T',
     1' ',' ','O','R',' ',' ','B','L','O','O','D','F','L','O','W',
     1' ',' ','L','/','M','I','N',' '/
      DATA IHEAD2/' ',' ','R','E','T','E','N','T','I','O','N',
     1' ',' ',' ',' ',' ','A','N','D',' ',' ',' ',' ',' ','E','X',
     1'C','R','E','T','I','O','N',' ',' ',' '/
      DATA IBORDE/'.'/
      DATA IBLANK/' '/
      DATA INAM1/'V','E','N','T','I','L','A','T','I','O','N',
```

```
      1' ','=' ,' ','O'/
      DATA INAM2/'B','L','O','O','D','F','L','O','W',' ',' ',
      1' ','=',' ','*'/
      DATA LRUN/'0','1','2','3','4','5','6','7','8','9'/
      IWID=61
      RWID=IWID-1
      IF(IM.EQ.2) GO TO 20
      DO 10 I=7,21
      IGRAPH(I,3)=INAM1(I-6)
   10 IGRAPH(I,5)=INAM2(I-6)
   20 CONTINUE
      IF(KRUN.LT.10) IGRAPH(55,3)=LRUN(KRUN+1)
      IF(KRUN.LT.20.AND.KRUN.GE.10) IGRAPH(55,3)=LRUN(2)
      IF(KRUN.LT.20.AND.KRUN.GE.10) IGRAPH(56,3)=LRUN(KRUN-9)
      IF(KRUN.LT.30.AND.KRUN.GE.20) IGRAPH(55,3)=LRUN(3)
      IF(KRUN.LT.30.AND.KRUN.GE.20) IGRAPH(56,3)=LRUN(KRUN-19)
      IPOINT=ISYMBO
      YL=YMAX
      YS=0.0
      MATRIX=IWID*LINES
      A=LINES-1
      YSCALE=(YL-YS)/A
      XL=1000.0
      XS=0.0001
      IF(LOGX) 30,50,30
   30 DO 40 I=1,LAST
   40 X(I)=ALOG10(XX(I))
      XS=ALOG10(XS)
      XL=ALOG10(XL)
      XSCALE=(XL-XS)/RWID
   50 IF(LOGY) 60,80,60
   60 DO 70 I=1,LAST
   70 Y(I)=ALOG10(Y(I))
      YS=ALOG10(YS)
      YL=ALOG10(YL)
   80 IF(NO-1) 120,90,120
   90 CONTINUE
      MATRIX=IWID*LINES
      DO 100 I=1,LINES
      DO 100 J=1,IWID
  100 IGRAPH(J,I)=IBLANK
      XSCALE=(XL-XS)/RWID
      A=LINES-1
      YSCALE=(YL-YS)/A
      DO 110 I=1,LINES
```

```
      IGRAPH(1,I)=IBORDE
110   IGRAPH(IWID,I)=IBORDE
120   DO 170 I=1,LAST
      IF(XL-X(I)) 170,130,130
130   IF(X(I)-XS) 170,140,140
140   IF(YL-Y(I)) 170,150,150
150   IF(Y(I)-YS) 170,160,160
160   IX=(X(I)-XS)/XSCALE + 1.5
      IY=(Y(I)-YS)/YSCALE + 0.5
      IY=LINES-IY
      IGRAPH(IX,IY)=IPOINT
170   CONTINUE
      IF(NO-MOST) 180,190,180
180   RETURN
190   CONTINUE
      WRITE(3,320)
      YES=YL+YSCALE
      DO 200 I=1,LINES
      YES=YES-YSCALE
      IF(IM.EQ.1) IHEAD=IHEAD1(I)
      IF(IM.EQ.2) IHEAD=IHEAD2(I)
200   WRITE(3,330)YES,IHEAD,(IGRAPH(J,I),J=1,IWID)
      WRITE(3,340)
      ZX(1)=0.0
      IF(IM.EQ.2) ZX(1)=0.00013
      ZX(2)=0.0013
      ZX(3)=0.013
      ZX(4)=0.13
      ZX(5)=1.0
      ZX(6)=10.0
      ZX(7)=100.0
      ZX(8)=1000.0
      WRITE(3,350)(ZX(K),K=1,8)
      IF(LOGX) 250,220,250
220   IF(LOGY) 230,310,230
230   WRITE(3,240)
240   FORMAT(26H Y IS PLOTTED ON LOG SCALE)
      GO TO 310
250   IF(LOGY) 260,280,260
260   WRITE(3,270)
270   FORMAT(34H X AND Y ARE PLOTTED ON LOG SCALES)
      GO TO 310
280   CONTINUE
      IF(IM.EQ.1) WRITE(3,290)
      IF(IM.EQ.2) WRITE(3,300)
```

```
290    FORMAT(/21X'VENTILATION - PERFUSION  RATIO,  LOG SCALE'///)
300    FORMAT(/20X'BLOOD:GAS PARTITION COEFFICIENT,  LOG SCALE'/////)
310    RETURN
320    FORMAT(11H   Y VALUES 3X,1H.,7X,1H.,7X,1H.,7X,1H.,7X,1H.,
      17X,1H.,7X,1H.,7X,1H.)
330    FORMAT(1H F8.3,1X,A1,1X,101A1)
340    FORMAT(14X,1H.,7X,1H.,7X,1H.,7X,1H.,7X,1H.,7X,1H.,8X,1H.,7X,
      11H.)
350    FORMAT(F17.4,F8.3,F8.2,F7.1,F8.1,F8.1,F10.1,F9.1)
       RETURN
       END

       SUBROUTINE BLOOD(PO2,PCO2,O2C,CO2C)
C
C      calls the Kelman blood gas subotuines for O2, CO2 and pH
C
       COMMON/OXY1/HB,HCRIT,TEMP,APH,BPH,APCO2,BPCO2,KOUNT,DP50,PIO2,SO2
       PH1=PH(PCO2,0.0)
       Y=0.003*HB*(1.0-SATURA(PO2,PCO2,PH1)/100.0)
       PH2=PH(PCO2,Y)
       SATRN=SATURA(PO2,PCO2,PH2)
       O2C=0.0139*HB*SATRN + SO2*PO2
       CO2C=CO2CON(PCO2,PH2,SATRN)
       RETURN
       END
       FUNCTION PH(PCO2,Y)
C
C      Kelman pH routine
C
       COMMON/OXY1/HB,HCRIT,TEMP,APH,BPH,APCO2,BPCO2,KOUNT,DP50,PIO2,SO2
       IF(PCO2.LT.0.001) PCO2=0.001
       IF(APH-1.0) 10,10,20
 10    PH=7.59 + Y - 0.2741*ALOG(PCO2/20.0)
       GO TO 30
 20    PH=BPH+Y+(APH-BPH)*ALOG(PCO2/BPCO2)/ALOG(APCO2/BPCO2)
 30    RETURN
       END
       FUNCTION CO2CON(PCO2,PHE,SATN)
C
C      Kelman CO2 routine
C
       COMMON/OXY1/HB,HCRIT,TEMP,APH,BPH,APCO2,BPCO2,KOUNT,DP50,PIO2,SO2
       P=7.4-PHE
       PK=6.086+0.042*P+(38.0-TEMP)*(0.00472+0.00139*P)
       SOL=0.0307  +  0.00057*(37.0-TEMP)  +  0.00002*(37.0-TEMP)*
       (37.0-TEMP)
```

```
      DOX=0.59+0.2913*P-0.0844*P*P
      DR=0.664+0.2275*P-0.0938*P*P
      DDD=DOX+(DR-DOX)*(1.-SATN/100.0)
      CP=SOL*PCO2*(1.0+10.0**(PHE-PK))
      CCC=DDD*CP
      CO2CON=(HCRIT*CCC*0.01 + (1.0-HCRIT*0.01)*CP)*2.22
      RETURN
      END
      FUNCTION SATURA(PO2,PCO2,PHE)
C
C     Kelman O2Hb curve routine
C
      COMMON/OXY1/HB,HCRIT,TEMP,APH,BPH,APCO2,BPCO2,KOUNT,DP50,
      PIO2,SO2
      KOUNT=KOUNT+1
      A1=-8532.229
      A2=2121.401
      A3=-67.07399
      A4=935960.9
      A5=-31346.26
      A6=2396.167
      A7=-67.10441
      B=0.43429*ALOG(40.0/PCO2)
      X=PO2*10.0**(0.024*(37.0-TEMP)+0.4*(PHE-7.4)+0.06*B)
      X=26.8*X/(26.8+DP50)
      IF(X-10.0) 10,20,20
 10   SAT=0.003683*X + 0.000584*X*X
      GO TO 30
 20   SAT=(X*(X*(X*(X+A3)+A2)+A1))/(X*(X*(X*(X+A7)+A6)+A5)+A4)
 30   SATURA=100.0*SAT
      RETURN
      END
      SUBROUTINE SAMEO2(PPO,PPCO,ARTO2C,CO2CT2)
C
C     SAMEO2 and FNDTEN are a pair of routines that allow you to
      compute PO2 and PCO2
C     from O2 and CO2 concentrations
C
      COMMON/OXY1/HB,HCRIT,TEMP,APH,BPH,APCO2,BPCO2,KOUNT,DP50,
      PIO2,SO2
      E=0.0
      F=PIO2+10.0
 10   G=(E+F)/2.0
      CALL BLOOD(G,PPCO,O2CNT2,CO2CT2)
      A10=ABS(O2CNT2-ARTO2C)
```

```
      IF(A10-0.001) 50,50,20
 20   IF(O2CNT2-ARTO2C) 30,50,40
 30   E=G
      GO TO 10
 40   F=G
      GO TO 10
 50   CONTINUE
      PPO=G
      RETURN
      END
      SUBROUTINE FNDTEN(PPO,PPCO,ARTO2C,ARTCO2)
      DIMENSION PI(4)
      PI(1)=10.0
      PI(2)=1.0
      PI(3)=0.1
      PI(4)=0.01
      PPCO=0.0
      DO 20 K=1,4
 10   PPCO=PPCO+PI(K)
      CALL SAMEO2 (PPO,PPCO,ARTO2C,CO2CT2)
      IF(CO2CT2-ARTCO2) 10,20,20
 20   PPCO=PPCO-PI(K)
      RETURN
      END
      SUBROUTINE DETERM(U,F,G,DET)
C
C     subotuine used in the algorithm that finds the alveolar PO2
      and PCO2 in every VA/Q compartment
C
      DIMENSION U(3),F(3),G(3),W(3)
      I=1
      J=2
      K=3
 10   W(I)=U(I)*(F(J)*G(K) - F(K)*G(J))
      IF(I-3) 20,50,50
 20   IF(I-1) 30,30,40
 30   I=2
      J=3
      K=1
      GO TO 10
 40   I=3
      J=1
      K=2
      GO TO 10
 50   DET=0.0
```

```
      DO 60 I=1,3
 60   DET=DET+W(I)
      RETURN
      END
      SUBROUTINE SUMUP(KRUN)
C
C     routine that summs up the contributions of all 50 compart-
      ments for O2 and CO2
C
      COMMON/INERT/NVAQS,VT,QT,NGASES,FACT,Z,VQLO,VQHI
      COMMON/OXY1/HB,HCRIT,TEMP,APH,BPH,APCO2,BPCO2,KOUNT,DP50,
      PIO2,SO2
      COMMON/OXY2/PB,FIO2,PICO2,FICO2,VS,QS,SKEW,GVO2,GVCO2,TOL,
     + PVO2,PVCO2,FMVO2,FMVCO2,PVN2,ALPHA,SHUNT,QR,GVAQ,FVQ,PAO2,
     + PACO2,O2CON,CCO2,PO2,PCO2,FVO2,FVCO2,RZ,RM,ARTO2C,ARTCO2,AMO2C
      COMMON/VQ/V(500),Q(500),QQ(500),VAQ(500)
      COMMON/BOHR/DM,VC,BETA,RNT,DDLO2,DDLCO2,PMAO2,PMACO2,PO2B,PCO2B,
     +IBOHR,CAIII,IPRN,ARTPM,IWARN,IWARN2
      COMMON/O2ARAY/OO2CON(500),OCCO2(500),FVQQ(500),PN22(500),
     +PCO22(500),PO22(500),RZZ(500),PBO2(500),PBCO2(500)
      CALL BLOOD(PVO2,PVCO2,FMVO2,FMVCO2)
      KNT=0
      DM1=0.020*GVO2
      DM2=0.025*GVO2
      DM3=0.030*GVO2
C
C     ALTERNATIVE FIRST THREE GUESSES COULD BE:
C
C     0.8*QT
C     1.0*QT
C     1.5*QT
C
 10   KNT=KNT+1
      IF(KNT.EQ.1) DM=DM1
      IF(KNT.EQ.2) DM=1.2*DM1
      IF(KNT.EQ.3) DM=DM3
 20   CC=0.0
      DD=0.0
      DLO2=0.0
      DLCO2=0.0
      IWARN2=0
      DO 30 I=1,NVAQS
      FVQQ(I)=0.0
      PO22(I)=0.0
      PCO22(I)=0.0
```

```
      PN22(I)=0.0
      RZZ(I)=0.0
      PBO2(I)=0.0
      PBCO2(I)=0.0
      OO2CON(I)=0.0
      OCCO2(I)=0.0
30    CONTINUE
      IF(IBOHR.EQ.2) WRITE(*,40)KRUN
40    FORMAT(//,3X'#'5X'PAO2'6X'PaO2'6X'PACO2'5X'PaCO2       SET:'I5/)
      DO 80 I=2,NVAQS-1
      IF(IBOHR.EQ.2.AND.Q(I).LE.1.E-6) GO TO 80
      GVAQ=VAQ(I)
      FI=FIO2+FICO2
      SUM=ABS(FI-1.0)
      IF(SUM.LE.0.0001) GO TO 60
      CALL VQSOLN
      IF (IWARN.EQ.1) IWARN2=1
      IF(IBOHR.EQ.2) WRITE(*,50)I,PAO2,PO2B,PACO2,PCO2B
50    FORMAT(' ',I3,4F10.2)
      GO TO 70
60    CALL PUREO2
70    CONTINUE
      PBO2(I)=PO2B
      PBCO2(I)=PCO2B
      IF(IBOHR.EQ.2) CALL BLOOD(PO2B,PCO2B,O2CON,CCO2)
      FVQQ(I)=FVQ
      PO22(I)=PAO2
      PCO22(I)=PACO2
      PN22(I)=PIO2/FIO2-PAO2-PACO2
      RZZ(I)=RZ
      OO2CON(I)=O2CON
      OCCO2(I)=CCO2
      CC=CC+OO2CON(I)*Q(I)
      DD=DD+OCCO2(I)*Q(I)
      DLO2=DLO2+DDLO2*Q(I)
      DLCO2=DLCO2+DDLCO2*Q(I)
80    CONTINUE
      FVQQ(1)=0.0
      PO22(1)=PVO2
      PCO22(1)=PVCO2
      PBO2(1)=PVO2
      PBCO2(1)=PVCO2
      RZZ(1)=0.0
      PN22(1) = PIO2/FIO2-PVO2-PVCO2
      OO2CON(1) = FMVO2
```

```
      OCCO2(1) = FMVCO2
      FVQQ(NVAQS) = 30000.0
      PO22(NVAQS) = PIO2
      PCO22(NVAQS) = PICO2
      PN22(NVAQS) = PIO2/FIO2-PIO2-PICO2
      RZZ(NVAQS) = 0.0
      OO2CON(NVAQS) = 0.0
      OCCO2(NVAQS) = 0.0
      Q(NVAQS) = 0.0
      CC=CC+OO2CON(1)*Q(1)
      DD=DD+OCCO2(1)*Q(1)
      ARTO2C = CC/QT
      ARTCO2=DD/QT
      DLO2=DLO2/(QT-Q(1))
      DLCO2=DLCO2/(QT-Q(1))
      FVO2=10.0*QT*(ARTO2C-FMVO2)
      FVCO2=10.0*QT*(FMVCO2-ARTCO2)
      CALL FNDTEN(PP1,PP2,ARTO2C,ARTCO2)
      PO2=PP1
      PCO2=PP2
      PO2ER=PO2-PMAO2
      IF(IBOHR.EQ.2) WRITE(3,90)KNT,DM,PO2ER
      IF(IBOHR.EQ.2) WRITE(*,90)KNT,DM,PO2ER
 90   FORMAT(' ITN',I3,';  DM=',F8.2,'; & PO2 ERROR=',F7.2)
      IF (IBOHR.EQ.1) GO TO 140
      IF(ABS(PO2ER).LT.0.1) GO TO 140
      IF(KNT.EQ.2) GO TO 100
      IF(KNT.EQ.3) GO TO 110
      IF(KNT.GT.3) GO TO 120
      IF(PO2ER.GT.-5.0.AND.PO2ER.LT.0.0) GO TO 99
      IF(PO2ER.LT.-10.0) DM=1.2*DM
      IF(PO2ER.LT.-10.0) GO TO 20
      IF(PO2ER.LT.-5.0) DM=1.1*DM
      IF(PO2ER.LT.-5.0) GO TO 20
      IF(PO2ER.GT.0.0) DM=0.8*DM
      IF(PO2ER.GT.0.0) GO TO 20
 99   DM1=DM
      ERR1=PO2ER
      GO TO 10
100   DM2=DM
      ERR2=PO2ER
      GO TO 10
110   CONTINUE
      IF(PO2ER.LE.-10.0) DM=4.0*DM
      IF(PO2ER.GT.-10.0.AND.PO2ER.LE.-5.0) DM=2.0*DM
```

```
      IF(PO2ER.GT. -5.0.AND.PO2ER.LE.-2.0) DM=1.50*DM
      IF(PO2ER.GT. -2.0.AND.PO2ER.LT. 0.0) DM=1.1*DM
      IF(PO2ER.LT.0.0) GO TO 20
      IF(PO2ER.GE.(ARTPM-0.5)) DM=0.8*DM
      IF(PO2ER.GE.(ARTPM-0.5)) GO TO 20
      DM3=DM
      ERR3=PO2ER
      GO TO 130
120   DM1=DM2
      ERR1=ERR2
      DM2=DM3
      ERR2=ERR3
      DM3=DM
      ERR3=PO2ER
130   T1=(ERR1-ERR2)/(DM1-DM2)
      T2=(ERR1-ERR3)/(DM1-DM3)
      A=(T1-T2)/(DM2-DM3)
      B=T1 - A*(DM1+DM2)
      C=ERR1 - B*DM1 - A*DM1*DM1
      DM=(SQRT(B*B - 4.0*A*C) - B)/(2.0*A)
      GO TO 10
140   CONTINUE
      RETURN
      END
      SUBROUTINE PUREO2
C
C     this is a special routine for the case when FIO2 = 1.00 when the
      usual
C     logic for solving the O2/CO2 equations does not work because
      there is no Nitrogen
C
      COMMON/OXY1/HB,HCRIT,TEMP,APH,BPH,APCO2,BPCO2,KOUNT,DP50,
      PIO2,SO2
      COMMON/OXY2/PB,FIO2,PICO2,FICO2,VS,QS,SKEW,GVO2,GVCO2,TOL,
     + PVO2,PVCO2,FMVO2,FMVCO2,PVN2,ALPHA,SHUNT,QR,GVAQ,FVQ,PAO2,
     + PACO2,O2CON,CCO2,PO2,PCO2,FVO2,FVCO2,RZ,RM,ARTO2C,ARTCO2,
       AMO2C
      TOL1=0.001
      PCO21=PICO2
      PCO22=PICO2+PIO2
 10   PACO2=(PCO21+PCO22)/2.0
      PAO2=PIO2+PICO2-PACO2
      CALL BLOOD(PAO2,PACO2,O2CON,COCON)
      R1=(O2CON-FMVO2)/(FMVCO2-COCON)
      CON1=1.0 + FICO2*(R1-1.0)
```

```
      FVQ=8.63*(FMVCO2-COCON)*CON1/(PACO2-PICO2)
      DIFF=ABS(GVAQ-FVQ)
      IF(DIFF.LE.TOL1) GO TO 40
      IF(GVAQ-FVQ) 20,40,30
 20   PCO21=PACO2
      GO TO 10
 30   PCO22=PACO2
      GO TO 10
 40   CONTINUE
      RZ=1.0/R1
      CCO2=COCON
      PO2=PAO2
      PCO2=PACO2
      RETURN
      END
      SUBROUTINE FNDMVP(KRUN,X,Y)
C
C     this routine is the master routine for solving the VA/Q equa-
      tion (ie that finds the
C     alveolar PO2 and PCO2 for a given VA/Q ratio
C
C     it also has the ability to search for the mixed venous PO2 and
      PCO2 thatwould allow the
C     VO2 and VCO2 to be matched to measured values if you choose
      that option (ie, TOL=20)
C
C     when TOL is as usual, 99000, mixed venous PO2 and PCO2 are
      calcu;lated without searching
C     simply by using the Fick principle (CvO2= CaO2 - VO2/QT),
      ditto for CO2
C     and then calling SAMEO2/FNDTEN to convert those concentrations
      to partial pressures.
C
      DIMENSION X(4),Y(4),F(4),G(4),U(4)
      DIMENSION VINSP(500)
      COMMON/VQ/V(500),Q(500),QQ(500),VAQ(500)
      COMMON/INERT/NVAQS,VT,QT,NGASES,FACT,Z,VQLO,VQHI
      COMMON/OXY1/HB,HCRIT,TEMP,APH,BPH,APCO2,BPCO2,KOUNT,DP50,
      PIO2,SO2
      COMMON/OXY2/PB,FIO2,PICO2,FICO2,VS,QS,SKEW,GVO2,GVCO2,TOL,
     + PVO2,PVCO2,FMVO2,FMVCO2,PVN2,ALPHA,SHUNT,QR,GVAQ,FVQ,PAO2,
     + PACO2,O2CON,CCO2,PO2,PCO2,FVO2,FVCO2,RZ,RM,ARTO2C,ARTCO2,AMO2C
      COMMON/BOHR/DM,VC,BETA,RNT,DDLO2,DDLCO2,PMAO2,PMACO2,PO2B,
      PCO2B,
     +IBOHR,CAIII,IPRN,ARTPM,IWARN,IWARN2
```

```
      COMMON/O2ARAY/OO2CON(500),OCCO2(500),FVQQ(500),PN22(500),
     +PCO22(500),PO22(500),RZZ(500),PBO2(500),PBCO2(500)
      IF(ARTPM.LE.1.0.AND.IBOHR.EQ.2) WRITE(3,1)
  1   FORMAT('  NOT  ATTEMPTING  BOHR  INTEGRATION  BECAUSE',/,'
      PREDICTED',
     1' ARTERIAL PO2 IS LESS THAN 1 TORR MORE THAN MEASURED')
      IF(ARTPM.LE.1.0.AND.IBOHR.EQ.2) RETURN
      IF(IBOHR.EQ.2) WRITE(3,5)
  5   FORMAT(//,19X,'***** BOHR INTEGRATION RESULTS *****'//)
      DO 10 I=1,NVAQS
      Q(I)=QQ(I)
 10   CONTINUE
      PVN2=PIO2/FIO2-PIO2-PICO2
      FBTPS=(273.0+TEMP)*PB*FIO2/(273.0*PIO2)
C       WRITE(3,20)
 20   FORMAT(/,8X'ITERATION'6X'PVO2'5X'PVCO2'5X'+VO2'4X'+VCO2')
      ITER=0
      NNN=3
 30   DO 70 N=1,NNN
      PVO2=X(N)
      PVCO2=Y(N)
      CALL SUMUP(KRUN)
      F(N)=FVO2-GVO2
      G(N)=FVCO2-GVCO2
      IF(NNN.EQ.3) WRITE(3,20)
      WRITE(3,40)ITER,X(N),Y(N),F(N),G(N)
 40   FORMAT(11X,I3,6X,F7.2,2X,F8.2,2F9.1)
      IF(ABS(F(N))-TOL) 50,50,70
 50   IF(ABS(G(N))-TOL) 60,60,70
 60   CONTINUE
      GO TO 190
 70   CONTINUE
      DO 80 N=1,3
      U(N)=1.0
 80   CONTINUE
      CALL DETERM(U,F,G,DET1)
      DO 90 N=1,3
 90   U(N)=X(N)
      NFLAG=0
100   CALL DETERM(U,F,G,DET2)
      IF(NFLAG-1) 110,120,120
110   X(4)=DET2/DET1
      GO TO 130
120   Y(4) = DET2/DET1
130   IF(NFLAG-1) 140,160,160
```

```
140   DO 150 N=1,3
150   U(N)=Y(N)
      NFLAG=1
      GO TO 100
160   DO 170 N=1,2
      J=4-N
      X(J)=X(J-1)
      Y(J)=Y(J-1)
      F(J)=F(J-1)
      G(J)=G(J-1)
170   CONTINUE
      X(1)=X(4)
      Y(1)=Y(4)
      NNN=1
      ITER=ITER+1
      IF(ITER-10) 180,190,190
180   GO TO 30
190   CONTINUE
      PVVO2=X(N)
      PVVCO2=Y(N)
      CALL BLOOD(PVVO2,PVVCO2,FVVO2,FVVCO2)
      IF(IBOHR.EQ.1) WRITE(3,195)
195   FORMAT(/2X'N'7X'VA'7X'Q'6X'VA/Q'3X'PO2'3X'PCO2'3X'O2CON',
     12X'CO2CON'3X'VINSP'3X'RQ')
      IF(IBOHR.EQ.2) WRITE(3,200)
200   FORMAT(/2X'N'5X'VA/Q'3X'PAO2'3X'PaO2'2X'PACO2'2X
     1'PaCO2'2X'O2CON'2X'CO2CON'3X'RQ')
      PAN2=0.0
      DO 210 I=1,NVAQS-1
      IF(IBOHR.EQ.2.AND.I.EQ.1) GO TO 210
      IF(IBOHR.EQ.2.AND.Q(I).LE.1.0E-6) GO TO 210
      VINSP(I)=V(I)*PO22(I)/PIO2 + 8.63*Q(I)*(OO2CON(I)-FMVO2)/PIO2
      IF(IBOHR.EQ.1) WRITE(3,215)I,V(I),Q(I),VAQ(I),PO22(I),PCO22(I),
     1OO2CON(I),OCCO2(I),VINSP(I),RZZ(I)
      IF(IBOHR.EQ.2) WRITE(3,220)I,VAQ(I),PO22(I),PBO2(I),
     1PCO22(I),PBCO2(I),OO2CON(I),OCCO2(I),RZZ(I)
      PAN2=PAN2+Q(I)*PN22(I)
210   CONTINUE
215   FORMAT(I3,F10.4,F9.4,F8.4,4F7.2,F8.2,F6.2)
220   FORMAT(I3,F9.4,6F7.2,F6.2)
      I=NVAQS
      VINSP(I) = V(NVAQS)
      IF(IBOHR.EQ.1) WRITE(3,230)I,V(I),Q(I),PO22(I),PCO22(I),
     1OO2CON(I),OCCO2(I),VINSP(I),RZZ(I)
230   FORMAT(I3,F10.4,F9.4,'   INF    ',4F7.2,F8.2,F6.2)
```

```
      PAN2=PVN2*(1.0-QR/QT) + PAN2/QT
      CAN2=ALPHA*PAN2
      CVN2=ALPHA*PVN2
      VN2=10.0*QT*(CAN2-CVN2)
      WRITE(3,240)PAN2,PVN2,VN2
240   FORMAT(/11X,' MIXED ARTERIAL PN2 '21X,'=',F9.1,/,
     111X,' MIXED VENOUS   PN2 '21X,'=',F9.1,/,
     211X,' N2 UPTAKE, ML/MIN  '21X,'=',F9.1/)
      RETURN
      END
      SUBROUTINE WRITE
C
C     routine to summarize and print output findings
C
      COMMON/VQ/V(500),Q(500),QQ(500),VAQ(500)
      COMMON/INERT/NVAQS,VT,QT,NGASES,FACT,Z,VQLO,VQHI
      COMMON/OXY1/HB,HCRIT,TEMP,APH,BPH,APCO2,BPCO2,KOUNT,DP50,PIO2,SO2
      COMMON/OXY2/PB,FIO2,PICO2,FICO2,VS,QS,SKEW,GVO2,GVCO2,TOL,
     + PVO2,PVCO2,FMVO2,FMVCO2,PVN2,ALPHA,SHUNT,QR,GVAQ,FVQ,PAO2,
     + PACO2,O2CON,CCO2,PO2,PCO2,FVO2,FVCO2,RZ,RM,ARTO2C,ARTCO2,AMO2C
      COMMON/BOHR/DM,VC,BETA,RNT,DDLO2,DDLCO2,PMAO2,PMACO2,PO2B,PCO2B,
     +IBOHR,CAIII,IPRN,ARTPM,IWARN,IWARN2
      COMMON/O2ARAY/OO2CON(500),OCCO2(500),FVQQ(500),PN22(500),
     +PCO22(500),PO22(500),RZZ(500),PBO2(500),PBCO2(500)
      IF(IBOHR.EQ.2) GO TO 160
      AA=0.0
      BB=0.0
      EE=0.0
      FFF=0.0
      DO 10 I=1,NVAQS
      AA=AA+V(I)*PO22(I)
      BB=BB+V(I)*PCO22(I)
      EE=EE+V(I)*PN22(I)
      FFF=FFF+Q(I)*PN22(I)
 10   CONTINUE
      ALVPO2=AA/VT
      ALVPCO=BB/VT
      ALVPN2=EE/VT
      ARTPN2=FFF/QT
      O2IN=FVO2
      CO2OUT=FVCO2
      OVERAL=CO2OUT/O2IN
      RMEAS=GVCO2/GVO2
      WRITE(3,20)ALVPO2,ALVPCO,O2IN,CO2OUT,OVERAL,RMEAS
 20   FORMAT(12X,'MIXED EXPIRED  PO2                    =',F10.2,/,
```

```
     112X,'MIXED EXPIRED  PCO2                          =',F10.2,/,
     112X,'OXYGEN UPTAKE                                =',F10.2,/,
     112X,'CARBON DIOXIDE OUTPUT                        =',F10.2,/,
     112X,'PREDICTED R                                  =',F10.2,/,
     112X,'MEASURED R                                   =',F10.2,/)
      CALL FNDTEN(ARTPO2,ARTPCO,ARTO2C,ARTCO2)
      O2DIF=ALVPO2-ARTPO2
      CO2DIF=ARTPCO-ALVPCO
      DIFN2=ARTPN2-ALVPN2
      WRITE(3,30)ARTPO2,ARTPCO,ARTO2C,ARTCO2,O2DIF,CO2DIF,DIFN2
 30   FORMAT(12X,'ARTERIAL PO2                                 =',F10.2,/,
     112X,'ARTERIAL PCO2                                =',F10.2,//,
     112X,'ARTERIAL O2 CONTENT                          =',F10.2,/,
     112X,'ARTERIAL CO2 CONTENT                         =',F10.2,/,
     112X,'MIXED ALVEOLAR-ARTERIAL PO2 DIFFERENCE  =',F10.2,/,
     112X,'MIXED ALVEOLAR-ARTERIAL PCO2 DIFFERENCE =',F10.2,/,
     112X,'MIXED ALVEOLAR-ARTERIAL PN2 DIFFERENCE  =',F10.2)
      IF(OVERAL.LE.0.0) RETURN
      GRADMP=0.0
      ARTPM=ARTPO2-PMAO2
      CONPM=ARTO2C-AMO2C
      DO 140 II=1,2
      IF(II.EQ.1) WRITE(3,40)
      IF(II.EQ.2) WRITE(3,50)
 40   FORMAT(/'  IDEAL CALCULATIONS USING PREDICTED VO2 & VCO2:'/)
 50   FORMAT(/'  IDEAL CALCULATIONS USING  MEASURED VO2 & VCO2:'/)
      P1=0.0
      P2=1.2*PVCO2
      IRCNT=0
 60   PACO2=(P1+P2)/2.0
      IRCNT=IRCNT+1
      IF(II.EQ.1) RM=OVERAL
      IF(II.EQ.2) RM=RMEAS
      IF(II.EQ.2) ARTPO2=PMAO2
      FI=FIO2+FICO2
      SUM=ABS(FI-1.0)
      IF(SUM.LE.0.0001) GO TO 70
      GO TO 80
 70   PAO2=PIO2-PACO2+PICO2
      GO TO 90
 80   PAO2=PIO2*RM+PACO2*FIO2*(1.0-RM)+PICO2-PACO2
      PAO2=PAO2/(RM+FICO2*(1.0-RM))
 90   CONTINUE
      CALL BLOOD(PAO2,PACO2,O2CONE,CCO2E)
      IF(O2CONE.EQ.FMVO2) GO TO 110
```

```
      BLOODR=(FMVCO2-CCO2E)/(O2CONE-FMVO2)
      DIFF=ABS(RM-BLOODR)
      IF(DIFF.LE.0.001) GO TO 110
      IF(IRCNT.GT.50) GO TO 110
      IF(BLOODR.GT.RM) GO TO 100
      P2=PACO2
      GO TO 60
100   P1=PACO2
      GO TO 60
110   CONTINUE
      CON1=1.0+FICO2*(RM-1.0)
      GVAQ=8.63*(FMVCO2-CCO2E)*CON1/(PACO2-PICO2)
      PO2IDE=PAO2
      PCO2ID=PACO2
      DEALDI=PO2IDE-ARTPO2
      DEALO2=O2CONE
      IF(II.EQ.1) GRADMP=GRADMP-DEALDI
      IF(II.EQ.2) GRADMP=GRADMP+DEALDI
      ALVDS=100.0*(PCO2ID-ALVPCO)/(PCO2ID-PICO2)
      QSQT=0.0
      IF(DEALO2.EQ.FMVO2) GO TO 120
      QSQT=100.0*(DEALO2-ARTO2C)/(DEALO2-FMVO2)
      IF(II.EQ.2) QSQT=100.0*(DEALO2-AMO2C)/(DEALO2-FMVO2)
120   WRITE(3,130)PO2IDE,PCO2ID,DEALDI,DEALO2,ALVDS,QSQT,KOUNT,GVAQ
130   FORMAT(12X,'IDEAL ALVEOLAR PO2                    =',F10.2,/,
     112X'IDEAL ALVEOLAR PCO2                          =',F10.2,/,
     112X'IDEAL ALVEOLAR-ARTERIAL PO2 DIFFERENCE       =',F10.2,/,
     112X'IDEAL O2 CONTENT                             =',F10.2,/,
     112X'PHYSIOLOGIC DEAD SPACE PCT                   =',F10.2,/,
     112X'VENOUS ADMIXTURE PCT                         =',F10.2,/,
     112X'NUMBER OF TIMES SATURA CALLED                =',I10,/,
     112X'IDEAL VA/Q                                   =',F10.2)
140   CONTINUE
      WRITE(3,150) ARTPM,CONPM,GRADMP
150   FORMAT(/,
     1'  PREDICTED - MEASURED ARTERIAL PO2                    ='F10.2,/,
     2'  PREDICTED - MEASURED ARTERIAL CONTENT                ='F10.2,/,
     3'  MEASURED - PREDICTED ALVEOLAR-ARTERIAL GRADIENT      ='F10.2//)
      GO TO 220
160   CONTINUE
      IF(ARTPM.LE.1.0) GO TO 220
      AA=0.0
      BB=0.0
      EE=0.0
      FFF=0.0
```

```
      DO 180 I=1,NVAQS
      AA=AA+V(I)*PO22(I)
      BB=BB+V(I)*PCO22(I)
      EE=EE+V(I)*PN22(I)
      FFF=FFF+Q(I)*PN22(I)
180   CONTINUE
      ALVPO2=AA/VT
      ALVPCO=BB/VT
      ALVPN2=EE/VT
      ARTPN2=FFF/QT
      O2IN=FVO2
      CO2OUT=FVCO2
      OVERAL=CO2OUT/O2IN
      RMEAS=GVCO2/GVO2
      WRITE(3,190) ALVPO2,ALVPCO,O2IN,CO2OUT,OVERAL,RMEAS
190   FORMAT(12X,'MIXED EXPIRED  PO2                        =',F10.2,/,
     112X,'MIXED EXPIRED  PCO2                       =',F10.2,/,
     112X,'OXYGEN UPTAKE                             =',F10.2,/,
     112X,'CARBON DIOXIDE OUTPUT                     =',F10.2,/,
     112X,'PREDICTED R                               =',F10.2,/,
     112X,'MEASURED R                                =',F10.2,/)
      CALL FNDTEN(ARTPO2,ARTPCO,ARTO2C,ARTCO2)
      O2DIF=ALVPO2-ARTPO2
      CO2DIF=ARTPCO-ALVPCO
      DIFN2=ARTPN2-ALVPN2
      WRITE(3,200) ARTPO2,ARTPCO,ARTO2C,ARTCO2,O2DIF,CO2DIF,DIFN2
200   FORMAT(12X,'ARTERIAL PO2                              =',F10.2,/,
     112X,'ARTERIAL PCO2                             =',F10.2,//,
     112X,'ARTERIAL O2 CONTENT                       =',F10.2,/,
     112X,'ARTERIAL CO2 CONTENT                      =',F10.2,/,
     112X,'MIXED ALVEOLAR-ARTERIAL PO2 DIFFERENCE    =',F10.2,/,
     112X,'MIXED ALVEOLAR-ARTERIAL PCO2 DIFFERENCE   =',F10.2,/,
     112X,'MIXED ALVEOLAR-ARTERIAL PN2 DIFFERENCE    =',F10.2)
      NT=RNT
      WRITE (3,210) DDLO2,DDLCO2,NT
210   FORMAT(/12X,'DLO2  by Bohr Integration               =',F10.2,/,
     112X,'DLCO2 set to 5*DLO2                       =',F10.2,/,
     212X,'NUMBER OF STEPS USED                      =',I10,///)
      IF (IWARN2.EQ.0) GOTO 220
      WRITE(3,215)
      WRITE(*,215)
215   FORMAT(' Possible truncation error: try smaller step size')
220   CONTINUE
      RETURN
      END
```

```
      SUBROUTINE VQSOLN
C
C     this is the subroutine that actually solves the VA/Q equation
C     it is called by subroutine SUMUP once for every VA/Q compartment
C
C     VQSOLN will call the Bohr integration subroutine (BOHRI) when
      asking
C     for a diffusing capacity estimate has been requested
C     at the time of running the program and selecting run options
C
      DIMENSION X(4),Y(4),G(4),U(4),F(4)
      COMMON/INERT/NVAQS,VT,QT,NGASES,FACT,Z,VQLO,VQHI
      COMMON/OXY1/HB,HCRIT,TEMP,APH,BPH,APCO2,BPCO2,KOUNT,DP50,PIO2,SO2
      COMMON/OXY2/PB,FIO2,PICO2,FICO2,VS,QS,SKEW,GVO2,GVCO2,TOL,
     + PVO2,PVCO2,FMVO2,FMVCO2,PVN2,ALPHA,SHUNT,QR,GVAQ,FVQ,PAO2,
     + PACO2,O2CON,CCO2,PO2,PCO2,FVO2,FVCO2,RZ,RM,ARTO2C,ARTCO2,AMO2C
C     COMMON/VQ/V(500),Q(500),QQ(500),VAQ(500)
      COMMON/BOHR/DM,VC,BETA,RNT,DDLO2,DDLCO2,PMAO2,PMACO2,PO2B,PCO2B,
     + IBOHR,CAIII,IPRN,ARTPM,IWARN,IWARN2
C       COMMON/O2ARAY/OO2CON(500),OCCO2(500),FVQQ(500),PN22(500),
C     +PCO22(500),PO22(500),RZZ(500),PBO2(500),PBCO2(500)
      TOL1=0.001
      Y(1)=PVCO2*6.0/(6.0 + GVAQ)
      Y(2)=Y(1) + 5.0
      Y(3)=Y(2)
      IF(GVAQ.LE.0.55) GO TO 20
      IF(GVAQ.GE.10.0)GO TO 10
      X(1) = PIO2-30.0
      X(2) = PIO2-60.0
      X(3) = PIO2-30.0
      GO TO 30
 10   CONTINUE
      X(1)=PIO2
      X(2)=0.95*PIO2+0.05*PVO2
      X(3) = X(1)
      GO TO 30
 20   CONTINUE
      X(1) = PVO2 + 0.1
      X(2)=X(1)+10.0
      X(3)=X(1)
 30   CONTINUE
      ITER=0
      NNN=3
 40   DO 90 N=1,NNN
      CALL BOHRI(X(N),Y(N),O2CON,CCO2)
```

```
      IF(PVN2.EQ.0.0) GO TO 50
C     N2 EXCHANGE INCORPORATED
      RZ=(FMVCO2-CCO2)/(O2CON-FMVO2)
      PAO2=X(N)
      PACO2=Y(N)
      FAO2=PAO2*FIO2/PIO2
      FACO2=PACO2*FIO2/PIO2
      FVN2=PVN2*FIO2/PIO2
      C1=1.0-FAO2-FACO2
      C2=C1-FVN2
      C3=1.0-FIO2-FICO2
      B1=GVAQ*C1/C3
      B2=8.63*ALPHA*C2/C3
      O2CON1=FMVO2 + (PIO2*(B1+B2) - PAO2*GVAQ)/8.63
      CCO21=FMVCO2 - (PACO2*GVAQ - PICO2*(B1+B2))/8.63
      GO TO 60
 50   CONTINUE
C     N2 EXCHANGE IGNORED
      ABAB=Y(N)*(1.0-FIO2) - FICO2*(PIO2/FIO2 - X(N))
      AAAA=ABAB*GVAQ/(8.63*(1.0-FIO2-FICO2))
      BBAA=PIO2-X(N)*(1.0-FICO2) - FIO2*Y(N)
      AABB=Y(N)*(1.0-FIO2) - PICO2 + FICO2*X(N)
      RZ=AABB/BBAA
      O2CON1 = FMVO2 + AAAA/RZ
      CCO21 = FMVCO2-AAAA
 60   CONTINUE
      F(N) = O2CON-O2CON1
      G(N) = CCO2-CCO21
      IF(ABS(F(N))-TOL1) 70,70,90
 70   IF(ABS(G(N))-TOL1) 80,80,90
 80   CONTINUE
      PAO2=X(N)
      PACO2=Y(N)
      GO TO 210
 90   CONTINUE
      DO 100 N=1,3
100   U(N)=1.0
      CALL DETERM(U,F,G,DET1)
      DO 110 N=1,3
110   U(N)=X(N)
      NFLAG=0
120   CALL DETERM(U,F,G,DET2)
      IF(NFLAG-1) 130,140,140
130   IF(DET1.NE.0.0) X(4) = DET2/DET1
      IF(DET1.EQ.0.0) X(4)=PAO2 + 1.0
```

```
      GO TO 150
140   IF(DET1.NE.0.0) Y(4) = DET2/DET1
      IF(DET1.EQ.0.0) Y(4)=PACO2+1.0
150   IF(NFLAG-1) 160,180,180
160   DO 170 N=1,3
170   U(N)=Y(N)
      NFLAG=1
      GO TO 120
180   DO 190 N=1,2
      J=4-N
      X(J)=X(J-1)
      Y(J)=Y(J-1)
      F(J)=F(J-1)
      G(J)=G(J-1)
190   CONTINUE
      X(1)=X(4)
      Y(1)=Y(4)
      NNN=1
      ITER=ITER+1
      IF(ITER-20) 200,210,210
200   GO TO 40
210   CONTINUE
      IF(PICO2.GT.0.0) GO TO 220
      FVQ=8.63*(FMVCO2-CCO2)/PACO2
      GO TO 230
220   CONTINUE
      D1=(O2CON-FMVO2)/PIO2 + (FMVCO2-CCO2)/PICO2
      D2=PACO2/PICO2-PAO2/PIO2
      FVQ=8.63*D1/D2
230   CONTINUE
      PO2=PAO2
      PCO2=PACO2
      RETURN
      END
      SUBROUTINE FTEN(X,Y,OXO2,COCON,PO2,PCO2)
C     DOUBLE PRECISION U,F,G,DET1,DET2
      DIMENSION X(4),Y(4),F(4),G(4),U(4)
      COMMON/OXY1/HB,HCRIT,TEMP,APH,BPH,APCO2,BPCO2,KOUNT,DP50,PIO2,SO2
      DO 10 KK=1,4
      F(KK)=0.
      G(KK)=0.
      U(KK)=0.
 10   CONTINUE
      ITER=0
      TOL=.001
```

```
      NNN=3
20    DO 50 N=1,NNN
      PH1=PH(Y(N),0.0)
      Y1=.003*HB*(1.-SATURA(X(N),Y(N),PH1)/100.)
      PH2=PH(Y(N),Y1)
      SATRN=SATURA(X(N),Y(N),PH2)
      FNN=.0139*HB*SATRN+SO2*X(N)-OXO2
      GNN=CO2CON(Y(N),PH2,SATRN)-COCON
      F(N)=FNN
      G(N)=GNN
      IF(ABS(FNN)-TOL)30,30,50
30    IF(ABS(GNN)-TOL)40,40,50
40    PO2=X(N)
      PCO2=Y(N)
      GO TO 170
50    CONTINUE
      DO 60 N=1,3
      U(N)=1.
60    CONTINUE
      CALL DETER(U,F,G,DET1)
      DO 70 N=1,3
      U(N)=X(N)
70    CONTINUE
      NFLAG=0
      PNO2=X(1)
      PNCO2=Y(1)
80    CALL DETER(U,F,G,DET2)
      IF(NFLAG-1)90,100,100
90    IF(DET1.EQ.0.) X(4)=PNO2*1.05
      IF(DET1.NE.0.) X(4)=DET2/DET1
      GO TO 110
100   IF(DET1.EQ.0.) Y(4)=PNCO2*.95
      IF(DET1.NE.0.) Y(4)=DET2/DET1
110   IF(NFLAG-1)120,140,140
120   DO 130 N=1,3
      U(N)=Y(N)
130   CONTINUE
      NFLAG=1
      GO TO 80
140   DO 150 N=1,2
      J=4-N
      X(J)=X(J-1)
      Y(J)=Y(J-1)
      F(J)=F(J-1)
      G(J)=G(J-1)
```

```
150     CONTINUE
        X(1)=X(4)
        Y(1)=Y(4)
        NNN=1
        ITER=ITER+1
        IF(ITER-30)160,170,170
160     GO TO 20
170     CONTINUE
        X(1)=PO2
        X(2)=PO2
        X(3)=PO2+5.
        Y(1)=PCO2
        Y(2)=PCO2-5.
        IF(Y(2).LE.1.0) Y(2)=1.0
        Y(3)=PCO2
        RETURN
        END
        SUBROUTINE BOHRI(PAAO2,PAACO2,O2CONE,CCO2E)
C
C       this routine does the Bohr integration along the capillary to
        calculate
C       separate alveolar and endcapillary PO2/PCO2 values for a trial O2
        diffusing capacity
C       in every VA/Q compartment when this option has been selected.
C
C       Obviously, when the Bohr integration has NOT been requested,
        VQSOLN will find
C       the alevolar PO2 and PCO2 values assuming they equal the
        endcapillary values
C
        DIMENSION X(4),Y(4),GRADO2(5),GRADCO(5)
        COMMON/INERT/NVAQS,VT,QT,NGASES,FACT,Z,VQLO,VQHI
        COMMON/OXY1/HB,HCRIT,TEMP,APH,BPH,APCO2,BPCO2,KOUNT,DP50,PIO2,SO2
        COMMON/OXY2/PB,FIO2,PICO2,FICO2,VS,QS,SKEW,GVO2,GVCO2,TOL,
       + PVO2,PVCO2,FMVO2,FMVCO2,PVN2,ALPHA,SHUNT,QR,GVAQ,FVQ,PAO2,
       + PACO2,O2CON,CCO2,PO2,PCO2,FVO2,FVCO2,RZ,RM,ARTO2C,ARTCO2,AMO2C
        COMMON/O2ARAY/OO2CON(500),OCCO2(500),FVQQ(500),PN22(500),
       +PCO22(500),PO22(500),RZZ(500),PBO2(500),PBCO2(500)
        COMMON/BOHR/DM,VC,BETA,RNT,DDLO2,DDLCO2,PMAO2,PMACO2,PO2B,PCO2B,
       + IBOHR,CAIII,IPRN,ARTPM,IWARN,IWARN2
        IWARN=0
        PAO2=PAAO2
        PACO2=PAACO2
        IF (DM.GT.9000.0.OR.IBOHR.EQ.1) GO TO 30
        DO 10 I=1,5
```

```
      GRADCO(I)=0.0
      GRADO2(I)=0.0
 10   CONTINUE
      O2CI=FMVO2
      CO2CI=FMVCO2
      X(1)=PVO2
      X(2)=PVO2
      X(3)=PVO2+5.
      Y(1)=PVCO2
      Y(2)=PVCO2-5.
      Y(3)=PVCO2
      X(4)=0.
      Y(4)=0.
      PASTPO=0.
      PASTPC=0.
      NT=RNT
      DDLO2=DM
      DDLCO2=5.0*DM
      DO 20 I=1,NT
      CALL CALC(X,Y,O2CI,CO2CI,FO21,FCO21)
      IF((FO21.EQ.9000.0).OR.(FCO21.EQ.9000.0)) GOTO 30
      CALL CALC(X,Y,O2CI+FO21/2.0,CO2CI+FCO21/2.0,FO22,FCO22)
      IF((FO22.EQ.9000.0).OR.(FCO22.EQ.9000.0)) GOTO 30
      CALL CALC(X,Y,O2CI+FO22/2.0,CO2CI+FCO22/2.0,FO23,FCO23)
      IF((FO23.EQ.9000.0).OR.(FCO23.EQ.9000.0)) GOTO 30
      CALL CALC(X,Y,O2CI+FO23,CO2CI+FCO23,FO24,FCO24)
      IF((FO24.EQ.9000.0).OR.(FCO24.EQ.9000.0)) GOTO 30
      O2CI=O2CI+(FO21+FO22*2.0+FO23*2.0+FO24)/6.0
      CO2CI=CO2CI+(FCO21+FCO22*2.0+FCO23*2.0+FCO24)/6.0
      CALL FTEN(X,Y,O2CI,CO2CI,PO2I,PCO2I)
      PASTPO=PO2I
      PASTPC=PCO2I
 20   CONTINUE
      IF (DM.LT.9000.) GO TO 40
C
C     COME HERE IF DM IS ESSENTIALLY INFINITE ....
C
 30   PO2I=PAO2
      PCO2I=PACO2
      IWARN=1
      CALL BLOOD(PO2I,PCO2I,O2CI,CO2CI)
 40   CONTINUE
      O2CONE=O2CI
      CCO2E=CO2CI
      PO2B=PO2I
```

```
      PCO2B=PCO2I
      RETURN
      END
      SUBROUTINE CALC(X,Y,O2CI,CO2CI,DO2CI,DCO2CI)
C
C     THIS SUBROUTINE CALCULATES DO2,DCO2 FROM OTHER INPUTS
C
      DIMENSION X(4),Y(4)
      COMMON/INERT/NVAQS,VT,QT,NGASES,FACT,Z,VQLO,VQHI
      COMMON/OXY1/HB,HCRIT,TEMP,APH,BPH,APCO2,BPCO2,KOUNT,DP50,PIO2,SO2
      COMMON/OXY2/PB,FIO2,PICO2,FICO2,VS,QS,SKEW,GVO2,GVCO2,TOL,
     + PVO2,PVCO2,FMVO2,FMVCO2,PVN2,ALPHA,SHUNT,QR,GVAQ,FVQ,PAO2,
     + PACO2,O2CON,CCO2,PO2,PCO2,FVO2,FVCO2,RZ,RM,ARTO2C,ARTCO2,AMO2C
      COMMON/O2ARAY/OO2CON(500),OCCO2(500),FVQQ(500),PN22(500),
     +PCO22(500),PO22(500),RZZ(500),PBO2(500),PBCO2(500)
      COMMON/BOHR/DM,VC,BETA,RNT,DDLO2,DDLCO2,PMAO2,PMACO2,PO2B,PCO2B,
     + IBOHR,CAIII,IPRN,ARTPM,IWARN,IWARN2
      O2C=(1.39*HB) + (SO2*PIO2)
      IF ((CO2CI.LT.0.0).OR.(O2CI.GT.O2C)) GOTO 10
      CALL FTEN(X,Y,O2CI,CO2CI,PO2I,PCO2I)
      IF ((PO2I.LE.PAO2+2.0).AND.(PCO2I.GE.PACO2-1.0)) GOTO 20
 10   IF (PO2I.GT.PAO2+2.0) DO2CI=9000.0
      IF (PCO2I.LT.PACO2-1.0) DCO2CI=9000.0
      IF (CO2CI.LT.0.0) DCO2CI=9000.0
      IF (O2CI.GT.O2C) DO2CI=9000.0
      GOTO 30
 20   CONTINUE
      DO2CI=(PAO2-PO2I)*DDLO2/(RNT*(QT-QS)*10.)
      DCO2CI=(PACO2-PCO2I)*DDLCO2/(RNT*(QT-QS)*10.)
 30   IF(IPRN.LE.0) GO TO 40
      WRITE(*,35)O2CI,CO2CI,PO2I,PCO2I
 35   FORMAT(' IN CALC, O2CI,CO2CI,PO2I & PCO2I ARE:',4F10.3)
      WRITE(*,37)PAO2,PO2I,DM,DDLO2
 37   FORMAT(' IN CALC, PAO2,PO2I,DM & DDLO2 ARE:',4F10.3)
 40   CONTINUE
      RETURN
      END
      SUBROUTINE DETER(U,F,G,DET)
C
C     THIS SUBROUTINE IS USED BY BOTH THE MATCHING PROCEDURES.
C
      DIMENSION U(4),F(4),G(4),W(4)
C       DOUBLE PRECISION U,F,G,W,D,DET
      DO 10 KK=1,4
 10   W(KK)=0.0
```

```
      I=1
      J=2
      K=3
20    W(I)=U(I)*(F(J)*G(K)-F(K)*G(J))
      IF(I-3)30,60,60
30    IF(I-1)40,40,50
40    I=2
      J=3
      K=1
      GO TO 20
50    I=3
      J=1
      K=2
      GO TO 20
60    D=0.
      DO 70 I=1,3
      D=D+W(I)
70    CONTINUE
      DET=D
      RETURN
      END
```

MIX
Papier aus verantwortungsvollen Quellen
Paper from responsible sources
FSC® C105338

If you have any concerns about our products, you can contact us on
ProductSafety@springernature.com

In case Publisher is established outside the EU, the EU authorized representative is:
Springer Nature Customer Service Center GmbH
Europaplatz 3, 69115 Heidelberg, Germany

Printed by Libri Plureos GmbH
in Hamburg, Germany